ADVANCED TOPICS
IN SCIENCE AND TECHNOLOGY IN CHINA

ADVANCED TOPICS
IN SCIENCE AND TECHNOLOGY IN CHINA

Zhejiang University is one of the leading universities in China. In Advanced Topics in Science and Technology in China, Zhejiang University Press and Springer jointly publish monographs by Chinese scholars and professors, as well as invited authors and editors from abroad who are outstanding experts and scholars in their fields. This series will be of interest to researchers, lecturers, and graduate students alike.

Advanced Topics in Science and Technology in China aims to present the latest and most cutting-edge theories, techniques, and methodologies in various research areas in China. It covers all disciplines in the fields of natural science and technology, including but not limited to, computer science, materials science, life sciences, engineering, environmental sciences, mathematics, and physics.

Maohong Yu
Guowei Ma
Jianchun Li

Structural Plasticity
Limit, Shakedown and Dynamic Plastic Analyses of Structures

With 238 figures

AUTHORS:

Prof. Maohong Yu,
MOE Key Lab for Strength and
Vibration of Structures
Department of Civil Engineering
Xi'an Jiaotong University
Xi'an, 710049, China
E-mail: mhyu@mail.xjtu.edu.cn

Dr. Jianchun Li,
School of Civil and Environmental
Engineering
Nanyang Technological University
Singapore
E-mail: jcli@ntu.edu.sg

Dr. Guowei Ma,
School of Civil and Environmental
Engineering
Nanyang Technological University
Singapore
E-mail:cgwma@ntu.edu.sg

ISBN 978-7-308-06168-1 **Zhejiang University Press, Hangzhou**
ISBN 978-3-540-88151-3 **Springer Berlin Heidelberg New York**
e-ISBN 978-3-540-88152-0 **Springer Berlin Heidelberg New York**

Series ISSN 1995-6819 Advanced topics in science and technology in China
Series e-ISSN 1995-6827 Advanced topics in science and technology in China

Library of Congress Control Number: 2008936089

**Co-published by Zhejiang University Press, Hangzhou and Springer-
Verlag GmbH Berlin Heidelberg**

Springer is a part of Springer Science+Business Media
springer.com

Cover design: Frido Steinen-Broo, EStudio Calamar, Spain
Printed on acid-free paper

Preface

Structural Plasticity: Limit, Shakedown and Dynamic Plastic Analyses of Structures is the second monograph on plasticity. The others are *Generalized Plasticity*(Springer Berlin Heidelberg, 2006) and *Computational Plasticity*(forthcoming Zhejiang University Press Hangzhou and Springer Berlin Heidelberg, 2009) with emphasis on the application of the unified strength theory.

Generalized Plasticity, the first monograph on plasticity in this series, covers both traditional plasticity for metals (non-SD materials) and plasticity for geomaterials (SD materials). It describes the unified slip line theory for plane strain problems and characteristics theory for plane stress and axisymmetric problems, as well as the unified fracture criterion for mixed cracks. *Generalized Plasticity* can be used for either non-SD materials or SD materials. The second one is *Structural Plasticity: Limit, Shakedown and Dynamic Plastic Analyses of Structures*, which deals with limit analysis, shakedown analysis and dynamic plastic analyses of structures using the analytical method. The third one is *Computational Plasticity*, in which numerical methods are applied. The advances in strength theories of materials under complex stress are summarized in the book *Unified Strength Theory and Its Applications* (Springer Berlin Heidelberg, 2004).

The elastic and plastic limit analysis and shakedown analysis for structures can provide a very useful tool for the design of engineering structures. Conventionally, the Tresca yield criterion, the Huber-von Mises yield criterion, the maximum principal stress criterion and the Mohr-Coulomb criterion are applied in elastic-plastic limit analysis and shakedown analysis of structures. However, the result from each of the criteria above is a single solution suitable only for one kind of material. Only one or two principal stresses are taken into account in the maximum principal stress criterion, the Tresca criterion and the Mohr-Coulomb criterion. In addition, the Huber-von Mises criterion is inconvenient to use because of its nonlinear mathematical expression.

In the last decade more general solutions of plastic limit analysis and shakedown analysis for structures with a new unified strength theory have been presented. A series of unified solutions using the unified strength theory have been given. Unified plastic limit solutions of structures were presented in the literature, including unified solutions for circular plates, annular plates, oblique plates, rhombus plates, rectangular plates and square plates, orthogonal circular plates, thin plates with a hole, rotating discs and cylinders. So did unified solutions for the shakedown limit of pressure vessels, circular plates and rotating discs and for the dynamic plastic behavior of circular plates under soft impact. These unified solutions encompass not only the Tresca solution and the Mohr-Coulomb solution as special cases, but also a series of new solutions. The Huber-von Mises solution can also be approximated by the unified solution. The unified solution is a systematical one covering all results from a lower result to an upper result. These results can be suitable for a wide range of materials and engineering structures.

As an example, the unified solution of the limit load for oblique plates ($\theta = \pi/3$, $l_1 = 2l_2$) is illustrated in Fig.0.1. It can be seen from the figure that the limit load q can be obtained for various oblique plates with different angles and length and for various materials with a different strength ratio in tension and in compression and for various failure criteria with different parameter b.

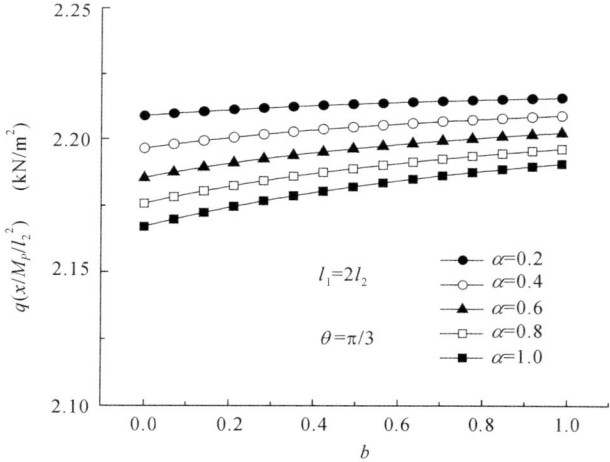

Fig. 0.1. Limit loads of oblique plate ($\theta = \pi/3, l_1 = 2l_2$) for different materials

The solution with $b = 0$ is the same as the solution of the Mohr-Coulomb material, and the solution with $b = 0$ and $\alpha = 1.0$ is the same for the Tresca material. The solution with $b = 1.0$ is for the generalized twin-shear criterion and the solution with $b = \alpha = 1.0$ is the solution of the twin-shear stress criterion. Other serial solutions between the single-shear theory (Tresca-Mohr-Coulomb theory) and the twin-shear theory are new solutions for different materials. Therefore the unified solution can be adopted for more materials and structures. It can be noted that all the solutions for the bearing capacity of structures with $b > 0$ are higher than those with the Tresca or Mohr-Coulomb criterion. The application of the unified solution is economical in the use of materials and energy. The other example is the determination of the limit pressure and thickness of pressure vessels in design. The relationship between limit pressure and wall thickness of a thin-walled vessel with the unified strength theory parameter b is shown in Fig.0.2 and Fig.0.3 respectively.

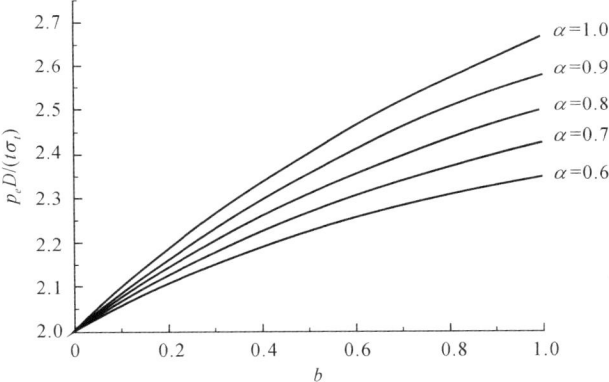

Fig. 0.2. Limit pressure versus unified strength theory parameter b

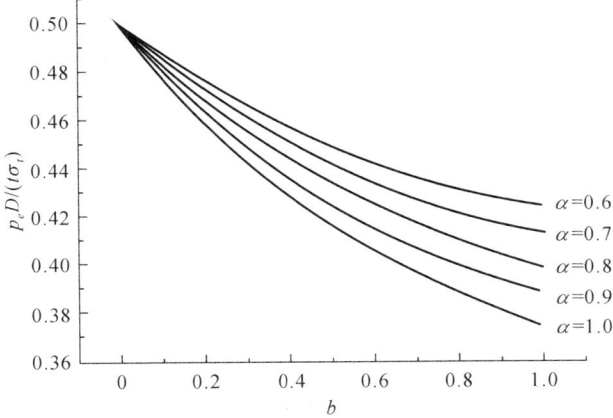

Fig. 0.3. Wall thickness versus unified strength theory parameter b

It can be seen that

- The conventional solution is a single solution ($b = 0$ in Fig.0.2 and Fig.0.3), which can be adopted only for one kind of material. The new solution is a unified solution, a serial solution, which can be adopted for more materials and structures.
- The solution for Tresca material ($b = 0$ and $\alpha = 1.0$) is identical to the solution for the Mohr-Coulomb material ($b = 0$ and $\alpha \neq 1.0$). It appears that the SD effect of materials ($\alpha \neq 1.0$ material) cannot be considered by the Mohr-Coulomb strength theory in this case.
- All the solutions for the bearing capacity of structures with $b > 0$ are higher than those for the Tresca-Mohr-Coulomb criterion. All the solutions for the wall thickness of a pressure vessel with $b > 0$ are lower than the solution using the conventional Tresca criterion or Mohr-Coulomb criterion.
- The applications of the unified strength theory and the unified solutions are more economical in the use of materials and the use of energy, leading to a reduction in environmental pollution.
- The wider application of the enhancement-factor concept on a global scale is, on the one hand, going to bring tremendous energy saving and pollution mitigation. It calls, on the other hand, for a theoretical support on which the concept can be based. Engineering practice in general has a desire to have a new strength theory, which should be more rational and more consistent with the experimental data than what can be achieved by using the Tresca-Mohr-Coulomb single-shear strength theory.

A series of the unified solutions for various structures are described in this book. It is organized as:

Chapters 2~4 give a brief introduction to the fundamental stress state, yield function and limit analysis theorem.

Chapters 5~9 deal with plastic limit analyses for circular plates, annular plates, oblique plates, rhombus plates, rectangular plates, square plates and cylinders by using the unified strength theory.

In Chapters 10, 11, the unified solutions of the dynamic plastic analysis of plates, and the limit velocity of rotating discs and cylinders are emphasized.

Penetration, wellbore analyses and orthogonal circular plates are presented in Chapters 12~14.

Chapters 15~17 are devoted to the shakedown theorem and shakedown analysis of pressure vessels, simply supported and clamped circular plates and rotating discs, using the unified strength theory. Brief summaries and references are given at the end of each chapter.

The unified strength theory and unified solutions provide a fundamental theory for the application of strength design of engineering structures. They can also be used for increasing the admissible loads or decreasing the cross-sections and the weight of structures. This results in a reduction in materials and energy consumption and a reduction in environmental pollution and the cost of structures.

The applications of the unified strength theory in plastic limit analysis and shakedown analysis for different structures are still developing. This book summarizes the research results obtained up to now. It is expected that the unified strength theory will have more and more applications in the future in addition to the plate and cylindrical structures discussed in this book. The applications of the unified strength theory in computational analysis are still growing. We hope that the Chinese idiom "Throwing out a brick to attract a piece of jade" becomes real and this book can serve as a solid brick.

The results of bearing capacity and shakedown loads obtained by using various yield criteria are very different. The results are influenced strongly by the selection of the yield criterion. We need to use a new efficient criterion. The straight-line segments on the unified strength theory make it convenient for analytical treatment of plasticity problems. The unified strength theory provides us with a very effective approach to studying the effect of yield criterion for various engineering problems. The serial results can be appropriate for most materials, from metallic materials to geomaterials.

Appreciation must be expressed for the support of the China Academy of Launch Vehicle Technology in Beijing, the Aircraft Strength Research Institute of China in Xi'an, the MOE Key Lab for Strength and Vibration at Xi'an Jiaotong University, Nanyang Technological University in Singapore, Springer and Zhejiang University Press.

I am pleased to acknowledge Dr. Guowei Ma, Dr. Jianchun Li, Dr. Fang Wang, Dr. Junhai Zhao, Dr. Xueying Wei, Dr. Yanping Wang, Dr. Shuanqiang Xu and others for their research in plastic analysis and design using the unified yield criterion and the unified strength theory. A series of unified solutions for various structures were presented. The pioneering work on the unified solution of the static and dynamic limit-bearing capacity of structures for non-SD materials using the unified yield criterion was performed by Dr. Ma during the period 1993 to 1999. The work on the unified solution of static and dynamic limit analysis of structures for SD materials was conducted by Drs. Li, Wei, Wang, Zhao, Fan and others. The work of unified solutions of shakedown for cylinders, rotating discs and circular plates was carried out by Prof. Xu Shuanqiang and Dr. Li.

I am also indebted to thank many other researchers for their research on the unified solution of structures in the fields of soil mechanics, rock mechanics, concrete mechanics and computational mechanics. I would like to express sincere thanks to Academician Shen ZJ, Academician Yang XM, Academician Chen SY, Academician Shen ZY, Academician Chen HQ, Prof. Fan SC at Nanyang Technological University, Singapore, Prof. Jiang MJ, Prof. Zhao JH, Prof. Fan W, Prof. Fung XD, Prof. Zhou XP, Prof. Yang XL, Prof. Liao HJ, Prof. Zhu XR, Prof. Chen CF, Dr. Zhang LY, Dr. Zhang SQ, et al. In addition, acknowledgement is made to Prof. Bingfeng Yu, Vice President of the Science and Technology Council, IIR and Director of Xi'an Jiaotong University Library and Dr. Maozheng Yu, Professor of the School of Energy and Power Engineering, Xi'an Jiaotong University for their kind discussions and suggestions. I would like to express my sincere thanks to the editors of Springer and Zhejiang University Press for their excellent editorial work on our manuscript.

Maohong Yu

Singapore

Contents

1

Introduction

1.1 Background

Plasticity is one of the underlying principles in the design of structures, especially metal and reinforced concrete structures. Numerous textbooks and monographs on structural plasticity and plastic design have been published since the 1950s (Baker et al., 1956; Baker, 1956; Neal, 1956; Heyman, 1958; 1971; Hodge, 1959; 1963; Horne, 1964; 1978; Baker and Heyman, 1969; Save and Massonnet, 1972; Chen, 1975; 1982; Morris and Randall, 1979; Horne and Morris, 1981; Zyczkowski, 1981; König and Maier, 1981; König, 1987; Mrazik et al., 1987; Save et al., 1997; Nielsen, 1999). The European Recommendations for the design of steelwork and reinforced concrete structures apply widely the plastic behavior of materials (Horne and Morris, 1981).

The advantages of the plastic method for structural analysis were discussed by Massonnet, Beedle, Heyman and Chen as follows:"The method of plastic design represents reality better than the conventional elastic method; it must lead to better proportioned and more economical structures"(Heyman, 1960). "For plastic design to represent reality means that the collapse load computed from plastic theory can be closely observed in practice" (Heyman, 1960).

"Engineers and research workers have been stimulated to study the plastic strength of steel structures and its application to design for three principal reasons: (a) it has a more logical design basis; (b) it is more economical in the use of steel, and (c) it represents a substantial saving of time in the design" (Beedle, 1960).

"The calculation of load-carrying capacity by use of the limit theorems is much easier than the calculation of stress. Answers obtained are not only physically more meaningful but also simpler. The simplicity of limit analysis opens the way to limit design, to direct design as contrasted with the trial-and-error procedure normally followed in conventional design." "The estimation of the collapse load is of great value, not only as a simple check for a more refined analysis, but also as a basis for engineering design" (Chen, 1982).

The elastic and plastic limit loads are two important indicators for the overall robustness of a structure to resist external loads. Load-carrying capacity of a structure usually refers to the plastic limit load, which is also called collapse load.

The plastic limit analysis gives a straightforward approximation of the maximum strength and shakedown load of structures. It simplifies the analysis of structures and also helps to derive a cost-effective design.

The elastic and plastic limit loads of structures are often used for design and safety evaluation purposes. Although advanced computer software using numerical methods have been applied widely in engineering practices, analytical solutions still play an important role because they can provide explicit forms of the limit loads, stress and bending moment fields and deformation field of structures, which enables a preliminary structure design and validation of numerical results. The elastic and plastic limit solutions are simple to derive, especially for plate and rotationally symmetrical solids, whose governing equations contain only a few variables.

Conventionally the Tresca and Mohr-Coulomb criteria are adopted to derive the plastic limit load for structures because they have piecewise linear mathematical expressions, which makes the integration of the governing differential equations tractable. The plastic limit loads based on the Tresca criterion for metallic material structures and the Mohr-Coulomb criterion for geomaterial structures have been well applied in design and safety evaluation practice. However, these strength criteria ignore the influence of the intermediate principal stress on the material strength and may lead to improper or over-conservative design of structures.

On the other hand, another widely used criterion, the Huber-von Mises yield criterion has a nonlinear mathematical expression, which renders its application unstraightforward. The effect of different yield criteria on the estimation of the load-bearing capacity of structures has not been fully explored. Most of the textbooks on the plastic limit analysis introduce only solutions based on the Tresca criterion. However, these solutions give a lower bound of the plastic limit load of structures. They may underestimate the load-bearing capacity of structures, which will be addressed in the following chapters. To achieve a better design it is necessary to investigate accurately the plastic limit load of structures.

A unified strength theory was brought up and developed by Yu (1991; 1992; 2004), which is formulated by piecewise linear expression. The theory is based on the assumption that material yielding is dominated by the two larger principal shear stresses (or twin-shear stresses). A parameter b ranging from 0 to 1 is applied on the second principal shear stress, which adjusts the relatively low effect of the second principal shear stress compared to that of the first one. The parameter b reflects different material strength behaviors indicated by different relative shear strengths, which are supported by much experimental evidence. The unified strength theory can represent or approximate the most prevailing yield and strength criteria, such as the Tresca criterion, the Huber-von Mises criterion, the Mohr-Coulomb criterion, and the twin-shear stress criterion. Using the unified strength theory, the plastic limit load of structures made of different materials can be derived conveniently in a unified manner. This book presents the plastic limit and shakedown analyses of various plates and rotationally symmetrical structures based on the unified strength theory. The derived solutions demonstrate the effect of different yield and strength criteria on the plastic limit loads of structures.

This book is one of a series of books on the fundamentals, developments, and applications of the unified strength theory (Yu, 2004; Yu et al., 2006).

1.2 Unification of Yield and Strength Criteria

In general, a yield criterion refers to a simpler form of a strength criterion that is applied for metallic materials which are ductile and have identical tensile and compressive uniaxial strengths. A strength criterion is more general and applicable to both metallic and non-metallic materials, which may exhibit strength difference in tension and compression. Non-metallic materials, such as rock, concrete, and soils, are also referred as strength difference or SD materials. Correspondingly, the uniaxial tensile and compressive strengths of metallic materials are identified as those of non-SD materials. The unified strength theory consists of a unified strength theory, which is applicable to the SD materials, and a unified yield criterion to the non-SD materials. The unified yield criterion is a specific form of the unified strength theory when the strength difference in uniaxial tension and compression can be ignored.

For stable and isotropic metal materials, the Tresca criterion (or the single-shear criterion) is the lower bound yield criterion, and the twin-shear yield criterion (Yu, 1961; 1983) is the upper bound according to the convexity condition. The lowest and the highest load carrying capacities of metal structures are calculated with respect to these two criteria. Some nonlinear yield criteria, whose geometrical graphs lie between the two surfaces defined by the single-shear criterion and the twin-shear criterion, have also been proposed. However, their nonlinear mathematical expressions make the derivation of the closed-form solutions of plastic limit loads for structures very complicated.

The Yu unified yield criterion bounded by the single-shear criterion and the twin-shear criterion has the advantage of simple expression in giving the unified solution of the load-bearing capacity of structures. A series of different solutions can be derived by choosing the unified yield criterion parameter b from 0 to 1. It can be applied to various non-SD materials. The plastic limit loads estimated by the single-shear criterion (Tresca criterion), the Huber-von Mises criterion and the twin-shear stress criterion are special cases or close approximations of the solutions based on the unified yield criterion with specific values of strength parameter b.

The unified strength theory, which is a straightforward extension of the twin-shear yield criterion (Yu, 1961; 1983) and twin-shear strength theory (Yu et al., 1985), also has a piecewise linear mathematical expression. It gives the plastic limit solutions for structures of SD materials. The plastic limit load based on the Tresca criterion or the Mohr-Coulomb criterion is one of the specific forms of the solutions given by the unified strength theory. A series of new solutions can be derived by varying the unified strength theory parameter b from 0 to 1.

A detailed description of the unified strength theory can be found in the companion volumes of *Unified Strength Theory and Its Applications* and *General Plasticity* published by Springer in 2004 and 2006. A paper entitled "Remarks on Model of Mao-Hong Yu" was made by Altenbach and Kolupaev (2008). Reviews of "Unified Strength Theory and Its Applications" were made by Shen (2004) and Teodorescu (2006). The comments on the unified strength theory were made by Fan and Qiang (2001) and Zhang et al. (2001).

1.3 Plastic Limit Analysis

The traditional estimation of the load-carrying capacity of a structure under static loading was based on the "local" permissible stress condition. More realistic approaches must take plastic deformation into account. The simplest estimation of the load-bearing capacity is furnished by concept of the limit-carrying capacity. The related ideas date back to the 18th century (Gvozdev, 1938; Prager et al., 1951; Zyczkowski, 1981). The theorems of limit analysis which provide upper and lower bounds to the true collapse load were first presented by Gvozdev (1938) and independently proved by Hill (1951) for the rigid-perfectly plastic materials, and by Drucker et al. (1952) for the elastic-perfectly plastic materials.

The circular plate is an important structural element in many branches of engineering. Reliable prediction of the load-bearing capacity of the circular plate is crucial in achieving an optimal structural design. Previous studies by other researchers have shown that the limit analysis is an effective measure in exploring the strength behaviors of a circular plate in the plastic limit state (Hodge, 1963). Hopkins and Prager (1953), Zaid (1958), and Hodge (1963) investigated the load-bearing capacity of circular and annular plates with limit analysis theorems and proved that an exact plastic limit solution for an axisymmetrical plate can be analytically derived if the material of the plate satisfies a linear yield criterion. Ghorashi (1994) derived the plastic limit solutions for circular plates subjected to arbitrary-rotationally symmetric loading. However, most of the reported results have been derived in terms of the maximum stress yield criterion or the Tresca criterion that takes account of the effect of only one or two principal stresses in the three dimensional stress space. Solutions in terms of these yield criteria may not reflect the real characteristics of a circular plate in the plastic limit state. Hopkins and Wang (1954) investigated the load-bearing capacities of circular plates with respect to the Huber-von Mises criterion and a parabolic criterion by iterative method. The percentage differences of the plastic limit loads with respect to the Huber-von Mises criterion and the Tresca criterion are approximately 8% and 10% for a simply supported and a clamped circular plate, respectively. These observations indicate that the plastic limit loads in terms of different yield criteria are different and the difference varies with the variation of the constraint conditions of the plate.

Limit analysis of structures applies only if the loading magnitude is less than the plastic collapse force. With impact or explosive blast loading the structures may be subjected to an intense but short duration pressure or force pulse that exceeds the plastic collapse force. The response of circular plates to pulse loading was presented by Florence (1966; 1977), Youngdagl (1971; 1987), Li and Jones (1994), Li and Huang (1989), etc. The dynamic plastic behavior of beams has been investigated in detail by Stronge and Yu (1993). However, it is difficult to get the analytical solution for plate and shell in a dynamic plastic deformation state because of the complicated constitutive formulation. On the other hand, the dynamic analytical solution for a circular plate is much simpler because of the axisymmetry. Exact theoretical solutions to the dynamic response of a rigid, perfectly plastic, simply supported circular plate have been explored initially by Hopkins and Prager (1954). In the past forty years a number of studies (Jones, 1980; 1989; Florence, 1966; 1977; Symonds, 1979; Li and Jones, 1994; Liu and Jones, 1996) have been done on this subject by introducing various boundaries, loading conditions, and plastic flow assumptions for a circular plate. So far all these studies are based on the Tresca yield criterion or the maximum stress yield criterion. The influence of different yield criteria on the dynamic plastic behavior of circular plates (Florence, 1966) has not been addressed. Recently the influence of the transverse shear force on the final central deflection of circular plates has attracted some attention (Liu and Stronge, 1996; Youngdagl, 1971; 1987; Shen and Jones, 1993; Woodward, 1987). Jones and Oliveira (1980) analyzed a simply supported circular plate subjected to an impulsive velocity uniformly distributed over the entire plate. Dynamic plastic behavior of annular plates with transverse shear effects was studied by Lellep and Torn (2006).

Circular plates are sometimes strengthened in the radial direction or the circumferential direction with stiffeners, which induces orthotropic yield moments. Material orthotropy can also arise from the cold forming process, which results in different yield strengths in different directions. Orthotropic yield criteria for those plates have been suggested by many researchers (Sawczuk, 1956; Markowitz and Hu, 1965; Save et al., 1997). They are mainly modifications of the Tresca criterion. Olszak and Sawczuk (1960) investigated the plastic limit behavior of an orthotropic circular plate in terms of the modified Tresca criterion. The results have been extended to various loading cases by Markowitz and Hu (1965). The plastic limit solution given in these studies satisfies both a statically admissible moment field and kinematically admissible velocity fields and thus, is an exact solution.

Many monographs concerning the limit analysis of a structure were published. It can be seen in Baker (1956), Baker, Horne and Heyman (1956), Neal (1956), Hodge (1959; 1963), Horne (1964), Baker and Heyman (1969), Heyman (1971), Save and Massonnet (1972), Chen (1975; 1981; 1998), Horne (1978), Morris and Randall (1979), Horne and Morris (1981), Zyczkowski (1981), Xu and Liu (1985), Mrazik, Skaloud and Tochacek (1987), Xiong (1987), Save, Massonnet and Saxce (1997), Huang and Zhen (1998), Nielsen (1999). The number of papers devoted to problems of the limit carrying capacity of various structures is enormous. The Tresca criterion, Huber-von Mises criterion and maximum normal stress criterion are used to obtain the plastic limit of structures in most books and papers. The literature concerning the plastic analysis of a structure was summarized by Zyczkowski (1981).

The twin-shear yield criterion was used for limit analysis of a structure. Some solutions were presented by Li (1988) and Huang and Zeng (1989). Fourteen problems using the twin-shear criterion were collected in the monograph of Huang and Zeng (1998). Application of the twin-shear strength theory in the strength calculation of gun barrels was reported by Liu, Ni, Yan et al. (1998). The twin-shear criterion was also used in axisymmetric identification of a semi-infinite medium (Zhao, Xu, and Yang et al., 1998) and mathematical solutions for forming mechanics of continuum (Zhao, 2004). The calculation of stable loads of strength-differential thick cylinders and spheres by the twin-shear strength theory was reported by Ni, Liu, and Wang (1998).

The unified yield criterion was used for plastic analysis of structures of non-SD materials by Ma, He and Yu (1993), Ma and He (1994) and Ma, Yu, Iwasaki et al. (1994; 1995). Following these results, Ma et al. presented a series of unified solutions for non-SD materials. The unified solutions of structures for SD materials were presented by Li and Yu (2000b), Wei and Yu (2001; 2002), Wang and Yu (2002; 2005), Xu and Yu (2004), and others.

1.4 Plastic Limit Analysis of Rotating Solids

Rotating discs are used widely as important structural elements in mechanical engineering. In a structural design procedure, it is inevitable that we must estimate the angular velocity and the stress distribution of a rotating disc in a fully plastic state. The theoretical study of a rotating disc was presented by Nadai and Donnell (1929), and since then numerous works involving plastic collapse speeds (Heyman, 1958; Lenard and Haddow, 1972), thermal effects (Thompson, 1946; Gamer and Mack, 1985), acceleration effects (Reid, 1972), and variable thickness influence (Güven, 1992; 1994), etc., have been reported. Most of them employed the Tresca criterion which results in an over-conservative estimation of the load-bearing capacity of a structure because it does not take account of the effect of the second intermediate principal stress on material yielding (Hill, 1950). As Gamer (1983; 1985) pointed out, when the Tresca criterion and its associated flow rule is used, the displacement across the elasto-plastic interface of a rotating disc is not continuous, and negative circumferential plastic strain is derived in the disc center area where the stresses are tensile. To solve this problem, Gamer (1984) suggested an additional strain-hardening region at the center area of the disc. The idea was extended by Güven (1994) in investigating the plastic limit of angular velocity of a solid rotating disc with variable thickness. Calladine (1969), on the other hand, explained that the singularities in the strain increments at the center could be interpreted as a tendency for the disc to "thin" so much as to produce a small hole very quickly. From a mathematical point of view, the deficiencies in the Tresca solution can be avoided by applying a non-associated flow rule, e.g., combining the Tresca criterion with a Levy-Mises flow rule, or by applying the Huber-von Mises criterion and its associated flow rule (Rees, 1999). The latter must resort to a numerical iteration method because of the nonlinear expression of the Huber-von Mises criterion.

The limit of angular velocity of a rotating disc based on the Tresca criterion always gives the lowest estimation. The effect of different yield criterion on the limit of angular velocity of a rotating disc with variable thickness has not been well studied because the solution based on the Huber-von Mises criterion can only be obtained with numerical iteration (Rees, 1999).

1.5 Shakedown Analysis of Structures

The concept and method of shakedown analysis were first brought up in the 1930s and widely explored in the 1950s. The most significant milestones in shakedown theory of elasto-plastic structures are the pioneering works by Bleich (1932), Melan (1936), and Koiter (1953; 1960). They brought up two crucial shakedown theorems, namely the static shakedown theorem (also called the Melan's theorem, the first shakedown theorem, or the lower bound shakedown theorem), and the dynamic shakedown theorem (also called the Koiter's theorem, the second shakedown theorem or the upper bound shakedown theorem). The later developments of shakedown analysis can be categorized into the static and the dynamic shakedown analysis methods. The shakedown theory has constituted a well-established branch of plasticity theory. The bound to shakedown loads was discussed by Zouain and Silveira (2001).

In recent years, the shakedown analysis of an elasto-plastic structure has gradually attracted attention in engineering due to the requirements of modern technologies such as nuclear power plants, the chemical industry, aeronautical and astronautical technologies. The shakedown theory has been applied successfully in a number of engineering problems such as the construction of nuclear reactors, highways and railways, and employed as one of the tools for structural design and safety assessment in some design standards, rules, and regulations (König, 1987; Polizzotto, 1993; Feng et al., 1993; 1994; Maier, 2001). Shakedown analysis of the shape-memory-alloy structures is presented by Feng and Sun (2007). A new unified solution of the shakedown of cylinder and rotating disc for non-SD materials and SD materials is given by Xu and Yu (2005). Xu and Yu (2005) also give a shakedown analysis of a thick-walled spherical shell of material with different strengths in tension and compression.

1.6 Plastic Limit Analysis Based on the Unified Strength Theory

There have been significant developments in plastic limit analyses of plates and rotationally symmetrical solids based on the unified strength theory in recent years. This book mainly updates these developments which have considerable potential in extending the derived solutions to various other structural forms.

The yield loci of the unified strength theory cover all the traditional convex yield criteria. The Tresca yield criterion, the Huber-von Mises yield criterion, the twin-shear yield criterion and a series of other new linear yield criteria are special cases or approximations of the unified strength theory. It provides a new approach to the study of the load-carrying capacities of structures in a unified manner.

Ma et al. (1993; 1994; 1995a; 1995b; 1995c) derived a unified plastic limit solution to a circular plate under uniform and partially uniform load. Ma et al. (1998) gave a unified solution to simply supported circular plates and clamped circular plates in terms of the Yu's unified yield criterion. Applications of the unified yield criterion to unified plastic limit analysis of circular plates under arbitrary load were reported by Ma et al. (1998; 1999; 2001). The unified solutions of the limit speed of disc and cylinder using the unified yield criterion were derived by Ma et al. (1994; 1995a; 1995b; 1995c) for non-SD materials. The solutions in terms of maximum principal stress criterion, the Tresca yield criterion, the Huber-von Mises criterion, and the twin-shear yield criterion are all special cases or close approximations to the solutions using the unified yield criterion.

The unified solutions of the plastic limit and shakedown load to plate, cylinder, and limit speed of rotating disc and cylinder for SD materials have been reported by Wang and Fan (1998), Li and Yu (2001), Wei and Yu (2002), Wang and Yu (2002; 2005), Xu and Yu (2004; 2005). It will be described in this book, too.

A new orthotropic yield criterion, which is an extension of the unified yield criterion, has been suggested by Ma et al. (2002). The new orthotropic yield criterion is applicable to the plastic limit analysis of orthotropic plates.

The results derived from the present study show the influence of different yield criteria, which is helpful in achieving optimized design of structures and in validating the numerical models in plastic analysis.

1.7 Summary

Various single yield criteria are used for the limit analysis, shakedown analysis and dynamic plastic analysis of structures. The solution is a single result adapted for one kind of material. Owing to the development of the unified strength theory, a series of unified solutions of limit analysis, shakedown analysis and dynamic plastic analysis of structures were presented during the last decade.

The unified strength theory is an accumulation of serial yield criteria adapted for non-SD materials and SD materials. The serial criteria cover all areas between the lower bound (single-shear criterion, Tresca-Mohr-Coulomb (1864; 1900)) and upper bound (twin-shear criteria, Yu (1961; 1983; 1985)). It is well known that all the yield criteria of the unified strength theory are piecewise linear with the attendant simplification of the analytical solution of the structure. The application of the unified strength theory gives not only a single solution, but also a series of solutions. The serial solution for a structure is referred to the unified solution, which can be adopted for more materials and structures.

References

Altenbach H, Kolupaev VA (2008) Remarks on Model of Mao-Hong Yu. The Eighth International Conference on Fundamentals of Fracture (ICFF VIII), Zhang TY, Wang B, Feng XQ (eds.) 270-271

Baker AL (1956) Ultimate load theory applied to the design of reinforced and prestressed concrete frames. Concrete Publ., London

Baker JF, Heyman J (1969) Plastic design of frames: fundamentals. Cambridge University Press, London

Baker JF, Horne MR, Heyman J (1956) The steel skeleton. Vol. II, Cambridge University Press, London

Beedle LS (1960) On the application of plastic design. In: Lee EH, Symonds PS (eds.) Plasticity: Proceedings of the Second Symposium on Naval Structural Mechanics, Pergamon Press, Oxford, 538-567

Bleich H (1932) Uber die Bemessung statisch unbestimmter Stahl-tragwerke unter Bauingenieur. 13:261-267

Calladine CR (1969) Engineering plasticity. Pergamon, Oxford

Chen WF (1975) Limit analysis and soil plasticity. Elsevier, Amsterdam

Chen WF (1982) Plasticity in reinforced concrete. McGraw-Hill, New York

Chen WF (1998) Concrete plasticity: past, present and future. In: Yu MH, Fan SC (eds.) Strength Theory: Applications, Developments and Prospects for the 21st Century, Science Press, Beijing, New York, 7-48

Chen WF, Saleeb AF (1981) Constitutive equations for engineering materials. Vol.1, Elasticity and Modeling; Vol.2, Plasticity and Modelling, Wiley, New York

Drucker DC, Prager W, Greenberg HJ (1952) Extended limit design theorems for continuous media. Quart. Appl. Math., 9:381-392

Fan SC, Qiang HF (2001) Normal high-velocity impact concrete slabs—a simulation using the meshless SPH procedures. Computational Mechanics—New Frontiers for New Millennium, Valliappan S, Khalili N (eds.) Elsevier, Amsterdam, pp 1457-1462

Feng XQ, Liu XS (1993) Factors influencing shakedown of elastoplastic structures. Adv. Mech., 23(2):214-222 (in Chinese)

Feng XQ, Sun QP (2007) Shakedown analysis of shape-memory-alloy structures. Int. J. Plasticity, 23(2):183-206

Feng XQ, Yu SW (1994) An upper bound on damage of elastic-plastic structures at shakedown. Int. J. Damage Mech., 3:277-289

Florence AL (1966a) Clamped circular plates under central blast loading. Int. J. Solids Struct., 2:319-335

Florence AL (1966b) Clamped circular rigid-plastic plates under blast loading. J. Appl. Mech. 33:256-260

Florence AL (1977) Response of circular plates to central pulse loading. Int. J. Solids Struct., 13:1091-1102

Gamer U (1983) Tresca's yield condition and the rotating disk. J. Appl. Mech. 50:676-678

Gamer U (1984) Elastic-plastic deformation of the rotating solid disk. Ingenieure Archiv 54:345-354

Gamer U (1985) Stress distribution in the rotating elastic-plastic disk. J. Appl. Math. Mech (ZAMM), 65:T136

Gamer U, Mack W (1985) Thermal stress in an elastic-plastic disk exposed to a circular heat source. J. Appl. Math. Phys. (ZAMP), 36:568-580

Ghorashi M (1994) Limit analysis of circular plates subjected to arbitrary rotational symmetric loading. Int. J. Mech. Sci., 36(2):87-94

Güven U (1992) Elastic-plastic stresses in a rotating annular disk of variable thickness and variable density. Int. J. Mech. Sci., 34(2):133-138

Güven U (1994) The fully plastic rotating solid disk of variable thickness. J. Appl. Math. Mech (ZAMM), 74(1):61-65

Gvozdev AA (1938) The determination of the value of the collapse load for statically indeterminate systems undergoing plastic deformation. In: Proceedings of the Conference on Plastic DeformationsAkademiia Nauk SSSR, Moscow, 1938, pp. 19-33 (Russian original); Translated into English by R.M. Haythornthwaite, Int. J. MechSci. 1960, 1(4):322-355

Heyman J (1958) Plastic design of rotating discs. Proc. Inst. Mech. Eng., 172:531-539

Heyman J (1960) Progress in plastic design. In: Plasticity: Lee EH, Symonds PS (eds.) Proceedings of the Second Symposium on Naval Structural Mechanics, Pergamon Press, Oxford, 511-537

Heyman J (1971) Plastic design of frames: applications. Cambridge University Press, London

Hill R (1950) The mathematical theory of plasticity. Clarendon, Oxford

Hill R (1951) On the state of stress in a plastic rigid body at the yield point. Phi. Mag. 42:868-875

Hodge PG Jr (1959) Plastic analysis of structures. McGraw-Hill, New York

Hodge PG Jr (1963) Limit analysis of rotationally symmetric plates and shells. Prentice-Hall, New Jersey

Horne MR (1964) The plastic Design of columns. BCSA Pub. No. 23

Horne MR (1978) Plastic theory of structures. Pergamon Press, Oxford

Horne MR, Morris LJ (1981) Plastic design of low-rise frames. Collins, London

Hopkins HG, Prager W (1953) The load carrying capacities of circular plates. J. Mech. Phys. Solids. 2:1-13

Hopkins HG, Prager W (1954) On the dynamics of plastic circular plates. J. App. Math. Phys. (ZAMP), 5:317-330

Hopkins HG, Wang AJ (1954) Load carrying capacities for circular plates of perfectly-plastic material with arbitrary yield condition. J. Mech. Phys. Solids, 3:117-129

Huang WB, Zeng GP (1989) Solving some plastic problems by using the Twin shear stress criterion. Acta Mechanica Sinica 21(2), 249-256 (in Chinese, English abstract)

Huang WB, Zeng GP (1998) Analyses of difficult questions in elasticity and plasticity. Higher Education Press, Beijing

Jones N (1989) Structural impact. Cambridge University Press

Jones N, De Oliveira JG (1980) Dynamic plastic response of circular plates with transverse shear and rotatory inertia. J. Appl. Mech. 47(1):27-34

Koiter WT (1953) Stress-strain relations, uniqueness and variational theorems for elastic-plastic materials with a singular yield surface. Quart. Appl. Math. 11:350-354

Koiter WT (1960) General theorems for elastic-plastic solids. North-Holland Publ. Co., Amsterdam

König JA (1987) Shakedown of Elastic-Plastic Structures. Elsevier, Amsterdam

König JA, Maier G (1981) Shakedown analysis of elastic-plastic structures: a review of recent developments. Nuclear Engineering and Design 66:81-95

Lellep J, Torn K (2006) Dynamic plastic behavior of annular plates with transverse shear effects. Int. J. Impact Eng. 32:601-623

Lenard L, Haddow JB (1972) Plastic collapse speeds for rotating cylinders. Int J. Mech. Sci. 14:285-292

Li JC, Yu MH (2000a) Unified limit solution for metal oblique plates. Chinese J. Mech. Eng., 36(8):25-28 (in Chinese, English abstract)

Li JC, Yu MH, Fan SC (2000b) A unified solution for limit load of simply supported oblique plates, rhombus plates, rectangle plates and square plates. China Civil Engineering Journal, 33(6):76-89 (in Chinese, English abstract)

Li QM, Huang YG (1989) Dynamic plastic response of thin circular plates with transverse shear and rotatory inertia subjected to rectangular pulse loading. Int. J. Impact Eng. 8:219-228

Li QM, Jones N (1994) Blast loading of fully clamped circular plates with transverse shear effects. Int. J. Solids Struct., 31(14):1861-1876

Li YM (1988) Elastoplastic limit analyses with a new yield criterion (twin-shear yield criterion). J. Mech. Strength, 10(3):70-74 (in Chinese, English abstract)

Liu D, Stronge WJ (1996) Deformation of simply supported circular plate by central pressure pulse. Int. J. Solids Struct., 33(2):283-299

Liu J, Jones N (1996) Shear and bending response of a rigid-plastic circular plate struck transversely by a mass. Mech. Struct. & Mach., 24(3):361-388

Liu XQ, Ni XH, Yan S et al (1998) Application of the twin shear strength theory in strength-calculation of gun barrels. Strength Theory: Applications, Developments and Prospects for the 21st Century. Yu MH, Fan SC (eds.) Science Press, New York, Beijing, 1039-1042

Ma GW, Gama BA, Gillespie JW Jr (2002) Plastic limit analysis of cylindrically orthotropic circular plates. Composite Structures, 55:455-466

Ma GW, Hao H, Iwasaki S (1999a) Plastic limit analysis of a clamped circular plate with unified yield criterion. Structural Engineering and Mechanics, 7(5):513-525

Ma GW, Hao H, Iwasaki S (1999b) Unified plastic limit analysis of circular plates under arbitrary load. J. Appl. Mech., ASME, 66(2): 568-570

Ma GW, Hao H, Iwasaki S, Miyamoto Y, Deto H (1998) Plastic behavior of circular plate under soft impact. In Proc. of International Symposium on Strength Theories: Application & Developments. Yu MH, Fan SC (eds.) Science Press, New York, Beijing, 957-962

Ma GW, Hao H, Miyamoto Y (2001) Limit angular velocity of circular disc using unified yield criterion. Int. J. Mech. Sci., 43(5):1137-1153

Ma GW, He LN (1994) Unified solution to plastic limit of simply supported circular plate. Mechanics and Practice, 16(6):46-48 (in Chinese, English abstract)

Ma GW, He LN, Yu MH (1993) Unified solution of plastic limit of circular plate under portion uniform loading. The Scientific and Technical Report of Xi'an Jiaotong University, 1993-181 (in Chinese)

Ma GW, Iwasaki S, Miyamoto Y, Deto H (1998) Plastic limit analysis of circular plates with respect to the unified yield criterion. Int. J. Mech. Sci., 40(10):963-976

Ma GW, Iwasaki S, Miyamoto Y, Deto H (1999c) Dynamic plastic behavior of circular plate using unified yield criterion. Int. J. Solids Struct. 36(22):3257-3275

Ma GW, Yu MH, Iwasaki S, Miyamoto Y (1994) Plastic analysis of circular plate on the basis of twin shear stress criterion. In Proceedings of International Conferences on Computational Method on Structural & Geotechnical Engineering. III, Hong Kong, 930-935

Ma GW, Yu MH, Iwasaki S, Miyamoto Y (1995a) Elasto-plastic response of axisymmetrical solids under impulsive loading. Acta Mechanica Solida Sinca 8(Special Issue):641-645

Ma GW, Yu MH, Iwasaki S, Miyamoto Y, Deto H (1995b) Unified plastic limit solution to circular plate under portion uniform loading. J. Struct. Eng., JSCE, 41A:385-392

Ma GW, Yu MH, Iwasaki S, Miyamoto Y, Deto H (1995c) Unified elastoplastic solution to rotating disc and cylinder. J. Struct. Eng., JSCE. 41A:79-85

Maier G (2001) On some issues in shakedown analysis. J. Appli. Mech. 68:799-807

Markowitz J, Hu LW (1965) Plastic analysis of orthotropic circular plates. Proc. ASCE, J. Eng. Mech. Div. 90(EM5):251-259

Melan E (1936) Theorie statisch unbestimmter Systeme. In: Prelim. Publ. 2nd Congr. Intern. Assoc. Bridge and Structural Eng. Berlin, 43-64

Melan E (1936) Theorie statisch unbestimmter Systeme aus ideal plastischem Baustoff. Sitz. Ber. Akad. Wiss. Wien, II.a,145:195-218

Morris LJ, Randall AL (1979) Plastic design. Constrado Pub

Mrazik A, Skaloud M, Tochacek M (1987) Plastic design of steel structures. Ellis Horwood, Chichester, New York

Nadai A, Donnell LN (1929) Stress distribution in rotating disks of ductile material after the yield point has been reached. Trans. ASME, 51:173-181

Neal BG (1956) The plastic method of structural analysis. Chapman and Hall (2nd ed. Wiley, New York, 1963)

Ni XH, Liu XQ, Liu YT, Wang XS (1998) Calculation of stable loads of strength-differential thick cylinders and spheres by the twin shear strength theory. In: Yu MH, Fan SC (eds.) Strength Theory: Applications, Developments and Prospects for the 21st Century. Science Press, New York, Beijing, 1043-1046

Nielsen MP (1999) Limit analysis and concrete plasticity (2nd ed.) CRC Press, Boca Raton, London

Olszak W, Sawczuk A (1960) Theorie de la capacite portante des constructions nonhomogenes et orthotropes (Load bearing capacity of nonhomogeneous and orthotropic structures). Institut Technique du Batiment et des Travaux Publics-Annales, 113(149):517-536

Polizzotto C (1993) On the conditions to prevent plastic shakedown of structures: part 1-Theory. ASME, J. Appl. Mech. 60:15-19

Polizzotto C (1993) On the conditions to prevent plastic shakedown of structures: part 2-The plastic shakedown limit load. ASME, J. Appl. Mech. 60:20-25

Polizzotto C (1993) A study on plastic shakedown of structures: part 1 - Basic properties. ASME, J. Appl. Mech. 60:318-323

Polizzotto C (1993) A study on plastic shakedown of structures: part 2 - Theorems. ASME, J. Appl. Mech. 60:324-330

Prager W, Hodge PG Jr (1951) Theory of perfectly plastic solids. John & Wiley, New York

Rees DWA (1999) Elastic-plastic stresses in rotating discs by von Mises and Tresca. J. Appl. Math. Mech. (ZAMM) 79(4):281-288

Reid SR (1972) On the influence of acceleration stresses on the yielding of discs with uniform thickness. Int. J. Mech. Eng. Sci. 14(11):755-763

Save MA (1985) Limit analysis of plates and shells: Research over two decades. J. Struct. Mech., 13:343-370

Save MA, Massonnet CE (1972) Plastic analysis and design of plates, shells and disks. North-Holland Publ. Co., Amsterdam

Save MA, Massonnet CE, de Saxce G (1997) Plastic limit analysis of plates, shells and disks. Elsevier, Amsterdam

Sawczuk A (1965) Note on plastic expansion of irradiated spherical shells. Nucl. Struct. Eng., 1(2):155-158

Shen WQ, Jones N (1993) Dynamic response and failure of fully clamped circular plates under impulsive loading. Int. J. Impact Eng. 13:259-278

Shen ZJ (2004) Reviews to "Unified strength theory and its applications". Advances in Mechanics, 34(4):562-563 (in Chinese)

Stronge WJ, Yu TX (1993) Dynamic models for structural plasticity. Springer-Verlag, London

Symonds PS, Wierzbicki T (1979) Membrane mode solutions for impulsively loaded circular plates. J. Appl. Mech., 46(1):58-64

Teodorescu PP (2006) Review of "Unified strength theory and its applications. Springer, Berlin, 2004". Zentralblatt MATH 2006, Cited in Zbl. Reviews, 1059.74002 (02115115)

Thompson AS (1946) Stresses in rotating discs at high temperature. J. Appl. Mech. 13:A45-A52

Wang F, Fan SC (1998) Limit pressures of thick-walled tubes using different yield criteria. In: Yu MH, Fan SC (eds.) Strength Theory: Applications, Developments and Prospects for the 21st Century. Science Press, New York, Beijing, 1047-1052

Wang YB, Li JH, Wei XY, Yu MH (2005) Analysis of high-velocity tungsten rod on penetrating brittle target. Chinese J. High Pressure Physics, 19(3):257-263 (in Chinese)

Wang YB, Yu MH et al. (2005) Dynamic plastic response to circular plate based on unified strength theory. Int. J. Impact Eng. 31(1):25-40

Wang YB, Yu MH, Li JC et al. (2002a) Unified plastic limit analysis of metal circular plates subjected to border uniformly distributed loading. J. Mechanical Strength, 24:305-307 (in Chinese, English abstract)

Wang YB, Yu MH, Wei XY (2002b) Unified plastic limit analyses of circular plates. Engineering Mechanics, 19:84-88 (in Chinese, English abstract)

Wang YB, Zhang XQ, Yu MH et al. (2005) Penetration analysis of long rod projectiles on rock targets. Chinese J. of Rock Mechanics and Engineering, 24(8):1301-1307 (in Chinese)

Wei XY, Yu MH (2001) Unified plastic limit of clamped circular plate with strength differential effect in tension and compression. Chinese Quart. Mechanics, 22:78-83 (in Chinese, English abstract)

Wei XY, Yu MH (2002a) Analysis of tungsten rods on penetrating ceramics at high velocity. Acta Armamentaria, 23: 167-170

Wei XY, Yu MH (2002b) Unified solutions for plastic limit of annular plate. J. Mech. Strength, 24:140-143 (in Chinese, English abstract)

Woodward RL(1987) A structural model for thin plate perforation by normal impact of blunt projectiles. Int. J. Impact Eng. 6:129-140

Xiong ZH (1987) Plastic Analysis of Structures . Traffic Press, Beijing (in Chinese)

Xu BY, Liu XS (1985) Plastic limit analysis of structures. China Building Industry Press, Beijing (in Chinese)

Xu SQ, Yu MH (2004) Unified analytical solution to shakedown problem of thick-walled cylinder. Chinese J. Mechanical Engineering, 40(9):23-27 (in Chinese, English abstract)

Xu SQ, Yu MH (2004) Elasto Brittle-plastic carrying capacity analysis for a thick walled cylinder under unified theory criterion. Chinese Quart. Mechanics 25(4):490-495

Xu SQ, Yu MH (2005) Shakedown analysis of thick-walled cylinders subjected to internal pressure with the unified strength criterion. Int. J. Pressure Vessels and Piping, 82(9):706-712

Xu SQ, Yu MH (2005) Shakedown analysis of thick-walled spherical shell of material with different strength in tension and compression. Machine Design & Manufacture, 2005(1):36-37

Youngdagl CK (1971) Influence of pulse shape on the final plastic deformation of a circular plate. Int. J. Solids Struct. 7:1127-1142

Youngdagl CK (1987) Effect of pulse shape and distribution on the plastic deformation of a circular plate. Int. J. Solids Struct. 23:1179-1189

Yu MH (1961a) General behavior of isotropic yield function. Res. Report of Xi'an Jiaotong University, Xi'an, China (in Chinese)

Yu MH (1961b) Plastic potential and flow rules associated singular yield criterion. Res. Report of Xi'an Jiaotong University. Xi'an, China (in Chinese)

Yu MH (1983) Twin shear stress yield criterion. Int. J. Mech. Sci., 25(1):71-74

Yu MH (1992) A new system of strength theory. Xi'an Jiaotong University Press, Xi'an (in Chinese)

Yu MH (2002) Advances in strength theories for materials under complex stress state in the 20th century. Applied Mechanics Reviews ASME, 55(3):169-218

Yu MH (2004) Unified strength theory and its applications. Springer, Berlin

Yu MH, He LN (1983) Non-Schmid effect and twin shear stress criterion of plastic deformation in crystals and polycrystalline metals. Acta Metallurgica Sinica 19(5):190-196 (in Chinese, English abstract)

Yu MH, He LN (1991) A new model and theory on yield and failure of materials under complex stress state. In: Mechanical Behavior of Materials-6, Pergamon Press, Oxford, 3:841-846

Yu MH, He LN, Song LY (1985) Twin shear stress theory and its generalization. Scientia Sinica (Sciences in China), English ed. Series A, 28(11):1174-1183

Yu MH, Ma GW, Qiang HF, et al. (2006) Generalized Plasticity. Springer, Berlin

Zaid M (1958) On the carrying capacity of plates of arbitrary shape and variable fixity under a concentrated load. J. Appl. Mech., ASME, 25:598-602

Zhang XS, Guan H, Loo YC (2001) UST failure criterion for punching shear analysis of reinforcement concrete slab-column connections. Computational Mechanics—New Frontiers for New Millennium. Valliappan S, Khalili N (eds.) Elsevier, Amsterdam, 299-304

Zhao DW (2004) Mathematical solutions for forming mechanics of continuum. North-East University Press, Shenyang (in Chinese)

Zhao DW, Xu JZ, Yang H, Li XH, Wang GD (1998) Application of twin shear stress yield criterion in axisymmetric identification of a semi-infinite medium. In: Yu MH, Fan SC (eds.) Strength Theory: Applications, Developments and Prospects for the 21st Century. Science Press, Beijing and New York, 1079-1084

Zouain N, Silveira JL (2001) Bounds to shakedown loads. Int. J. Solid Struct. 38:2249-2266

Zyczkowski M (1981) Combined loadings in the theory of plasticity. Polish Scientific Publishers, PWN and Nijhoff

2

Fundamental Concepts of Stress and Strain

2.1 Stress Components and Invariants

The mechanical behavior at a point of a solid can be represented by stress and strain components in three-dimensional space. Consider a generic point O of an elementary parallelepiped of a continuum referred to by orthogonal Cartesian axes x, y, z as shown in Fig.2.1. Each of the three faces in the reference planes is in general subjected to one normal stress and two shear stresses. The state of stress at O is thus characterized by these nine stress components.

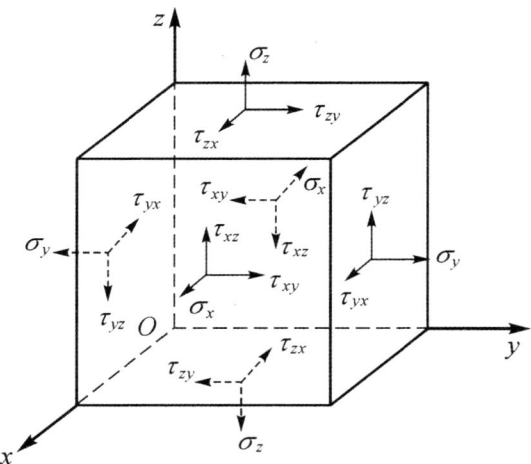

Fig. 2.1. Stress components

Rotational equilibrium of the parallelepiped about its axes yields equality of the shear stresses,

$$\tau_{xy} = \tau_{yx}, \tau_{yz} = \tau_{zy} \text{ and } \tau_{zx} = \tau_{xz}. \tag{2.1}$$

Thus there remain only six independent stress components, namely the normal stresses σ_x, σ_y, σ_z and the shear stresses τ_{xy}, τ_{yz}, τ_{zx}. Any stress tensor, which is usually denoted by $[\sigma]$, can be represented by these six components in the matrix form of

$$\sigma_{ij} = \begin{bmatrix} \sigma_x & \tau_{xy} & \tau_{xz} \\ \tau_{yx} & \sigma_y & \tau_{yz} \\ \tau_{zx} & \tau_{zy} & \sigma_z \end{bmatrix}, \tag{2.2a}$$

where x, y and z refer to orthogonal Cartesian axes.

It can be seen in the textbooks of the mechanics of materials, elasticity, mechanics of solids or plasticity that there exist three principal stresses σ_1, σ_2 and σ_3 and where the companion shear stresses are zero; and the state of stress at the element can be described as

$$\sigma_i = \begin{bmatrix} \sigma_1 & 0 & 0 \\ 0 & \sigma_2 & 0 \\ 0 & 0 & \sigma_3 \end{bmatrix}. \tag{2.2b}$$

An element of material subjected to principal stresses σ_1, σ_2 and σ_3 acting in mutually perpendicular directions is said to be in a state of triaxial stress or three-dimensional stress. If one of the principal stresses is equal to zero, it degrades to a plane stress state or a biaxial stress state. The triaxial or biaxial stress state is sometimes referred as polyaxial, multiaxial, or complex stress state. The principal planes are the planes on which the principal stresses occur. They are mutually perpendicular.

The principal stresses are the three roots of the equation of

$$\sigma^3 - (\sigma_x + \sigma_y + \sigma_z)\sigma^2 + (\sigma_x\sigma_y + \sigma_y\sigma_z + \sigma_z\sigma_x - \tau_{xy}^2 - \tau_{yz}^2 - \tau_{zx}^2)\sigma$$
$$\tag{2.3}$$
$$- (\sigma_x\sigma_y\sigma_z + 2\tau_{xy}\tau_{yz}\tau_{zx} - \sigma_x\tau_{yz}^2 - \sigma_y\tau_{zx}^2 - \sigma_z\tau_{xy}^2) = 0.$$

Eq.(2.3) can be rewritten in a simpler form of

$$\sigma^3 - I_1\sigma^2 + I_2\sigma - I_3 = 0, \tag{2.4}$$

where I_1, I_2 and I_3 are stress tensor invariants which have the following expressions:

$$I_1 = \sigma_x + \sigma_y + \sigma_z = \sigma_1 + \sigma_2 + \sigma_3, \tag{2.5a}$$

$$I_2 = \sigma_x \sigma_y + \sigma_y \sigma_z + \sigma_z \sigma_x - \tau_{xy}^2 - \tau_{yz}^2 - \tau_{zx}^2 = \sigma_1 \sigma_2 + \sigma_2 \sigma_3 + \sigma_3 \sigma_1, \tag{2.5b}$$

$$I_3 = \sigma_x \sigma_y \sigma_z + 2\tau_{xy}\tau_{yz}\tau_{zx} - \sigma_x \tau_{yz}^2 - \sigma_y \tau_{zx}^2 - \sigma_z \tau_{xy}^2 = \sigma_x \sigma_y \sigma_z = \sigma_1 \sigma_2 \sigma_3. \tag{2.5c}$$

The quantities I_1, I_2 and I_3 are independent of the direction of the axes chosen. The stresses σ_1, σ_2 and σ_3 are called principal stresses, and their directions principal directions. Each principal direction is orthogonal to the plane determined by the other two directions. The shear stresses associated with these stresses equal zero.

Plane stress states, when one principal stress vanishes, are frequently the case in practical engineering. Letting σ_3 be the vanishing principal stress, all the other stresses then lie in the plane (O, σ_1, σ_2) of the two other principal directions. Consider a plane surface element with the exterior normal coinciding with that of the (O, σ_1, σ_2) plane and the x-axis having an angle α (clockwise positive) with respect to the direction of the first principal stress. The principal stresses are related to the stress components σ_x, σ_y and τ_{xy} by

$$\sigma_1 = \frac{\sigma_x + \sigma_y}{2} + \sqrt{\frac{1}{4}(\sigma_x - \sigma_y)^2 + \tau_{xy}^2}, \tag{2.6a}$$

$$\sigma_2 = \frac{\sigma_x + \sigma_y}{2} - \sqrt{\frac{1}{4}(\sigma_x - \sigma_y)^2 + \tau_{xy}^2}, \tag{2.6b}$$

and the inclined angle α is given by

$$\tan 2\alpha = -\frac{2\tau_{xy}}{\sigma_x - \sigma_y}. \tag{2.7}$$

2.2 Deviatoric Stress Tensor and the Tensor Invariants

For convenient application in the study of strength theory and plasticity, the stress tensor is split into two parts; one is the deviatoric stress tensor S_{ij}, and the other the spherical stress tensor p_{ij}. There are

$$\sigma_{ij} = S_{ij} + p_{ij} = S_{ij} + \sigma_m \delta_{ij}. \tag{2.8}$$

The spherical stress tensor is the tensor whose components are $\sigma_m \delta_{ij}$, where σ_m is the mean stress, and

$$p_{ij} = \sigma_m \delta_{ij} = \sigma_m \begin{bmatrix} 1 & 0 & 0 \\ 0 & 1 & 0 \\ 0 & 0 & 1 \end{bmatrix} = \begin{bmatrix} \sigma_m & 0 & 0 \\ 0 & \sigma_m & 0 \\ 0 & 0 & \sigma_m \end{bmatrix}, \tag{2.9}$$

and

$$\sigma_m = (\sigma_x + \sigma_y + \sigma_z)/3 = (\sigma_1 + \sigma_2 + \sigma_3)/3 = I_1/3. \tag{2.10}$$

It is apparent that σ_m is not changed for all the orientations of the axes. Thus it is also referred as hydrostatic stress or spherical stress.

The deviatoric stress tensor S_{ij} can be written as

$$S_{ij} = \sigma_{ij} - p_{ij} = \sigma_{ij} - \sigma_m \delta_{ij}. \tag{2.11}$$

The invariants of the deviatoric stress tensor are denoted by J_1, J_2 and J_3, which have the form of

$$J_1 = S_1 + S_2 + S_3 = 0, \tag{2.12a}$$

$$J_2 = \frac{1}{2} S_{ij} S_{ij} = \frac{1}{6} \left[(\sigma_1 - \sigma_2)^2 + (\sigma_2 - \sigma_3)^2 + (\sigma_1 - \sigma_3)^2 \right], \tag{2.12b}$$

$$J_3 = |S_{ij}| = S_1 S_2 S_3, \tag{2.12c}$$

where S_1, S_2 and S_3 are three principal deviatoric stresses.

2.3 Principal Shear Stresses

The maximum shear stress acts on the plane bisecting the angle between the largest and the smallest principal stresses and it is equal to half of the difference between these principal stresses,

$$\tau_{\max} = \tau_{13} = \frac{1}{2}(\sigma_1 - \sigma_3). \tag{2.13}$$

It is also called the maximum principal shear stress. The shear stresses in the other two perpendicular planes are the intermediate or the minimum principal shear stresses. The three principal shear stresses τ_{13}, τ_{12} and τ_{23} can be obtained as

$$\tau_{13} = \frac{1}{2}(\sigma_1 - \sigma_3), \tau_{12} = \frac{1}{2}(\sigma_1 - \sigma_2), \tau_{23} = \frac{1}{2}(\sigma_2 - \sigma_3). \tag{2.14}$$

The companion normal stresses σ_{13}, σ_{12} and σ_{23} for the principal shear stresses have the form of

$$\sigma_{13} = \frac{1}{2}(\sigma_1 + \sigma_3), \sigma_{12} = \frac{1}{2}(\sigma_1 + \sigma_2), \sigma_{23} = \frac{1}{2}(\sigma_2 + \sigma_3). \tag{2.15}$$

The directions of the principal stresses and the principal shear stresses are illustrated schematically in Fig.2.2.

The three principal stresses, three principal shear stresses and three normal stresses acting on the principal shear stress plane can be illustrated by three stress circles referred to as the Mohr circles, as shown in Fig.2.3.

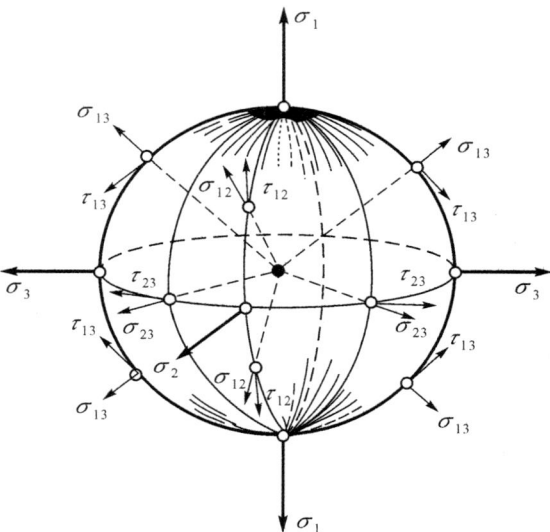

Fig. 2.2. Directions of the principal stresses and the principal shear stresses

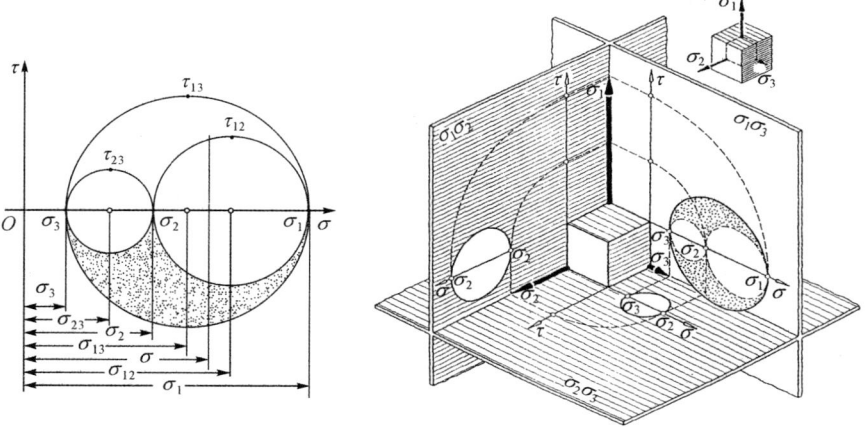

Fig. 2.3. Three principal stresses, three principal shear stresses and three normal stresses acting on the principal shear stress plane

The magnitudes of the normal and shear stresses in any plane are equal to the distance of the corresponding stress point on the stress circle to the origin of the referred coordinate frame. The three principal shear stresses are equal to the radii of three stress circles. Detailed descriptions of the stress circle can be found in Johnson and Mellor (1962), Kussmaul (1981), and Chakrabarty (1987) and other textbooks on plasticity or mechanics of materials.

2.4 Octahedral Shear Stress

If the normal of an oblique plane forms equal angles with the three principal axes, i.e.,

$$l = m = n = \pm\frac{1}{\sqrt{3}}, \qquad (2.16)$$

where l, m and n are the directional cosines of the principal planes, principal shear stress planes, and the octahedral plane respectively. The shear stresses acting on the octahedral plane are called octahedral shear stresses. The normal stress on this plane is accordingly referred as an octahedral normal stress σ_8 (or σ_{oct}) which equals the mean stress

$$\sigma_8 = \frac{1}{3}(\sigma_1 + \sigma_2 + \sigma_3) = \sigma_m. \qquad (2.17)$$

A tetrahedron similar to this one can be constructed in each of the four quadrants above/below the x-y plane. In the eight tetrahedral oblique planes the condition $l^2 = m^2 = n^2 = 1/3$ is satisfied. The tetrahedra can be differentiated from each other by the signs of l, m and n. Fig.2.4 shows various representative elements and the corresponding stress elements. The eight tetrahedral sections form octahedra as shown in Fig.2.4(e). The octahedral normal stress is given by Eq.(2.17), and the shear stress τ_8 (sometimes denoted as σ_{oct}) on the octahedral plane is

$$\begin{aligned}
\tau_8 &= \frac{1}{3}[(\sigma_1 - \sigma_2)^2 + (\sigma_2 - \sigma_3)^2 + (\sigma_3 - \sigma_1)^2]^{1/2} \\
&= \frac{1}{\sqrt{3}}[(\sigma_1 - \sigma_m)^2 + (\sigma_2 - \sigma_m)^2 + (\sigma_3 - \sigma_m)^2]^{1/2}.
\end{aligned} \qquad (2.18)$$

The directional cosines l, m and n of the principal planes, principal shear stress planes and octahedral plane, the normal stresses σ and shear stresses τ are listed in Table 2.1.

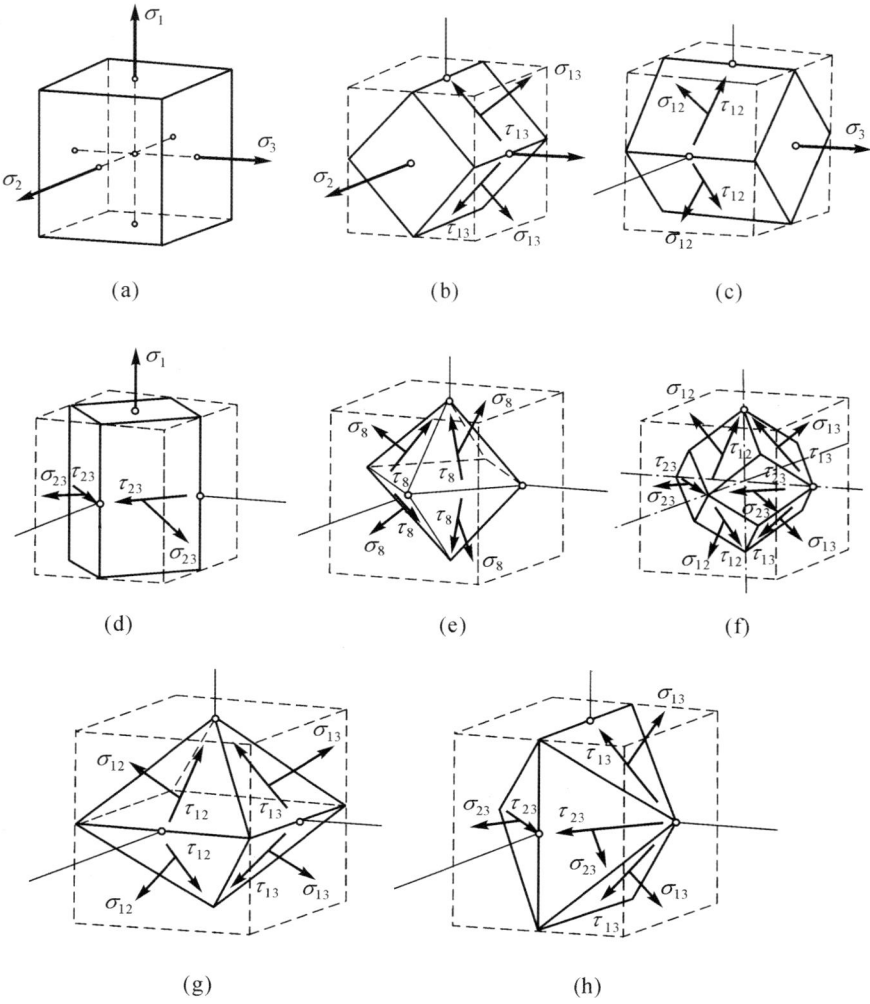

Fig. 2.4. Various polyhedral elements

2.5 Strain Components

When a continuum is deformed, a generic point experiences a displacement $\{U\}$ with components u, v, w with respect to Cartesian orthogonal axes x, y, z, respectively. For very small strains, the axial strains ε_x, ε_y, ε_z and shear strains γ_{xy}, γ_{yz}, γ_{zx} can be expressed by the displacement differentiation as follows:

Table 2.1. Directional cosines of the principal planes, the principal shear stress planes and the octahedral plane

	Principal planes			Principal shear stress planes			Octa. plane
$l =$	± 1	0	0	$\pm\frac{1}{\sqrt{2}}$	$\pm\frac{1}{\sqrt{2}}$	0	$\frac{1}{\sqrt{3}}$
$m =$	0	± 1	0	$\pm\frac{1}{\sqrt{2}}$	0	$\pm\frac{1}{\sqrt{2}}$	$\frac{1}{\sqrt{3}}$
$n =$	0	0	± 1	0	$\pm\frac{1}{\sqrt{2}}$	$\pm\frac{1}{\sqrt{2}}$	$\frac{1}{\sqrt{3}}$
$\sigma =$	σ_1	σ_2	σ_3	$\sigma_{12} = \frac{\sigma_1+\sigma_2}{2}$	$\sigma_{13} = \frac{\sigma_1+\sigma_3}{2}$	$\sigma_{23} = \frac{\sigma_2+\sigma_3}{2}$	σ_8
$\tau =$	0	0	0	$\tau_{12} = \frac{\sigma_1-\sigma_2}{2}$	$\tau_{13} = \frac{\sigma_1-\sigma_3}{2}$	$\tau_{23} = \frac{\sigma_2-\sigma_3}{2}$	τ_8

$$\varepsilon_x = \frac{\partial u}{\partial x}, \quad \varepsilon_y = \frac{\partial v}{\partial y}, \quad \varepsilon_z = \frac{\partial w}{\partial z}, \tag{2.19}$$

$$\gamma_{xy} = \frac{\partial u}{\partial y} + \frac{\partial v}{\partial x}, \gamma_{yz} = \frac{\partial v}{\partial z} + \frac{\partial w}{\partial y}, \gamma_{xz} = \frac{\partial u}{\partial z} + \frac{\partial w}{\partial x}. \tag{2.20}$$

Six strain components ε_x, ε_y, ε_z, γ_{xy}, γ_{yz}, γ_{zx} can describe completely the state of strain at the considered point. Similar to the stress tensor, there also exist principal strains ε_1, ε_2, ε_3 with companion shear strains equal to zero. For a plane strain state, i.e., the third principal strain ε_3 vanishes, the principal strains can be expressed as follows:

$$\varepsilon_1 = \frac{\varepsilon_x + \varepsilon_y}{2} + \sqrt{\frac{1}{4}(\varepsilon_x - \varepsilon_y)^2 + (\gamma_{xy}/2)^2},$$
$$\varepsilon_2 = \frac{\varepsilon_x + \varepsilon_y}{2} - \sqrt{\frac{1}{4}(\varepsilon_x - \varepsilon_y)^2 + (\gamma_{xy}/2)^2}. \tag{2.21}$$

The principal direction is given by

$$\tan 2\alpha = -\frac{\gamma_{xy}}{\varepsilon_x - \varepsilon_y}. \tag{2.22}$$

Eq.(2.22) still holds when $\varepsilon_z \equiv \varepsilon_3 \neq 0$, provided ε_z is a principal strain.

2.6 Equations of Equilibrium

The following three differential equations of equilibrium in the directions of the coordinate axes are

$$\frac{\partial \sigma_x}{\partial x} + \frac{\partial \tau_{xy}}{\partial y} + \frac{\partial \tau_{xz}}{\partial z} + X = 0,$$

$$\frac{\partial \tau_{xy}}{\partial x} + \frac{\partial \sigma_y}{\partial y} + \frac{\partial \tau_{yz}}{\partial z} + Y = 0, \qquad (2.23)$$

$$\frac{\partial \tau_{xz}}{\partial x} + \frac{\partial \tau_{yz}}{\partial y} + \frac{\partial \sigma_z}{\partial z} + Z = 0,$$

where X, Y, Z are the components of the body force per unit volume. For a body in the equilibrium state, the variation in stresses is governed by the above equations of equilibrium.

2.7 Generalized Hooke's Law

Equations relating stress, strain, stress rate (increase of stress per unit time) and strain rate are called the constitutive equations, which are determined by the material properties under consideration. In the case of elastic solids the constitutive equations take the form of the generalized Hooke's law that involves stress and strain instead of the stress rate and strain rate.

In a general three-dimensional stress state, the generalized Hooke's law has the form of

$$\varepsilon_x = \frac{1}{E}[\sigma_x - \nu(\sigma_y + \sigma_z)], \qquad (2.24a)$$

$$\varepsilon_y = \frac{1}{E}[\sigma_y - \nu(\sigma_x + \sigma_z)], \qquad (2.24b)$$

$$\varepsilon_z = \frac{1}{E}[\sigma_z - \nu(\sigma_x + \sigma_y)], \qquad (2.24c)$$

$$\gamma_{xy} = \frac{1}{G}\tau_{xy}, \ \gamma_{yz} = \frac{1}{G}\tau_{yz}, \ \gamma_{xz} = \frac{1}{G}\tau_{xz}, \qquad (2.24d)$$

where E and ν are the modulus of elasticity and the Poisson's ratio, respectively; G is the modulus of rigidity. Only two of them are independent, and there is

$$G = \frac{E}{2(1+\nu)}.$$

Eqs.(2.24a) $\sim$ (2.24d) may be rewritten conversely,

$$\sigma_x = 2G\varepsilon_x + \lambda(\varepsilon_x + \varepsilon_y + \varepsilon_z),$$

$$\sigma_y = 2G\varepsilon_y + \lambda(\varepsilon_x + \varepsilon_y + \varepsilon_z),$$

$$\sigma_z = 2G\varepsilon_z + \lambda(\varepsilon_x + \varepsilon_y + \varepsilon_z), \tag{2.25}$$

$$\tau_{xy} = G\gamma_{xy}, \ \tau_{yz} = G\gamma_{yz}, \ \tau_{zx} = G\gamma_{zx},$$

where the constants G and λ are called Lame's constants,

$$\lambda = \frac{\nu E}{(1+\nu)(1-2\nu)}. \tag{2.26}$$

Another important elastic constant is called the bulk modulus of elasticity K, which defines the dilatation (volumetric strain) ε_v as the unit change in volume,

$$\varepsilon_v = \varepsilon_x + \varepsilon_y + \varepsilon_z, \tag{2.27}$$

with the hydrostatic component of stress, or spherical component of stress σ_m,

$$\sigma_m = \frac{1}{3}(\sigma_x + \sigma_y + \sigma_z), \tag{2.28}$$

such that

$$\varepsilon_v = \frac{1}{K}\sigma_m. \tag{2.29}$$

From the generalized Hooke's law, K is derived as

$$K = \frac{E}{[3(1-2\nu)]}. \tag{2.30}$$

2.8 Compatibility Equations

Eqs.(2.19) and (2.20) implicitly show that the strain components are functions of the three displacement components. Differentiate the first equation of Eq.(2.19) twice with respect to y and the second equation of Eq.(2.19) twice with respect to x and add the results,

$$\frac{\partial^2 \varepsilon_x}{\partial y^2} + \frac{\partial^2 \varepsilon_y}{\partial x^2} = \frac{\partial^3 u}{\partial y^2 \partial x} + \frac{\partial^3 v}{\partial x^2 \partial y}. \tag{2.31}$$

Differentiating the first equation of Eq.(2.20) with respect to x and y yields

$$\frac{\partial^2 \gamma_{xy}}{\partial x \partial y} = \frac{\partial^2}{\partial x \partial y}\left(\frac{\partial u}{\partial y} + \frac{\partial v}{\partial x}\right). \tag{2.32}$$

And since the order of differentiation for single-value, continuous functions is immaterial, thus,

$$\frac{\partial^2 \varepsilon_x}{\partial y^2} + \frac{\partial^2 \varepsilon_y}{\partial x^2} = \frac{\partial^2 \gamma_{xy}}{\partial x \partial y}.$$

Similarly, we can derive the following additional equations:

$$\frac{\partial^2 \varepsilon_x}{\partial y^2} + \frac{\partial^2 \varepsilon_y}{\partial x^2} = \frac{\partial^2 \gamma_{xy}}{\partial x \partial y}, \tag{2.33a}$$

$$\frac{\partial^2 \varepsilon_y}{\partial z^2} + \frac{\partial^2 \varepsilon_z}{\partial y^2} = \frac{\partial^2 \gamma_{yz}}{\partial y \partial z}, \tag{2.33b}$$

$$\frac{\partial^2 \varepsilon_z}{\partial x^2} + \frac{\partial^2 \varepsilon_x}{\partial z^2} = \frac{\partial^2 \gamma_{zx}}{\partial z \partial x}, \tag{2.33c}$$

$$2\frac{\partial^2 \varepsilon_x}{\partial y \partial z} = \frac{\partial}{\partial x}\left(-\frac{\partial \gamma_{yz}}{\partial x} + \frac{\partial \gamma_{xz}}{\partial y} + \frac{\partial \gamma_{xy}}{\partial z}\right), \tag{2.33d}$$

$$2\frac{\partial^2 \varepsilon_y}{\partial z \partial x} = \frac{\partial}{\partial y}\left(\frac{\partial \gamma_{yz}}{\partial x} - \frac{\partial \gamma_{xz}}{\partial y} + \frac{\partial \gamma_{xy}}{\partial z}\right), \tag{2.33e}$$

$$2\frac{\partial^2 \varepsilon_z}{\partial x \partial y} = \frac{\partial}{\partial z}\left(\frac{\partial \gamma_{yz}}{\partial x} + \frac{\partial \gamma_{xz}}{\partial y} - \frac{\partial \gamma_{xy}}{\partial z}\right). \tag{2.33f}$$

Eqs.(2.33a) $\sim$ (2.33f) are called Saint-Venant compatibility equations, or compatibility equations in terms of strain.

In total there are fifteen governing equations, including three equilibrium equations (Eq.(2.23)), six strain displacement relations (Eq.(2.19) and Eq.(2.20)), and six stress-strain relations (Eq.(2.24)) to solve the fifteen variables (six stress components σ_x, σ_y, σ_z, τ_{xy}, τ_{yz} and τ_{xz}, six strain components ε_x, ε_y, ε_z, γ_{xy}, γ_{yz} and γ_{xz}, and three displacements u, v, w). The compatibility equations are derived from the strain-displacement equations and therefore cannot be counted as governing equations. The compatibility equations will be satisfied automatically if the fifteen governing equations are satisfied.

2.9 Governing Equations for Plane Stress Problems

For plane stress problems the stress components are simplified as

$$\sigma_x = \sigma_x(x, y), \ \sigma_y = \sigma_y(x, y), \ \tau_{xy} = \tau_{xy}(x, y), \tag{2.34a}$$

$$\tau_{xz} = \tau_{yz} = \sigma_z = 0. \tag{2.34b}$$

The equilibrium equations become

$$\frac{\partial \sigma_x}{\partial x} + \frac{\partial \tau_{xy}}{\partial y} + X = 0, \tag{2.35a}$$

$$\frac{\partial \tau_{xy}}{\partial x} + \frac{\partial \sigma_y}{\partial y} + Y = 0, \tag{2.35b}$$

where the body forces X and Y are functions of x and y only, and Z equals zero. The strain-stress relations take the form of

$$\varepsilon_x = \varepsilon_x(x, y) = \frac{1}{E}[\sigma_x - \nu\sigma_y], \tag{2.36a}$$

$$\varepsilon_y = \varepsilon_y(x, y) = \frac{1}{E}[\sigma_y - \nu\sigma_x], \tag{2.36b}$$

$$\varepsilon_z = \varepsilon_z(x, y) = \frac{1}{E}[-\nu(\sigma_x + \sigma_y)], \tag{2.36c}$$

$$\gamma_{xy} = \gamma_{xy}(x, y) = \frac{1}{G}\tau_{xy}. \tag{2.36d}$$

The two shear strains γ_{xz} and γ_{yz} and the normal strain ε_z vanish. Finally the strain-displacement relations are simplified as

$$\varepsilon_x = \frac{\partial u}{\partial x}, \quad \varepsilon_y = \frac{\partial v}{\partial y}, \quad \gamma_{xy} = \frac{\partial u}{\partial y} + \frac{\partial v}{\partial x}. \tag{2.37}$$

There are eight equations in total to correlate the eight unknown quantities of σ_x, σ_y, τ_{xy}, ε_x, ε_y, γ_{xy}, u and v. Again, the governing equations can only be solved with specific stress and displacement boundary conditions.

2.10 Governing Equations in Polar Coordinates

For analysis of circular ring and plate, rotating disk, curved bar of narrow rectangular cross section with a circular axis, etc., it is advantageous to use polar coordinates. If the external forces are also rotationally symmetric, the stress state can be assumed to be the plane stress independent of the z-axis which is perpendicular to the polar coordinates plane. The position of a point in the middle plane of a plate is then defined by the distance from the origin O and the angle θ between r and a certain axis O_x fixed in the plane. Denoting σ_r and σ_θ as the normal stress component in the radial and circumferential directions respectively, and $\tau_{r\theta}$ the shear stress component, the equation of equilibrium takes the form of

$$\frac{\partial \sigma_r}{\partial r} + \frac{1}{r}\frac{\partial \tau_{r\theta}}{\partial \theta} + \frac{\sigma_r - \sigma_\theta}{r} + R = 0, \tag{2.38a}$$

$$\frac{1}{r}\frac{\partial \sigma_\theta}{\partial \theta} + \frac{\partial \tau_{r\theta}}{\partial r} + \frac{2\tau_{r\theta}}{r} + S = 0, \tag{2.38b}$$

where R and S are the components of body force per unit volume in the radial and tangential directions, respectively.

The corresponding stress components are derived as

$$\sigma_r = \frac{A}{r^2} + B(1 + 2\log r) + 2C, \tag{2.39a}$$

$$\sigma_\theta = -\frac{A}{r^2} + B(3 + 2\log r) + 2C, \tag{2.39b}$$

$$\tau_{r\theta} = 0, \tag{2.39c}$$

where A, B and C are constant that can be determined by boundary conditions.

Denoting the displacements in the radial and tangential directions as u_r and u_θ respectively, the strain components in the polar coordinates are derived as

$$\varepsilon_r = \frac{\partial u_r}{\partial r}, \ \varepsilon_\theta = \frac{u_r}{r} + \frac{1}{r}\frac{\partial u_\theta}{\partial \theta} \ \text{ and } \ \gamma_{r\theta} = \frac{1}{r}\frac{\partial u_r}{\partial \theta} + \frac{\partial u_\theta}{\partial r} - \frac{u_\theta}{r}. \tag{2.40}$$

The generalized Hooke's law is then expressed by

$$\varepsilon_r = \frac{1}{E}(\sigma_r - \nu\sigma_\theta), \tag{2.41a}$$

$$\varepsilon_\theta = \frac{1}{E}(\sigma_\theta - \nu\sigma_r), \tag{2.41b}$$

$$\gamma_{r\theta} = \frac{1}{G}\tau_{r\theta}. \tag{2.41c}$$

Thus, based on the equilibrium equations, strain-displacement relations, compatibility equations, and Hooke's law plus relative boundary conditions, the stress and displacement fields of the rotational symmetrical body can be solved. Detailed derivations can be referred to *Theory of Elasticity* by Timoshenko and Goodier (1970) and *Elasticity: Tensor, Dynamic and Engineering Approaches* by Chou and Pagano (1967).

2.11 Bending of Circular Plate

It is assumed that a circular plate is perfectly elastic, isotropic and homogeneous, and is subjected to a variable symmetrical transverse load. The plate is initially flat and of uniform thickness. Maximum deflection is relatively small with respect to thickness (no more than half the thickness). Deformation of the plate is symmetrical about the cylindrical axis. During deformation the straight lines in the plate initially parallel to the cylindrical axis remain straight but become inclined. All forces, loads and reactions are parallel to the cylindrical axis. Shear effect on bending is negligible.

The pertinent strain equations according to Hooke's law for plane stress and the geometrical relations are (Griffel, 1968)

$$\varepsilon_r = \frac{\sigma_r}{E} - \nu\frac{\sigma_\theta}{E} = y\frac{d\phi}{dr}, \quad \varepsilon_\theta = \frac{\sigma_\theta}{E} - \nu\frac{\sigma_r}{E} = y\frac{\phi}{r}, \tag{2.42}$$

where

$$\phi \approx -\frac{dw}{dr}$$

and w is the transverse deflection of the plate; y is the distance from the mid-plane to the considered point in the cross section of the plate. Solving Eq.(2.42) for the radial and tangential unit stresses,

$$\sigma_r = \frac{Ey}{1 - \nu^2}\left(\frac{d\phi}{dr} + \nu\frac{\phi}{r}\right), \tag{2.43a}$$

$$\sigma_\theta = \frac{Ey}{1 - \nu^2}\left(\frac{\phi}{r} + \nu\frac{d\phi}{dr}\right). \tag{2.43b}$$

Assuming that unit stresses are proportional to the distance from the mid-plane, the radial and tangential bending moments can be written as

$$M_r = \int_{-h/2}^{h/2} \sigma_r y dA = D\left[\frac{d\phi}{dr} + \nu\frac{\phi}{r}\right], \tag{2.44a}$$

$$M_\theta = \int_{-h/2}^{h/2} \sigma_\theta y dA = D\left[\frac{\phi}{r} + \nu\frac{d\phi}{dr}\right], \tag{2.44b}$$

where

$$D = \frac{EI}{1 - \nu^2} = \frac{Eh^3}{12(1 - \nu^2)}. \tag{2.45}$$

The equilibrium equation is then expressed as

$$\frac{dM_r}{dr} + \frac{M_r - M_\theta}{r} = -\int p(r)dr, \tag{2.46}$$

where $p(r)$ is the transverse pressure symmetrically distributed on the plate. The equilibrium equation in terms of the bending angle and radius is the derivative of Eq.(2.42a). Substituting Eq.(2.42a) and Eqs.(2.44a) and (2.42b) into Eq.(2.46),

$$\frac{\mathrm{d}^2\phi}{\mathrm{d}r^2} + \frac{1}{r}\frac{\mathrm{d}\phi}{\mathrm{d}r} - \frac{\phi}{r^2} = \frac{\mathrm{d}}{\mathrm{d}r}\left[\frac{1}{r}\frac{\mathrm{d}}{\mathrm{d}r}(r\phi)\right] = -\frac{1}{D}\int p(r)\mathrm{d}r. \qquad (2.47)$$

Integrating Eq.(2.47), we derive

$$\frac{\mathrm{d}}{\mathrm{d}\phi}(r\phi) = -\frac{r}{D}\int\left(\int p(r)\mathrm{d}r\right)\mathrm{d}r + C_1 r. \qquad (2.48)$$

Integrating Eq.(2.47) once again leads to

$$-\frac{\mathrm{d}w}{\mathrm{d}r} \approx \phi = -\frac{1}{rD}\int r\left[\int\left(\int p(r)\mathrm{d}r\right)\mathrm{d}r\right]\mathrm{d}r + \frac{1}{2}C_1 r + \frac{1}{r}C_2. \qquad (2.49)$$

Thus

$$w = \frac{1}{D}\int\frac{1}{r}\left\{\int r\left[\int\left(\int p(r)\mathrm{d}r\right)\mathrm{d}r\right]\mathrm{d}r\right\}\mathrm{d}r - \frac{C_1}{4}r^2 - C_2\ln r + C_3. \qquad (2.50)$$

The moments M_r and M_θ are then calculated by

$$M_r = -\int\left(\int p(r)\mathrm{d}r\right)\mathrm{d}r - (1-\nu)\frac{1}{r^2}\int r\left[\int\left(\int p(r)\mathrm{d}r\right)\mathrm{d}r\right]\mathrm{d}r \qquad (2.51a)$$
$$+ \frac{C_1 D}{2}(1+\nu) - \frac{C_2 D}{r^2}(1-\nu),$$

$$M_\theta = -(1-\nu)\frac{1}{r^2}\int r\left[\int\left(\int p(r)\mathrm{d}r\right)\mathrm{d}r\right]\mathrm{d}r - \nu\int\left(\int p(r)\mathrm{d}r\right)\mathrm{d}r \qquad (2.51b)$$
$$+ \frac{C_1 D}{2}(1+\nu) + \frac{C_2 D}{r^2}(1-\nu).$$

For a special case when $p = 0$ which corresponds to an unloaded part of the plate, the moments M_r and M_θ are derived as

$$M_r = \frac{C_1 D}{2}(1+\nu) - \frac{C_2 D}{r^2}(1-\nu), \qquad (2.52a)$$

$$M_\theta = \frac{C_1 D}{2}(1+\nu) + \frac{C_2 D}{r^2}(1-\nu). \qquad (2.52b)$$

When a uniformly distributed pressure is applied to the plate, the moment M_r and M_θ become

$$M_r = -\frac{5-\nu}{8}pr^2 + \frac{C_1 D}{2}(1+\nu) - \frac{C_2 D}{r^2}(1-\nu), \tag{2.53a}$$

$$M_\theta = -\frac{1+5\nu}{8}pr^2 + \frac{C_1 D}{2}(1+\nu) + \frac{C_2 D}{r^2}(1-\nu). \tag{2.53b}$$

The integral constants C_1, C_2 and C_3 are determined by boundary conditions. The derived equations can be applied to either a solid plate or an annular plate with axially symmetrical loads. It should be noted that the above equations are valid with respect to elastic deformations.

2.12 Summary

This chapter presents the fundamentals of solid mechanics. Some basic concepts with respect to the stress tensors, stress tensor invariants, deviatory stress tensors, deviatory stress tensor invariants, octahedral shear and normal stresses, principal stresses and principal shear stresses, strain and strain-rate components are introduced. Governing equations for general stress state solids, plane stress solids, rotationally symmetrical solids, and rotationally symmetrical plates are given.

It should be mentioned that only the governing equations in the elastic range of solids are considered. Based on the elastic solutions, by adopting a proper yield criterion the elastic limit load of the solid body can be derived. For elasto-plastic analysis and plastic limit analysis, a yield criterion and a relevant flow law should be applied. The following chapter will introduce conventional yield criteria and a unified strength criterion developed by Yu (1991; 1992; 2004).

References

Chakrabarty J (1987) Theory of plasticity. McGraw-Hill, New York

Chou PC, Pagano NJ (1967) Elasticity: tensor, dynamic, and engineering approaches. Van Nostrand, Princeton, New Jersey

Griffel W (1968) Plate formulas. Frederick Ungar Publishing Co., New York

Johnson W, Mellor PB (1962) Plasticity for mechanical engineers. D. Van Nostrand Co., London

Kussmaul K (1981) Festigkeitslehre I. MPA Stuttgart, University Stuttgart

Timoshenko SP, Goodier JN (1970) Theory of elasticity. McGraw-Hill, New York

Yu MH (1992) A new system of strength theory. Xi'an Jiaotong University Press, Xi'an (in Chinese)

Yu MH (2004) Unified strength theory and its applications. Springer, Berlin

Yu MH, He LN (1991) A new model and theory on yield and failure of materials under complex stress state. In: Mechanical Behavior of Materials-6, Pergamon Press, Oxford, 3:841-846

3

Yield Condition

3.1 Introduction

A yield condition or yield criterion describes a material failure in structural plasticity. It defines the threshold state of a material between elastic and plastic or brittle failure deformations. It is very important to adopt a proper yield criterion in the design of a structure. The estimated load-bearing capacity of structures may be significantly affected by the choice of different yield criteria. Great efforts have been devoted to the formulation of yield criteria. Many different yield criteria have been proposed during the past 100 years (Pisarenko and Lebedev, 1976; Zyczkowski, 1981; Chen, 1982; 1998; Yu, 2002b; 2004). Amongst them, the Tresca criterion (Tresca, 1864), the Huber-von Mises criterion (Huber, 1904; von Mises, 1913), and the twin-shear yield criterion (Yu, 1961; 1983) are three representative criteria which can be used for materials that have identical strength in tension and compression (non-SD materials), and the shear strengths of $\tau_Y=0.5\sigma_Y$, $\tau_y=0.577\sigma_y$ and $\tau_y=0.667\sigma_y$, respectively, where σ_y is the uniaxial yield strength. The Drucker-Prager criterion is an extension of the Huber-von Mises criterion. Although the Drucker-Prager criterion has been widely applied for non-metallic materials, it contradicts some experimental results for geomaterials. The Mohr-Coulomb strength criterion (Mohr, 1900) and the twin-shear strength criterion (Yu, 1985) give the lower and upper bounds of the convex yield curves. Generally, a yield criterion is suitable for a certain type of material only. Thus it is of significance to develop different yield criteria for different materials.

A new strength theory, which gives a new system of yield criteria, was proposed by Yu (Yu and He, 1991; Yu, 1992; 2004). It consists of a unified yield criterion for metallic materials and a unified strength theory for SD materials (strength difference in tension and compression). It encompasses the most commonly used yield criteria and a series of new yield criteria as special cases or linear approximation. The conventional yield criteria, the unified yield criterion and unified strength theory will be introduced in this chapter. A detailed description of the unified strength theory can be found in the book entitled *Unified Strength Theory and Its Applications* published by Springer in 2004.

3.2 Conventional Yield Criteria

3.2.1 Maximum Normal Stress Criterion

When a stress state is uniaxial tension or compression, the yield condition for most metals is

$$\sigma = \pm\sigma_Y, \tag{3.1}$$

where σ_Y is the uniaxial material strength. This criterion is also suitable for brittle materials whose failure is characterized by the sudden breakdown of stress with an increase in deformation. It is also sometimes used as a cut-off criterion for geomaterial in tensile stress state.

3.2.2 Maximum Shear Stress-based Criteria — Single-shear Theory

3.2.2.1 Tresca Criterion for Non-SD Materials

In a multiaxial stress state, material yields when a stress state satisfies a certain condition. The yield condition should satisfy Eq.(3.1) when the stress state is in uniaxial tension or compression. For metals, particularly mild steel, it has been observed that the plastic deformations largely consist of the slip in crystals. Hence, it is assumed that the maximum shear stress governs material yielding. In other words, when the maximum shear stress reaches a specific value and above, material will yield. This yield condition is the well-known Tresca yield criterion.

The mathematical modeling of the Tresca criterion has a very simple form of

$$\tau_{\max} = C. \tag{3.2}$$

It can be represented by the maximum and minimum principal stresses

$$\tau_{\max} = \tau_{13} = \frac{\sigma_1 - \sigma_3}{2} = C. \tag{3.3}$$

In the uniaxial tensile stress state, the yield shear stress is written as

$$\tau_{\max,y} = \frac{\sigma_Y - 0}{2} = \frac{\sigma_Y}{2}, \tag{3.4}$$

where σ_Y is the tensile stress at yield or the tensile strength. By combining

Eq.(3.2) with Eq.(3.4), the maximum shear stress criterion or the Tresca criterion is given by

$$\sigma_1 - \sigma_3 = \sigma_Y. \tag{3.5}$$

In the case of in-plane bending of beams, the stress state is defined by $\sigma_x = \sigma$, $\sigma_y = 0$, and $\tau_{xy} = \tau$, we have

$$\tau_{\max} = \frac{1}{2}\sqrt{\sigma^2 + 4\tau^2}, \tag{3.6}$$

and the Tresca yield criterion becomes

$$\sigma^2 + 4\tau^2 = \sigma_Y^2. \tag{3.7}$$

In plane orthogonal Cartesian coordinates $(O\sigma,\ O\tau)$, Eq.(3.7) can be represented by an ellipse.

The yield surface of the Tresca criterion in the three-dimensional stress space is shown in Fig.3.1. Its axis is equally inclined with respect to the coordinate axes. When one of the principal stresses vanishes, e.g., σ_3 equals 0, the yield surface degenerates to be a hexagon obtained by intersecting the prism with the plane of $\sigma_3 = 0$ (Fig.3.1). Then the yield condition becomes

$$\max\left[|\sigma_1|, |\sigma_2|, |\sigma_1 - \sigma_2|\right] = \sigma_Y. \tag{3.8}$$

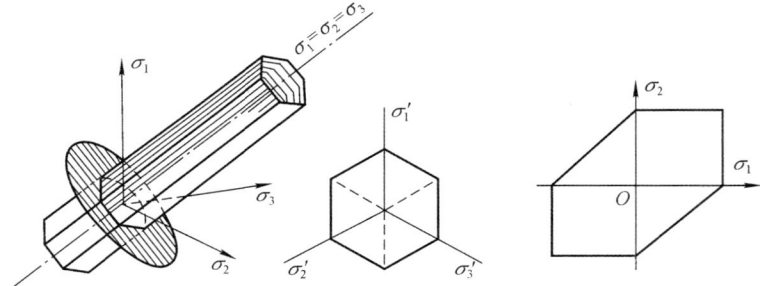

Fig. 3.1. Yield surface and yield loci in deviatoric plane and plane stress state of Tresca criterion

It is worth noting that the magnitude of the maximum shear stress is actually half of the (algebraic) difference of the two extreme principal stresses. Therefore the intermediate principal stress plays no role in the Tresca yield criterion. The Tresca yield criterion assumes that the uniaxial tensile and compressive strengths of the material are the same, thus it can only be used for non-SD materials.

3.2.2.2 Mohr-Coulomb Criterion for SD Materials

The Mohr-Coulomb criterion is an extension of the Tresca criterion. The mathematical modelling of the Mohr-Coulomb strength theory is expressed by

$$\tau_{13} + \beta\sigma_{13} = C. \tag{3.9}$$

The mathematical expression is

$$\sigma_1 - \alpha\sigma_3 = \sigma_t, \tag{3.10}$$

where β, C, α, and σ_t are material parameters, σ_t is the uniaxial tensile strength, α is the ratio of uniaxial tensile strength to compressive strength of the material, i.e. $\alpha = \sigma_t/\sigma_c$, where σ_c is the uniaxial compressive strength. In the uniaxial stress state, the Mohr-Coulomb criterion becomes the Tresca criterion when the material parameter β equals zero or the strength ratio α equals 1. In other words, the Mohr-Coulomb criterion can take into account the effect of the strength difference in tension and compression of materials (SD effect) by the introduction of extra material parameters into the Tresca criterion.

The yield surface of the Mohr-Coulomb criterion and the projections on the deviatoric plane and the yield loci in plane stress state are illustrated in Fig.3.2. The yield surface is pressure dependent from Fig.3.2.

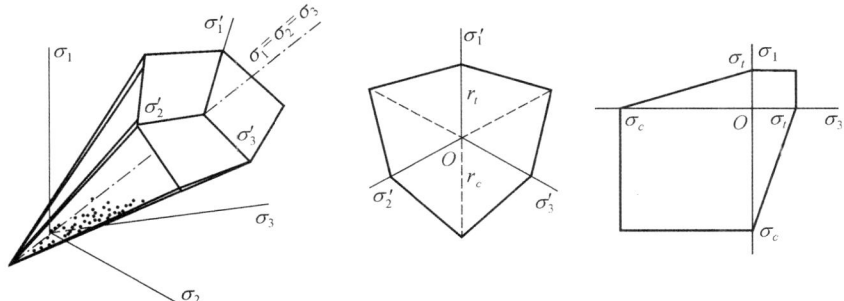

Fig. 3.2. Yield surface and yield loci in deviatoric plane and plane stress of Mohr-Coulomb criterion

The Mohr-Coulomb criterion has been widely used for non-metallic materials, especially for geomaterials such as rock, concrete, soil, etc., whose strength difference in tension and compression may not be ignored. And again the inefficiency of the Mohr-Coulomb criterion is that the effect of intermediate principal stress σ_2 on the material strength is ignored.

For geomaterials, the Mohr-Coulomb criterion sometimes takes the form of

$$\tau_f = c + \sigma_{nf} \tan \varphi, \tag{3.11}$$

where c and φ are the cohesion and the friction angle of the material, respectively. τ_f and σ_{nf} are the shear strength and the corresponding normal stress at failure, respectively. Eq.(3.11) is identical to Eq.(3.8) when the failure plane is parallel to the maximum shear stress.

There are many other extensions of the Tresca yield criterion. Among them, the single-shear strength theory assumes that the material yield is mainly dominated by the maximum shear stress and its associated normal stress.

3.2.2.3 The Mechanical Model of Mohr-Coulomb Criterion

The Mohr-Coulomb criterion can be introduced by using the mechanical model and mathematical modelling. It is convenient and clear. The key is to propose a reasonable model. A single-shear model (Yu, 1988) is proposed to introduce the Tresca criterion and the Mohr-Coulomb criterion as shown in Fig.3.3.

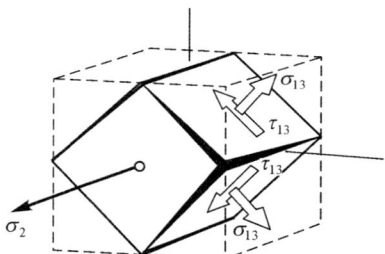

Fig. 3.3. Single-shear model of the Mohr-Coulomb criterion

The mathematical modelling and mathematical expressions of the Tresca criterion and the Mohr-Coulomb criterion can be introduced from the single-shear model. It is also seen from the single-shear model that the intermediate principal stress is not taken into account in the formulae of the Tresca criterion and the Mohr-Coulomb criterion.

3.2.3 Octahedral Shear Stress-based Criteria—Three-shear Theory

3.2.3.1 Huber-von Mises Criterion for Non-SD Materials

The Huber-von Mises criterion was proposed by Huber (1904), Hencky (1925), and von Mises (1913). It can also be introduced by using a regular octahedral model (Ros and Eichinger, 1926; Nadai, 1931), as shown in Fig.3.4.

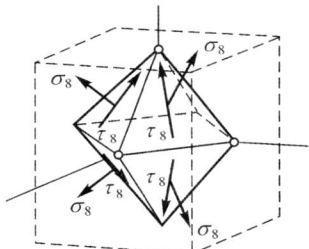

Fig. 3.4. Regular octahedral model

The octahedral shear stress criterion assumes that the material yield is governed by the octahedral shear stress. The mathematical modelling formula for the yield condition is

$$\tau_8 = C, \tag{3.12}$$

where τ_8 is the octahedral shear stress, and

$$\tau_8 = \frac{1}{3}\sqrt{(\sigma_1 - \sigma_3)^2 + (\sigma_1 - \sigma_2)^2 + (\sigma_2 - \sigma_3)^2}. \tag{3.13}$$

σ_1, σ_2, and σ_3 are the principal stresses and the yield condition can be rewritten in terms of the principal stresses as

$$\sigma_1^2 + \sigma_2^2 + \sigma_3^2 - \sigma_1\sigma_2 - \sigma_2\sigma_3 - \sigma_3\sigma_1 = \sigma_Y^2. \tag{3.14}$$

The yield surface is symmetrical regarding the three principal stresses. It implies that the effect of the three principal stresses on the material yielding is identical. When the stress components σ_x, σ_y, σ_z, τ_{xy}, τ_{yz} and τ_{xz} of the stress tensor are used, Eq.(3.14) becomes

$$\sigma_x^2 + \sigma_y^2 + \sigma_z^2 - \sigma_x\sigma_y - \sigma_y\sigma_z - \sigma_z\sigma_x + 3\tau_{xy}^2 + 3\tau_{yz}^2 + 3\tau_{zx}^2 = \sigma_Y^2. \tag{3.15}$$

The Huber-von Mises criterion is sometimes referred to as the maximum distortion strain energy criterion because the distortion strain energy is proportional to the square of the octahedral stress.

Eq.(3.14) can also be represented by the three principal shear stresses, i.e., $\tau_{13} = (\sigma_1 - \sigma_3)/2$, $\tau_{12} = (\sigma_1 - \sigma_2)/2$ and $\tau_{23} = (\sigma_2 - \sigma_3)/2$ in the form of

$$\tau_{13}^2 + \tau_{12}^2 + \tau_{23}^2 = C, \tag{3.16}$$

where C is a material constant. It is seen from Eq.(3.16) that the Huber-von Mises criterion includes the effects of the three principal shear stresses. It is thus categorized as a three-shear stress criterion (Yu, 1988; 2002).

The yield surface and yield loci in the deviatoric plane and plane stress state of the Huber-von Mises criterion are illustrated in Fig.3.5. In a plane stress state ($\sigma_3 = 0$), Eq.(3.14) can be rewritten as an equation of ellipse

$$\sigma_1^2 + \sigma_2^2 - \sigma_1\sigma_2 = \sigma_Y^2, \tag{3.17}$$

and the yield locus on the σ_1-σ_2 plane is shown in Fig.3.5, where the dotted line of the single-shear theory is also given for comparison.

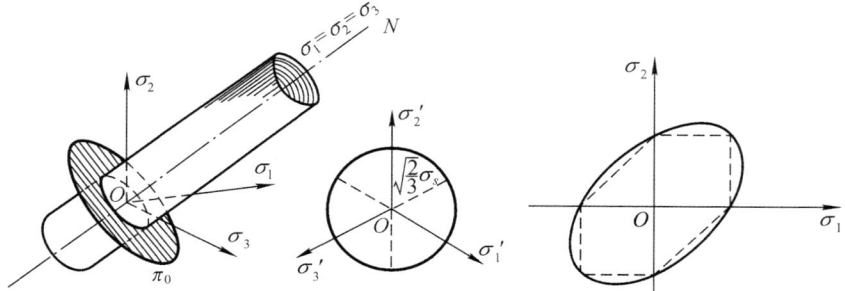

Fig. 3.5. Yield surface and yield loci in deviatoric plane and plane stress state of the Huber-von Mises criterion

In the particular plane stress state of bending beams with shear, where $\sigma_x = \sigma$, $\sigma_y = 0$, $\sigma_z = 0$, and τ is the shear stress, Eq.(3.14) becomes

$$\sigma^2 + 3\tau^2 = \sigma_Y^2. \tag{3.18}$$

From Eq.(3.14) the yield stress in pure shear is derived as

$$\tau_Y = \frac{\sigma_Y}{\sqrt{3}}, \tag{3.19}$$

while τ_Y has the magnitude of $\sigma_Y/2$ according to the Tresca condition.

Two yield conditions of the Tresca and Huber-von Mises criteria are the most widely-accepted yield criteria for metals. They are applicable to non-SD materials because the tensile and compressive strengths are assumed to be the same. In contrast to the uniaxial tensile strength criterion, these two criteria predict different shear yield strengths. The Tresca criterion is more suitable for a material with $\tau_Y = 0.5\sigma_Y$, while the Huber-von Mises criterion is more suitable for a material with $\tau_Y = 0.577\sigma_Y$. For materials with other shear-to-tensile strength ratios, different yield conditions are required.

3.2.3.2 Drucker-Prager Criterion for SD Materials

Drucker-Prager strength criterion is a simple adaptation of the Huber-von Mises criterion for material that has different tensile and compressive behavior by introducing an additional term to reflect the influence of a hydrostatic stress component on material failure. The expression of mathematical modeling for the Drucker-Prager criterion is

$$\tau_8 + \beta\sigma_8 = C, \tag{3.20}$$

where τ_8 and σ_8 are respectively the octahedral shear stress and the companion normal stress. β and C are material constants. The octahedral normal stress σ_8 is identical to the hydrostatic stress or the mean stress (σ_m) in magnitude,

$$\sigma_8 = \sigma_m = \frac{1}{3}(\sigma_1 + \sigma_2 + \sigma_3). \tag{3.21}$$

Eq.(3.20) is very similar to Eq.(3.9) for the Tresca criterion in that both of them have two material constants and can be applied to pressure-dependent materials or materials with the SD effect.

For geomaterials, the Drucker-Prager criterion takes the form of

$$k_1 I_1 + \sqrt{J_2} - k_2 = 0, \tag{3.22}$$

where I_1 is the first invariant of the stress tensor, J_2 is the second invariant of the deviatoric stress tensor. k_1 and k_2 are positive material constants. Eq.(3.22) is identical to Eq.(3.20) because the square root of J_2 and I_1 are proportional to τ_8 and σ_8, respectively. k_1 and k_2 are correlated to the cohesion c and the friction angle φ,

$$k_1 = \frac{2\sin\varphi}{\sqrt{3}(3 - \sin\varphi)}, \quad k_2 = \frac{6c\cos\varphi}{\sqrt{3}(3 - \sin\varphi)}. \tag{3.23}$$

The yield surface of the Drucker-Prager criterion is shown in Fig.3.6. Its yield locus in deviatoric plane is a circle.

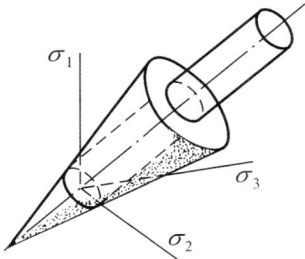

Fig. 3.6. Yield surface of the Drucker-Prager criterion

The Drucker-Prager criterion has been widely used in rock, concrete, and soil plasticity. The inefficiency of the criterion is that the projection of the

yield surface on the deviatoric plane is a circle. It has been criticized by
Zienkiewicz and Pande (1977), Chen (1982), Chen and Saleeb (1994), Chen
et al. (1994) on the grands that the Drucker-Prager criterion gives a very
poor approximation to the real failure conditions for rock, soil and concrete.

3.2.4 Twin-shear Stress-based Criterion—Twin-shear Theory

3.2.4.1 Twin-shear Model

It can be seen that the single-shear theory only takes the maximum principal
shear stress τ_{13} and the corresponding normal stress σ_{13} into account in the
yield of materials. The effect of the intermediate principal stress σ_2 is not
taken into account in the Tresca and the Mohr-Coulomb strength theory.
So they have obvious shortcomings in describing the realistic characteristics
of materials, since even if the value of σ_2 is zero, the other principal shear
stresses also reflect the effect of σ_2.

The principal stress state (σ_1, σ_2, σ_3) can be converted into the principal shear stress state (τ_{13}, τ_{12}, τ_{23}). The principal shear stress state can be
described by a rhombic dodecahedral element model as shown in Fig.3.7 (Yu
et al., 1985). It can also be used as a three-shear model.

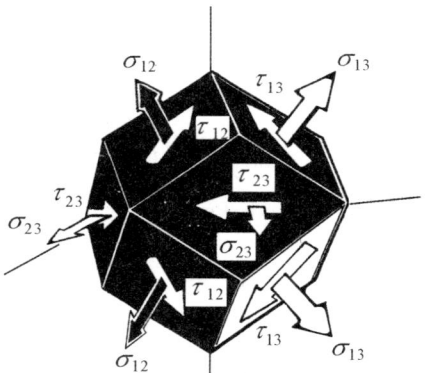

Fig. 3.7. Rhombic dodecahedral principal shear stress element

According to the definition of the principal shear stresses, only two of
them are independent because the maximum principal shear stress (τ_{13}) is
equal to the sum of the other two ($\tau_{12} + \tau_{23}$). So the idea of twin-shear is
developed (Yu, 1983; 1985; 1988).

3.2.4.2 Twin-shear Yield Criterion

A twin-shear yield criterion based on two principal shear stresses was developed by Yu (1961; 1983). The mathematical modeling is formulated in terms of the two larger principal shear stresses,

$$\tau_{13} + \tau_{12} = C \quad \text{when} \quad \tau_{12} \geqslant \tau_{23}, \tag{3.24a}$$

$$\tau_{13} + \tau_{23} = C \quad \text{when} \quad \tau_{12} \leqslant \tau_{23}, \tag{3.24b}$$

where τ_{13}, τ_{12}, τ_{23} are the three principal shear stresses, and there are $\tau_{13} = (\sigma_1 - \sigma_3)/2$, $\tau_{12} = (\sigma_1 - \sigma_2)/2$, and $\sigma_{23} = (\sigma_2 - \sigma_3)/2$, where σ_1, σ_2 and σ_3 are the principal stresses and satisfy the inequality $\sigma_1 \geqslant \sigma_2 \geqslant \sigma_3$, C is a material strength parameter to be determined by experiments.

The twin-shear stress yield criterion can be derived and expressed in terms of the principal stresses, as follows

$$\sigma_1 - \frac{1}{2}(\sigma_2 + \sigma_3) = \sigma_y, \quad \text{when} \quad \sigma_2 \leqslant \frac{1}{2}(\sigma_1 + \sigma_3), \tag{3.25a}$$

$$\frac{1}{2}(\sigma_1 + \sigma_2) - \sigma_3 = \sigma_y, \quad \text{when} \quad \sigma_2 \geqslant \frac{1}{2}(\sigma_1 + \sigma_3), \tag{3.25b}$$

where σ_Y is the uniaxial yield strength. The yield surface and yield loci in plane stress state and deviatoric plane of the twin-shear yield criterion in stress space are shown in Fig.3.8. The other two yield loci of the single-shear theory and three-shear theory in plane stress state are also given for comparison.

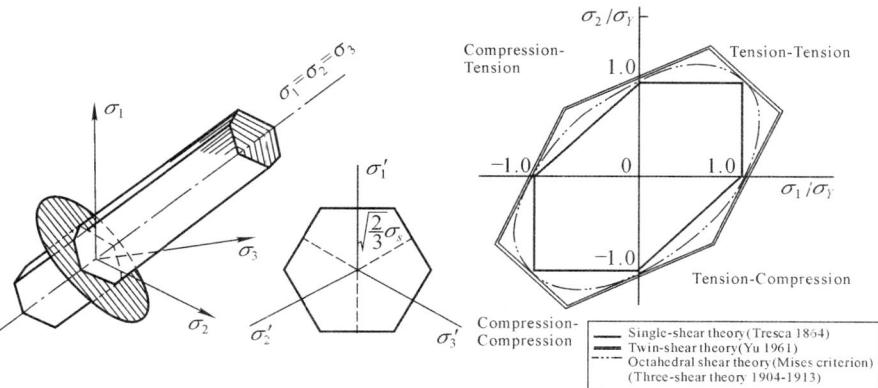

Fig. 3.8. Yield surface and yield loci in deviatoric plane and the plane stress state of the twin-shear yield criterion

It is seen that the expression of the twin-shear yield criterion is piecewise linear. In uniaxial tension and compression, the criterion is the same as other two yield criteria when the experimental uniaxial tension is considered. The Eq.(3.25a) of the twin-shear criterion is equivalent to the maximum deviatoric stress yield criterion (Haythornthwaite, 1961; Zyczkowski, 1981).

In the case of special plane stress state of normal stress with shear stress defined by $\sigma_x = \sigma$, $\sigma_y = 0$, and $\tau_{xy} = \tau$, the principal stresses are derived as

$$\sigma_1 = \frac{\sigma}{2} + \frac{1}{2}\sqrt{\sigma^2 + 4\tau^2}, \ \ \sigma_2 = 0 \text{ and } \sigma_3 = \frac{\sigma}{2} - \frac{1}{2}\sqrt{\sigma^2 + 4\tau^2}. \tag{3.26}$$

Thus, the twin-shear yield criterion has the form of

$$2\sigma^2 + 2\sigma\sigma_Y - 4\sigma_Y^2 + 9\tau^2 = 0. \tag{3.27}$$

According to Eq.(3.27), the shear yield strength is derived as

$$\tau_{\max} = \frac{2}{3}\sigma_Y. \tag{3.28}$$

The shear strength with respect to different yield criteria is compared in Table 3.1. It indicates that, even in a simple stress state, the predictions of material strength may be very different based on different yield criteria. Some materials that have larger shear strength may be more properly represented by the twin-shear yield criterion (Yu, 1961; 1983). The Tresca criterion always gives the lowest estimation of the material strength, while the twin-shear yield criterion gives the highest.

Table 3.1. Comparison of shear strength based on different criteria

Criteria	Tresca (1864)	Mises (1913)	Twin-shear (1961)
Shear strength (τ_{max})	$0.5\sigma_Y$	$0.577\sigma_Y$	$0.667\sigma_Y$

3.2.4.3 Twin-shear Strength Theory for SD Materials

The twin-shear yield criterion was extended to SD materials by Yu in 1985 by including the effect of the corresponding normal stresses of the two larger principal shear stresses. The principal shear stress state can be converted into the twin-shear stress state $(\tau_{13}, \tau_{12}; \sigma_{13}, \sigma_{12})$ or $(\tau_{13}; \tau_{23}; \sigma_{13}, \sigma_{23})$ as shown in Fig.3.9.

The mathematical modelling has the form of

$$\tau_{13} + \tau_{12} + \beta(\sigma_{13} + \sigma_{12}) = C, \quad \text{when} \quad \tau_{12} + \beta\sigma_{12} \geqslant \tau_{23} + \beta\sigma_{23}, \tag{3.29a}$$

$$\tau_{13} + \tau_{23} + \beta(\sigma_{13} + \sigma_{23}) = C, \quad \text{when} \quad \tau_{12} + \beta\sigma_{12} \leqslant \tau_{23} + \beta\sigma_{23}. \tag{3.29b}$$

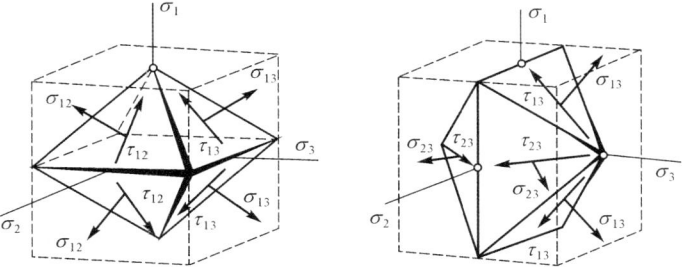

Fig. 3.9. Twin-shear model (orthogonal octahedral element)

The twin-shear strength theory can be obtained from combining Eqs.(3.29a) and (3.29b) and uniaxial tension condition and uniaxial compression condition. It can be expressed as follows:

$$\sigma_1 - \frac{\alpha}{2}(\sigma_2 + \sigma_3) = \sigma_t, \quad \text{when} \quad \sigma_2 \leqslant \frac{\sigma_1 + \alpha\sigma_3}{1 + \alpha}, \tag{3.30a}$$

$$\frac{1}{2}(\sigma_1 + \sigma_2) - \alpha\sigma_3 = \sigma_t, \quad \text{when} \quad \sigma_2 \geqslant \frac{\sigma_1 + \alpha\sigma_3}{1 + \alpha}. \tag{3.30b}$$

Compared to the single-shear theory and the three-shear theory, the Yu twin-shear theory is a completely new perspective for characterizing material yield and failure. It provides another option for describing the material strength besides the Tresca criterion, Mohr-Coulomb criterion and Huber-von Mises-Drucker-Prager criteria.

Fig.3.10 shows the yield surface and yield loci in the deviatoric plane and in the plane stress condition of the twin-shear strength criterion ($\alpha = 1/2$). The dotted lines represent the Mohr-Coulomb criterion for comparison.

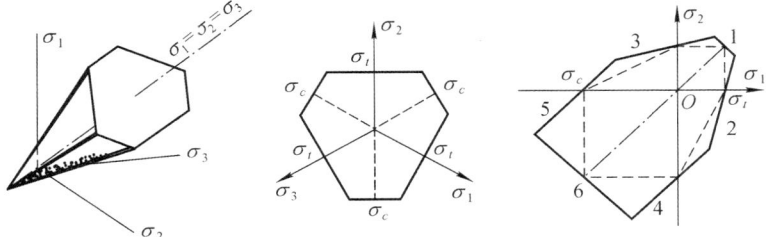

Fig. 3.10. Yield surface and yield loci in deviatoric plane and plane stress state of twin-shear strength theory ($\alpha = 1/2$)

3.3 Unified Yield Criterion for Metallic Materials (Non-SD Materials)

Further improvement has been made to the twin-shear yield criterion (Yu and He, 1991; Yu, 1992; 2004) based on the fact that the effects of the two larger principal shear stresses on yield of material may be different. A coefficient to the second principal shear stress is introduced. The mathematical modelling of the improved twin-shear yield criterion and the unified yield criterion becomes

$$\tau_{13} + b\tau_{12} = C, \quad \text{when} \quad \tau_{12} \geqslant \tau_{23}, \tag{3.31a}$$

$$\tau_{13} + b\tau_{23} = C, \quad \text{when} \quad \tau_{12} \leqslant \tau_{23}, \tag{3.31b}$$

where b is a strength parameter that reflects the relative influence of the intermediate principal shear stress. Yu (1991) suggested the range from 0 to 1 for b.

The expression of the unified yield criterion can be derived from the mathematical modelling and the experimental condition of axial yield test. We have

$$\sigma_1 - \frac{1}{1+b}(b\sigma_2 + \sigma_3) = \sigma_Y, \quad \text{when} \quad \sigma_2 \leqslant \frac{1}{2}(\sigma_1 + \sigma_3), \tag{3.32a}$$

$$\frac{1}{1+b}(\sigma_1 + b\sigma_2) - \sigma_3 = \sigma_Y, \quad \text{when} \quad \sigma_2 \geqslant \frac{1}{2}(\sigma_1 + \sigma_3), \tag{3.32b}$$

where σ_Y is the uniaxial yield strength.

The criterion is called unified yield criterion because, when b equals 0, the shape function of the criterion gives the lower limit of convex shape functions which is identical to that of the Tresca yield criterion; and when b equals 1, it gives the upper limit which is the twin-shear yield criterion. The Huber-von Mises yield criterion can be approximated by the unified yield criterion with $b = 0.5$. Any other convex shape functions can be obtained or approximated with different values of b. It is worth noting that the unified yield criterion unifies the convex shape functions in a simple formula.

Fig.3.11 shows the projections of the prism yield surface on the deviatoriy plane for the unified yield criterion. The unified yield criterion parameter b has its values $b=0$, $b=0.1$, $b=0.2$, $b=0.3$, $b=0.4$, $b=0.5$, $b=0.6$, $b=0.7$, $b=0.8$, $b=0.9$ and $b=1$ (Fig.3.11(a)); and $b=0$, $b=1/4$, $b=1/2$, $b=3/4$ and $b=1$ (Fig.3.11(b)), respectively.

It is worth indicating that the curved Huber-von Mises criterion is replaceable by the unified yield criterion with $b=1/2$ or $b = 1/(1 + \sqrt{3})$, as shown in Fig.3.11(b) and Fig.3.11(c). The unification of the yield criteria and their piecewise linear mathematical form make it possible to derive stress and deformation fields for different materials and structures in a unified manner.

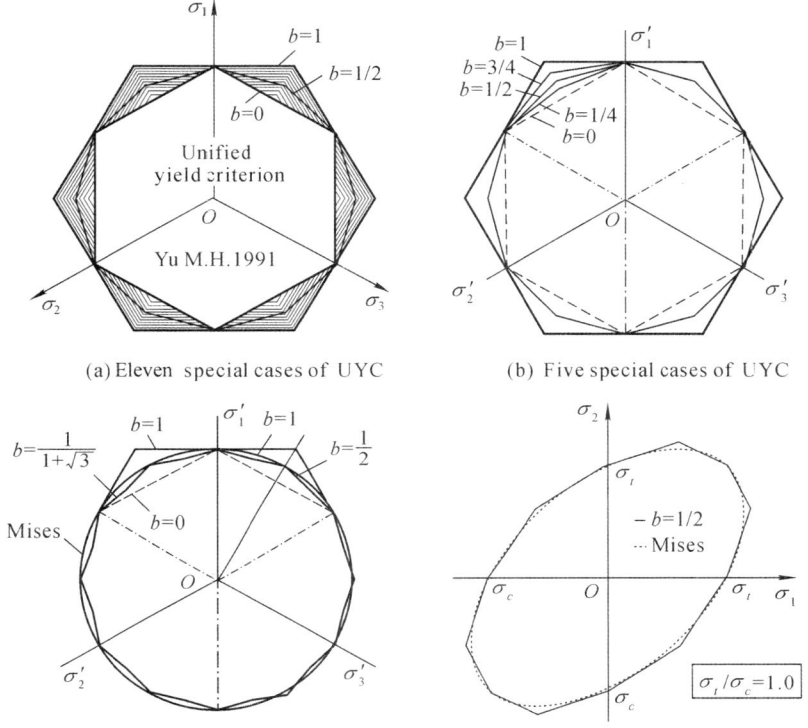

(a) Eleven special cases of UYC   (b) Five special cases of UYC

(c) Mises criterion and its linear approximation

Fig. 3.11. Yield loci of the unified yield criterion

However, the unified yield criterion can only be applied to materials which do not show apparent strength difference in tension and compression (non-SD material).

3.4 Unified Strength Theory for SD Materials

It is understood that a general yield/strength criterion for materials in a complex stress state should be a function of the three principal stresses,

$$F(\sigma_1, \sigma_2, \sigma_3) = 0. \tag{3.33a}$$

Alternatively, the yield function can be expressed by the three independent invariants I_1, J_2 and J_3

$$F(I_1, J_2, J_3) = 0. \tag{3.33b}$$

The maximum normal stress criterion assumes that the material yield or failure depends only on σ_1 or σ_3. The Tresca and Mohr-Coulomb criteria include only two principal stresses σ_1 and σ_3. The Huber-von Mises criterion considers only the effect of J_2. The Drucker-Prager criterion ignores J_3 (see Eq.(3.33b)). For a more generally applicable strength criterion, the effects of all the principal stresses σ_1, σ_2 and σ_3 or the three stress invariants I_1, J_2 and J_3 should be accounted for.

3.4.1 Mechanical Model of the Unified Strength Theory

It is clear that there are three principal shear stresses τ_{13}, τ_{12} and τ_{23} in the three-dimensional principal stress state σ_1, σ_2 and σ_3. However, only two principal shear stresses are independent variables among τ_{13}, τ_{12}, τ_{23} because the maximum principal shear stress equals the sum of the other two, that is,

$$\tau_{13} = \tau_{12} + \tau_{23}. \tag{3.34}$$

The mechanical model and mathematical modelling are important for establishing and understanding a new theory (Meyer, 1985; Tayler, 1986; Besseling and Giessen, 1994; Wu et al., 1999). The mechanical model is an abstraction, a formation of an idea or ideas that may involve the subject with special configurations. Mathematical modelling may involve relationships between continuous functions of space, time and other variations (Tayler, 1986; Meyer, 1985; Besseling and van der Giessen, 1994). Since there are only two independent principal shear stresses, the shear stress state can also be converted into the twin-shear stress state (τ_{13}, τ_{12}; σ_{13}, σ_{12}) or (τ_{13}, τ_{23}; σ_{13}, σ_{23}). This stress state corresponds to the twin-shear model proposed by Yu in 1961 and 1985. The eight sections that two groups of shear stresses act on consist of the orthogonal octahedral elements, so the twin-shear mechanical model can be obtained as shown in Fig.3.12.

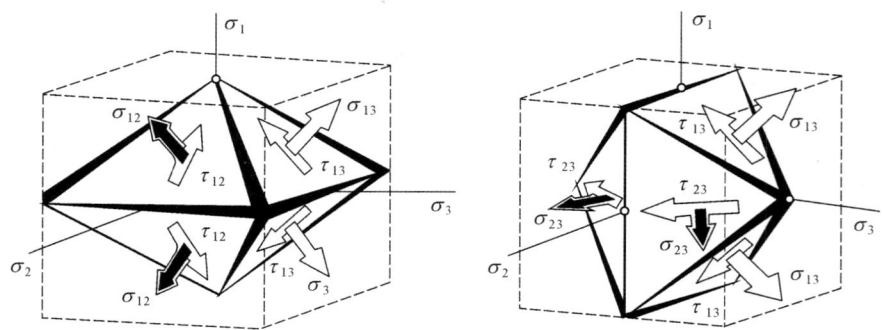

Fig. 3.12. Yu twin-shear model (orthogonal octahedral element)

By removing half of the orthogonal octahedral model, we can obtain a new pentahedron element, as shown in Fig.3.13. The relationship between the twin-shear stress and the principal stress σ_1 or σ_3 can be deduced from this element. Based on the orthogonal octahedral element and pentahedron element, the unified strength theory can be developed.

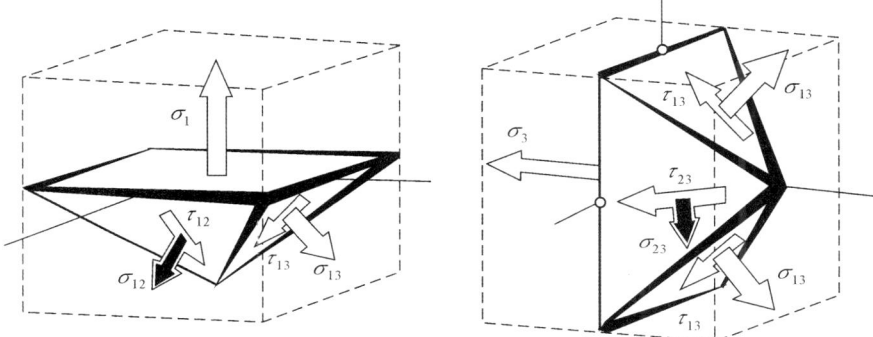

Fig. 3.13. Yu twin-shear model (pentahedron element)

The twin-shear orthogonal octahedral model is different from the regular octahedral model. The orthogonal octahedral model consists of two groups of four sections that are perpendicular to each other and acted on by the maximum shear stress τ_{13} and the intermediate principal shear stress τ_{12} or τ_{23}.

3.4.2 Mathematical Modelling of the Unified Strength Theory

The mathematical modelling of the unified strength theory has the expression of

$$F = \tau_{13} + b\tau_{12} + \beta(\sigma_{13} + b\sigma_{12}) = C, \quad \text{when} \quad \tau_{12} + \beta\sigma_{12} \geqslant \tau_{23} + \beta\sigma_{23}, \quad (3.35a)$$

$$F' = \tau_{13} + b\tau_{23} + \beta(\sigma_{13} + b\sigma_{23}) = C, \quad \text{when} \quad \tau_{12} + \beta\sigma_{12} \leqslant \tau_{23} + \beta\sigma_{23}, \quad (3.35b)$$

$$F'' = \sigma_1 = \sigma_2, \quad \text{when} \quad \sigma_1 > \sigma_2 > \sigma_3 > 0. \quad (3.35c)$$

The unified strength theory takes into account (1) the SD effect, (2) the hydrostatic stress effect, (3) the normal stress effect, (4) the effect of the intermediate principal stress, and (5) the effect of the intermediate principal shear stress.

The magnitudes of β and C can be determined by experimental results of uniaxial tensile strength σ_t and uniaxial compressive strength σ_c,

$$\sigma_1 = \sigma_t, \; \sigma_2 = \sigma_3 = 0,$$
$$\sigma_1 = \sigma_2 = 0, \; \sigma_3 = -\sigma_c. \tag{3.36}$$

So the material constants β and C can be determined by

$$\beta = \frac{\sigma_c - \sigma_t}{\sigma_c + \sigma_t} = \frac{1 - \alpha}{1 + \alpha}, \; C = \frac{2\sigma_c\sigma_t}{\sigma_c + \sigma_t} = \frac{2}{1 + \alpha}\sigma_t. \tag{3.37}$$

3.4.3 Mathematical Expression of the Unified Strength Theory

The mathematical expression can be derived from the mathematical modelling, uniaxial tension and uniaxial compression conditions as follows:

$$F = \sigma_1 - \frac{\alpha}{1 + b}(b\sigma_2 + \sigma_3) = \sigma_t, \quad \text{when} \quad \sigma_2 \leqslant \frac{\sigma_1 + \alpha\sigma_3}{1 + \alpha}, \tag{3.38a}$$

$$F' = \frac{1}{1 + b}(\sigma_1 + b\sigma_2) - \alpha\sigma_3 = \sigma_t, \quad \text{when} \quad \sigma_2 > \frac{\sigma_1 + \alpha\sigma_3}{1 + \alpha}, \tag{3.38b}$$

$$F'' = \sigma_1 = \sigma_2, \quad \text{when} \quad \sigma_1 > \sigma_2 > \sigma_3 > 0. \tag{3.38c}$$

Eq.(3.38c) is used only for the stress state of three tensile stresses. It is similar to the Mohr-Coulomb theory with tension cutoff suggested by Paul in 1961. It is expressed as Eq.(3.38c). The unified strength theory with the tension cutoff can be used for geomaterials.

When the uniaxial tensile and compressive strengths are identical, or $\alpha=1$, Eqs.(3.38a) and (3.38b) are simplified to Eqs.(3.32a) and (3.32b) for non-SD materials. Thus, the unified strength theory is applicable to both SD and non-SD materials. The widely used Mohr-Coulomb criterion for geomaterials is a specific form of the unified strength theory (Eqs.(3.38a) and (3.38b) with $b=0$).

3.4.4 Yield Surfaces and Yield Loci of the Unified Strength Theory

A series of yield surfaces of the unified strength theory can be illustrated in Fig.3.14 (drawn by Dr. Zhang, 2005). The inner yield surface is the same as the yield surface of the unified strength theory with $b=0$ or the yield surface of the Mohr-Coulomb strength theory, and the outer yield surface is the yield surface of the unified strength theory with $b=1$ or the yield surface of the twin-shear strength theory. The other surfaces are the yield surfaces of the unified strength theory with $0 < b < 1$.

Fig.3.15 shows the yield surfaces of three special cases with $b=0$, $b=1/2$ and $b=1$ (Zhang, 2005).

The projections of the unified strength theory on the deviatoric plane are illustrated in Fig.3.16. A series of yield loci on the deviatoric plane of unified yield criterion for non-SD materials ($\alpha =1$) is shown in Fig.3.17. The unified strength theory gives a series of yield and strength criteria and establishes the relationship among various failure criteria.

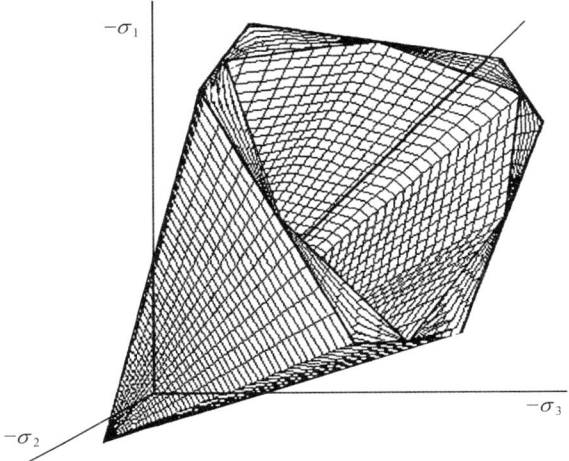

Fig. 3.14. Serial yield surfaces of the unified strength theory in stress space

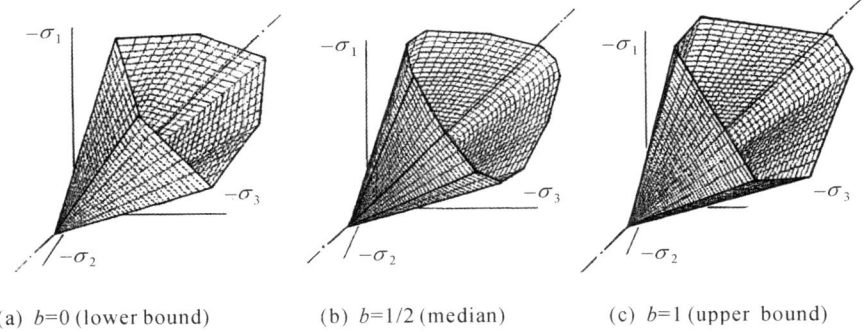

(a) $b=0$ (lower bound)   (b) $b=1/2$ (median)   (c) $b=1$ (upper bound)

Fig. 3.15. Three special cases ($b=0$, $b=1/2$ and $b=1$) of the unified strength theory

3.5 Significance of the Unified Strength Theory

The unified strength theory is introduced in the twin-shear orthogonal octa-hedral model (Fig.3.12) or twin-shear pentahedron element (Fig.3.13). These two models are spatial equipartition, which consists of completely filling a volume with polyhedra of the same kind. The combination of many twin-shear models can be used as a continuous body, as shown in Fig.3.18. It is worth noticing that the twin-shear model can be subjected to an affinity de-formation but remains a parallelehedron, which fills the space without gaps or overlapping.

The characteristics of the unified strength theory can be summarized as follows:

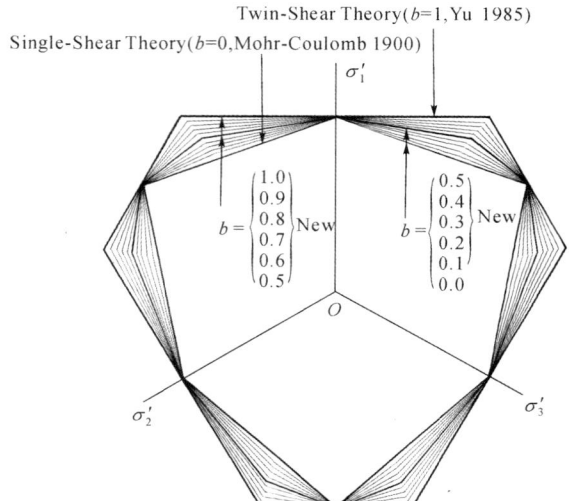

Fig. 3.16. Yield loci on deviatoric plane of unified strength theory for SD materials ($\alpha = 1/2$)

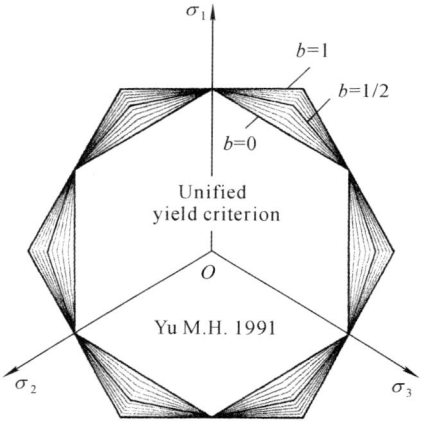

Fig. 3.17. Yield loci on deviatoric plane of unified yield criterion for non-SD materials ($\alpha = 1$)

(1) The unified strength theory is an assemblage of serial yield criteria adopted for non-SD materials and SD materials. The serial criteria cover the whole region between the lower bound (single-shear criterion, Tresca-Mohr-Coulomb, 1864; 1900) and upper bound (twin-shear criteria, Yu, 1961; 1985).

(2) Unified strength theory consists of various yield criteria by changing the parameter b. The single-shear stress-based criteria, the twin-shear stress-based criteria and some other strength criteria are special cases or the

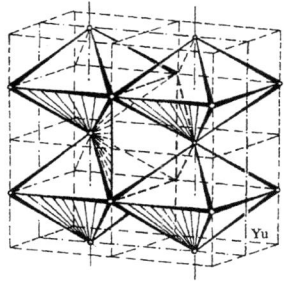

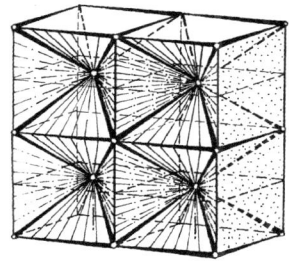

(a) Twin-shear orthogonal octahedral model   (b) Twin-shear pentahedron model

Fig. 3.18. Filling a space by twin-shear models

approximations of the unified strength theory as shown in Fig.3.19. A series of new criteria can also be obtained.

(3) All the yield criteria of the unified strength theory are piecewise linear, with the attendant simplification of the analytical solution of the structure which is well known. The application of the unified strength theory gives not only a single solution, but also a series of solutions. It is referred to as the unified solution adopted for more materials and structures.

(4) It is in good agreement with experimental results for various materials.

(5) The Yu unified strength theory can be used to built up unified elasto-plastic constitutive equations. It can be implemented in a finite element code in a unified manner. It is also convenient for elastic limit design, elasto-plastic and plastic limit analysis of structures because of its piecewise linear form.

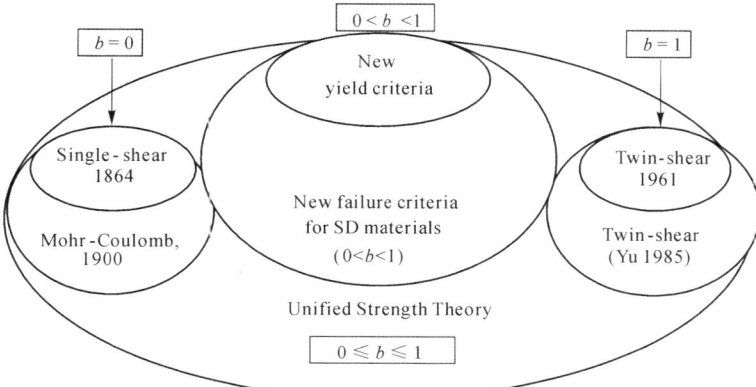

Fig. 3.19. Relationship among single-shear, twin-shear, and new criteria

In summary, the unified strength theory contains two families of yield criteria, namely yield criteria for non-SD materials ($\alpha=1$) and failure criteria for SD materials ($\alpha \neq 1$). The Tresca yield criterion and the twin-shear yield criterion are special cases of the unified yield criterion with $b=0$ and $b=1$, respectively. The Huber-von Mises criterion can be approximated by the unified yield criterion with $b=0.5$.

Fig.3.20 illustrates the relationship among the unified yield criterion, the unified strength criteria, the single-shear-based criteria, the twin-shear-based criteria and some new yield/strength criteria.

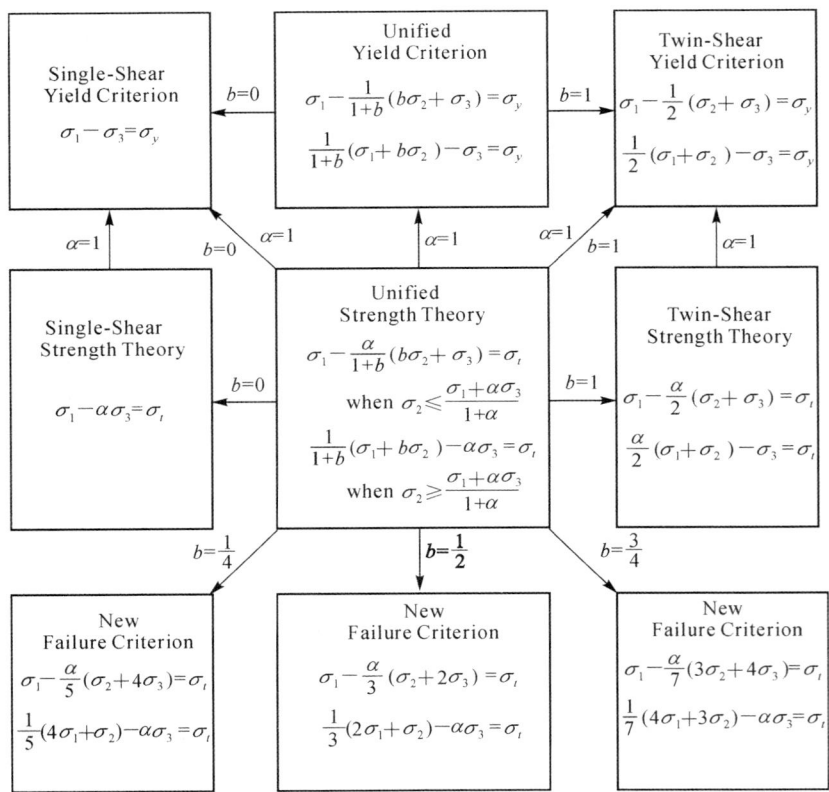

Fig. 3.20. Yu unified strength theory and its special cases

3.6 Unified Strength Theory in the Plane Stress State

The three-dimensional stress state is degenerated into a plane stress state if one of the three principal stresses is equal to zero. The unified strength theory in the plane stress state can be divided into three regions, as described in the following context.

3.6.1 $\sigma_1 \geqslant \sigma_2 > 0, \sigma_3 = 0$

The unified strength theory in the plane stress state in this region can be expressed as

$$\sigma_1 - \frac{\alpha b}{1+b}\sigma_2 = \sigma_t, \quad \text{when} \quad \sigma_2 \leqslant \frac{\sigma_1}{1+\alpha}, \tag{3.39a}$$

$$\frac{1}{1+b}\sigma_1 + \frac{b}{1+b}\sigma_2 = \sigma_t, \quad \text{when} \quad \sigma_2 > \frac{\sigma_1}{1+\alpha}. \tag{3.39b}$$

3.6.2 $\sigma_1 \geqslant 0, \sigma_2 = 0, \sigma_3 < 0$

In this region it becomes

$$-\frac{\alpha}{1+b}(b\sigma_2 + \sigma_3) = \sigma_t, \quad \text{when} \quad \sigma_2 \leqslant \frac{\alpha\sigma_3}{1+\alpha}, \tag{3.40a}$$

$$\frac{b}{1+b}\sigma_2 - \alpha\sigma_3 = \sigma_t, \quad \text{when} \quad \sigma_2 > \frac{\alpha\sigma_3}{1+\alpha}. \tag{3.40b}$$

Eqs.(3.39a) and (3.39b) to Eqs.(3.41a) and (3.41b) are the special cases of the unified strength theory (Eq.(3.38)) with one of the principal stresses zero. The yield loci in the plane stress state can be plotted in Fig.3.21. The yield loci are mirrored-symmetric on the line $\sigma_1 = \sigma_2$.

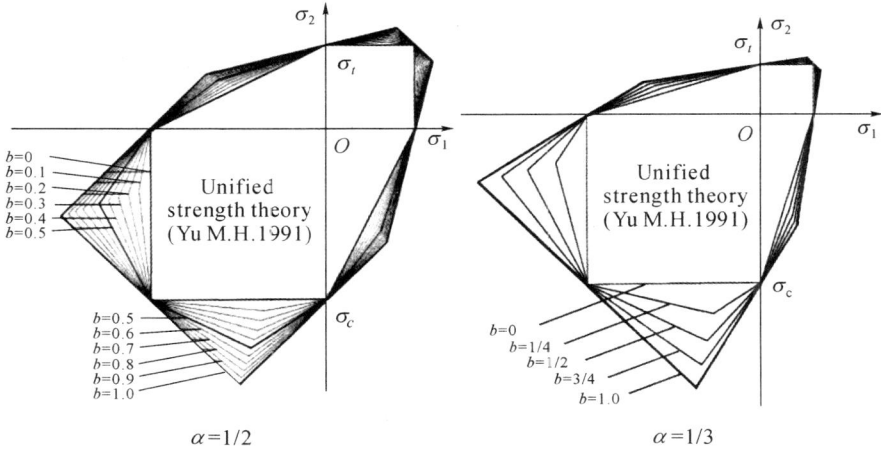

Fig. 3.21. Variation of yield loci of the UST in plane stress ($\sigma \neq 1$ materials)

Fig.3.22 illustrates the yield loci of the unified strength theory in the plane stress state for $\alpha = 1/2$ and $\alpha = 1/3$ materials. The unified strength theory parameter b varying from 0 to 1 gives a series of convex strength criteria in the (σ_1, σ_2) space. The unified strength theory with $b=0$ is the well-known Mohr-Coulomb strength theory (Yu, 2002; 2004).

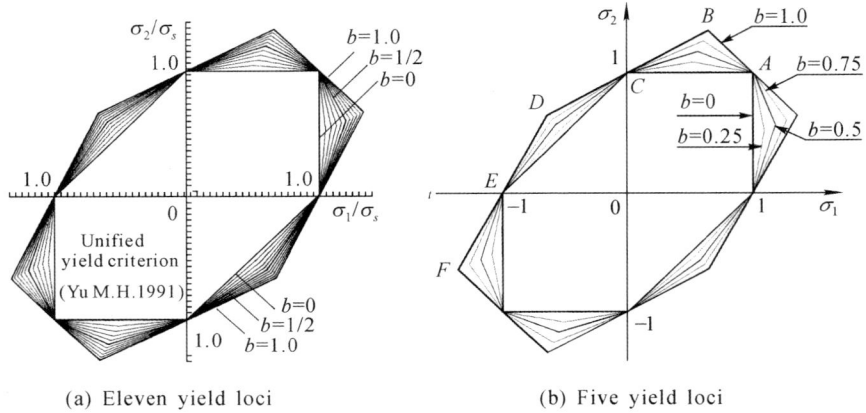

(a) Eleven yield loci   (b) Five yield loci

Fig. 3.22. Unified yield criterion in plane stress state

The equations of the twelve sides of the yield loci in Fig. 3.21 are expressed as

$$\sigma_1 - \frac{\alpha b}{1+b}\sigma_2 = \sigma_t, \qquad \sigma_2 - \frac{\alpha b}{1+b}\sigma_1 = \sigma_t, \qquad (3.41a)$$

$$\frac{1}{1+b}\sigma_1 + \frac{b}{1+b}\sigma_2 = \sigma_t, \qquad \frac{1}{1+b}\sigma_2 + \frac{b}{1+b}\sigma_1 = \sigma_t, \qquad (3.41b)$$

$$\sigma_1 - \frac{\alpha}{1+b}\sigma_2 = -\sigma_t, \qquad \sigma_2 - \frac{\alpha}{1+b}\sigma_1 = -\sigma_t, \qquad (3.41c)$$

$$\frac{1}{1+b}\sigma_1 - \alpha\sigma_2 = -\sigma_t, \qquad \frac{1}{1+b}\sigma_2 - \alpha\sigma_1 = -\sigma_t, \qquad (3.41d)$$

$$-\frac{\alpha}{1+b}(b\sigma_1 + \sigma_2) = \sigma_t, \qquad -\frac{\alpha}{1+b}(b\sigma_2 + \sigma_1) = \sigma_t, \qquad (3.41e)$$

$$\frac{b}{1+b}\sigma_1 - \alpha\sigma_2 = \sigma_t, \qquad \frac{b}{1+b}\sigma_2 - \alpha\sigma_1 = \sigma_t. \qquad (3.41f)$$

With $\alpha = 1$ for non-SD materials, the above equations are simplified as

$$\sigma_1 - \frac{b}{1+b}\sigma_2 = \pm\sigma_Y, \qquad \frac{b}{1+b}\sigma_1 - \sigma_2 = \pm\sigma_Y, \qquad (3.42a)$$

$$\frac{1}{1+b}\sigma_1 + \frac{b}{1+b}\sigma_2 = \pm\sigma_Y, \qquad \frac{b}{1+b}\sigma_1 + \frac{1}{1+b}\sigma_2 = \pm\sigma_Y, \qquad (3.42b)$$

$$\sigma_1 - \frac{1}{1+b}\sigma_2 = \pm\sigma_Y, \qquad \frac{1}{1+b}\sigma_1 - \sigma_2 = \pm\sigma_Y. \qquad (3.42c)$$

Fig.3.22(a) shows the serial yield loci of the unified yield criterion with respect to $b=0$, $b=0.1$, $b=0.2$, $b=0.3$, $b=0.4$, $b=0.5$, $b=0.6$, $b=0.7$, $b=0.8$, $b=0.9$, and $b=1$, respectively. It is seen that the criterion with $b=0$ is exactly the same as the Tresca criterion as shown in Fig.3.1. When $b=1$ it becomes the twin-shear yield criterion in plane stress condition (Fig.3.8). The unified yield criterion parameter b between 0 and 1 gives a series of new convex yield criteria bounded between the single-shear criterion (Tresca criterion) and the twin-shear yield criterion. Five special cases of the unified yield criterion are illustrated in Fig.3.22(b).

3.7 Summary

Each of the conventional yield criteria, including the Tresca yield criterion (1864), the Huber-von Mises yield criterion (1913), the twin-shear stress yield criterion (Yu, 1961a; 1983), or the maximum deviatoric stress yield criterion (Haythornthwaite, 1961), the Mohr-Coulomb criterion, and the twin-shear strength criterion (Yu et al., 1985) is applicable to a specific type of materials. Based on the concepts of the twin-shear models, a unified strength theory was developed by Yu (Yu and He, 1991; Yu, 1992; 2004). The unified strength theory is a completely new system of strength criteria. It embraces many well-established criteria in its special or approximating cases, such as the Tresca yield criterion, the Huber-von Mises yield criterion, the Mohr-Coulomb strength theory, the twin-shear yield criterion (Yu, 1961a), the twin-shear strength criterion (Yu et al., 1985), and the unified yield criterion (Yu and He, 1991; Yu, 1992). The unified strength theory forms an entire spectrum of convex and non-convex criteria, which can be used to describe many different types of engineering materials. The unified strength theory has a unified mechanical model, a unified mathematical model and a simple and unified mathematical expression that conform to the various experimental data. It is easy to be applied for both research and engineering design purposes because of its simple and linear form.

For more information readers can refer to Yu (1992; 2002; 2004), Yu et al. (1998b; 1999a; 2002; 2006). The Yu unified strength theory and the advances in strength theories in the twentieth century are summarized by Yu (2002; 2004). The theory has many connotations yet to be explored and its study has attracted much attention and many research efforts since 1998.

A paper entitled "Remarks on Model of Mao-Hong Yu" is made by Altenbach and Kolupaev (2008). Reviews of "Unified Strength Theory and Its Applications" were presented by Shen (2004) and Teodorescu (2006). The comments on the unified strength theory were presented by Fan and Qiang (2001) and Zhang et al. (2001).

3.8 Problems

Problem 3.1 Which criterion can be used for materials that have identical strength in tension and compression (non-SD materials), and the shear strengths of $\tau_Y = 0.5\sigma_Y$?

Problem 3.2 Which criterion can be used for materials that have identical strength in tension and compression (non-SD materials), and the shear strengths of $\tau_Y = 0.577\sigma_Y$?

Problem 3.3 Which criterion can be used for materials that have identical strength in tension and compression (non-SD materials), and the shear strengths of $\tau_Y = 0.667\sigma_Y$?

Problem 3.4 Can the Huber-von Mises criterion be used for SD materials (that have different strengths in tension and compression)?

Problem 3.5 Compare the four yield criteria of non-SD material, the yield loci in plane stress are shown in Fig.3.23.

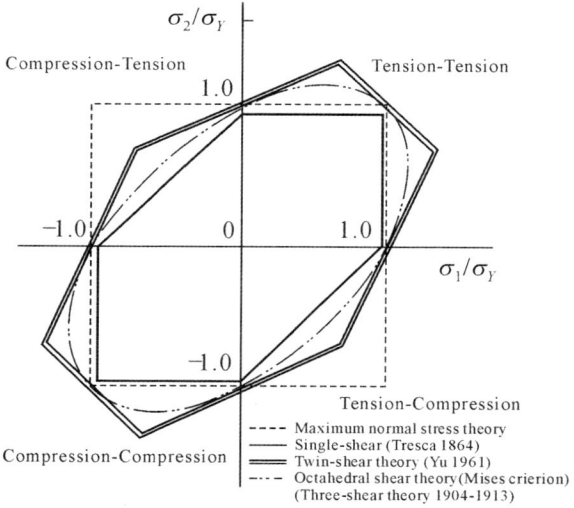

Fig. 3.23. The yield loci of four yield criteria in plane stress state

Problem 3.6 Introduce the well-known Mohr-Coulomb strength theory from the unified strength theory when $b=0$. Show the cross-sectional shapes of the unified strength theory when $b=0$ on the deviatory plane, on the meridian planes, and on the σ_1-σ_2 plane with $\sigma_3=0$.

Problem 3.7 Introduce a new failure criterion from the unified strength theory when $b=1/4$. Show the cross-sectional shapes of the unified strength theory when $b=1/4$ (new strength criterion) on the deviatory planes, on the meridian planes, and on the σ_1-σ_2 plane with $\sigma_3=0$.

Problem 3.8 Introduce a new failure criterion from the unified strength theory when $b=1/2$. Show the cross-sectional shapes of the unified strength theory when $b=1/2$ (new strength criterion) on the deviatory planes, on the meridian planes, and on the σ_1-σ_2 plane with $\sigma_3=0$.

Problem 3.9 Introduce a new failure criterion from the unified strength theory when $b=3/4$. Show the cross-sectional shapes of the unified strength theory when $b=3/4$ (new strength criterion) on the deviatory planes, on the meridian planes, and on the σ_1-σ_2 plane with $\sigma_3=0$.

Problem 3.10 Introduce the twin-shear strength theory from the unified strength theory when $b=1$. Show the cross-sectional shapes of the unified strength theory when $b=1$ (twin-shear strength theory) on the deviatory planes, on the meridian planes, and on the σ_1-σ_2 plane with $\sigma_3=0$.

Problem 3.11 Compare the unified strength theory when $b=1/2$ with the Drucker-Prager criterion.

Problem 3.12 Introduce a new failure criterion from the unified strength theory taking any value of b, and describe the characteristics of this criterion.

Problem 3.13 Introduce the unified strength theory in terms of stress invariant $F(I_1, J_2, \theta)$ and materials parameters σ_t and α.

Problem 3.14 Introduce the unified strength theory in terms of stress invariant $F(I_1, J_2, \theta)$ and materials parameters c and φ.

Problem 3.15 Introduce the unified strength theory by using the experimental condition of pure shear τ_0 and uniaxial tension strength σ_t.

Problem 3.16 Five kinds of yield loci of the unified strength theory (UST) are shown in Fig.3.24. Indicate the mathematical expressions of the parts AC and CB of the unified strength theory with $b=1$.

Problem 3.17 Five kinds of yield loci of the unified strength theory (UST) are shown in Fig.3.24. Indicate the mathematical expressions of the parts AC and CB of the unified strength theory with $b=3/4$.

Problem 3.18 Five kinds of yield loci of the unified strength theory (UST) are shown in Fig.3.24. Indicate the mathematical expressions of the parts AC and CB of the unified strength theory with $b=1/2$.

Problem 3.19 Five kinds of yield loci of the unified strength theory (UST) are shown in Fig.3.24. Indicate the mathematical expressions of the parts AC and CB of the unified strength theory with $b=1/4$.

Problem 3.20 Five kinds of yield loci of the unified strength theory (UST) are shown in Fig.3.24. Indicate the mathematical expressions of the parts AC and CB of the unified strength theory with $b=0$.

Problem 3.21 The yield equation in π-plane of the unified strength theory with $b=0$ (Mohr-Coulomb theory) is

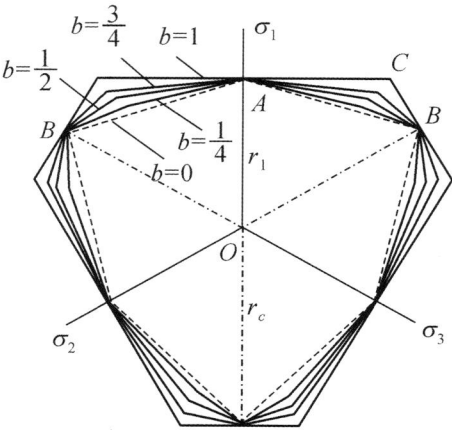

Fig. 3.24. Yield loci of UST in π-plane

$$F = F' = -\frac{\sqrt{2}}{2}ax + \frac{\sqrt{6}}{6}(2+\alpha)y + \frac{\sqrt{3}}{3}(1-\alpha)z = \sigma_t. \tag{3.43}$$

Draw the yield locus in π-plane.

Problem 3.22 A new failure criterion equation in π-plane of the unified strength theory with $b=1/4$ is

$$F = -\frac{3\sqrt{2}}{10}ax + \frac{\sqrt{6}}{6}(2+\alpha)y + \frac{\sqrt{3}}{3}(1-\alpha)z = \sigma_t, \tag{3.44a}$$

$$F' = -(\frac{1}{5}+\alpha)\frac{\sqrt{2}}{2}x + (\frac{7}{5}+\alpha)\frac{\sqrt{6}}{6}y + \frac{\sqrt{3}}{3}(1-\alpha)z = \sigma_t. \tag{3.44b}$$

Draw the yield locus in π-plane.

Problem 3.23 A new failure criterion equation in π-plane of the unified strength theory with $b=1/2$ is

$$F = -\frac{\sqrt{2}}{6}ax + \frac{\sqrt{6}}{6}(2+\alpha)y + \frac{\sqrt{3}}{3}(1-\alpha)z = \sigma_t, \tag{3.45a}$$

$$F' = -(\frac{1}{3}+\alpha)\frac{\sqrt{2}}{2}x + (1+\alpha)\frac{\sqrt{6}}{6}y + \frac{\sqrt{3}}{3}(1-\alpha)z = \sigma_t. \tag{3.45b}$$

Draw the yield locus in π-plane.

Problem 3.24 A new failure criterion equation in π-plane of the unified strength theory with $b=3/4$ is

$$F = -\frac{\sqrt{2}}{14}ax + \frac{\sqrt{6}}{6}(2+\alpha)y + \frac{\sqrt{3}}{3}(1-\alpha)z = \sigma_t, \quad (3.46a)$$

$$F' = -(\frac{3}{7}+\alpha)\frac{\sqrt{2}}{2}x + (\frac{5}{7}+\alpha)\frac{\sqrt{6}}{6}y + \frac{\sqrt{3}}{3}(1-\alpha)z = \sigma_t. \quad (3.46b)$$

Draw the yield locus in π-plane.

Problem 3.25 A new failure criterion equation in π-plane of the unified strength theory with $b=1$ (the twin-shear strength theory) is

$$F = \frac{\sqrt{6}}{6}(2+\alpha)y + \frac{\sqrt{3}}{3}(1-\alpha)z = \sigma_t, \quad (3.47a)$$

$$F' = -(\frac{1}{2}+\alpha)\frac{\sqrt{2}}{2}x + (\frac{1}{2}+\alpha)\frac{\sqrt{6}}{6}y + \frac{\sqrt{3}}{3}(1-\alpha)z = \sigma_t. \quad (3.47b)$$

Draw the yield locus in π-plane.

Problem 3.26 Show the cross-sectional shapes of the unified strength theory when $b=1$ and $\alpha=1/3$ (new strength criterion) on the meridian planes and on the σ_1-σ_2 plane with $\sigma_3=0$.

Problem 3.27 Show the cross-sectional shapes of the unified strength theory when $b=3/4$ and $\alpha=1/3$ (new strength criterion) on the meridian planes and on the σ_1-σ_2 plane with $\sigma_3=0$.

Problem 3.28 Five kinds of yield loci of the unified yield criterion when $\alpha = \sigma_t/\sigma_c = 1$ and in plane stress state are shown in Fig.3.25. These yield equations and yield loci of the unified yield criterion of $\alpha = \sigma_t/\sigma_c = 1$ materials for any value of parameter b can be obtained. For example, the 12 yield equations of the unified yield criterion under the plane stress state when $b=1/2$ can be given as follows. The yield loci of this yield criterion are illustrated in Fig.3.25.

$$f_{1,7} = \sigma_1 - \frac{1}{3}\sigma_2 = \pm\sigma_y; f_{2,8} = 2\sigma_1 + \sigma_2 = \pm\sigma_y, \quad (3.48a)$$

$$f_{3,9} = \frac{1}{3}(\sigma_1 + 2\sigma_2) = \pm\sigma_y; f_{4,10} = \frac{1}{3}\sigma_1 - \sigma_2 = \mp\sigma_y, \quad (3.48b)$$

$$f_{5,11} = \frac{2}{3}\sigma_1 - \sigma_2 = \pm\sigma_y; f_{6,12} = \sigma_1 - \frac{2}{3}\sigma_2 = \mp\sigma_y. \quad (3.48c)$$

Write out the 12 yield equations of the unified yield criterion under the plane stress state when $b = 0$, $b = 1/4$, $b = 3/4$, and $b = 1$, respectively.

Problem 3.29 The unified yield criterion in plane stress state can be divided into three cases as follows.

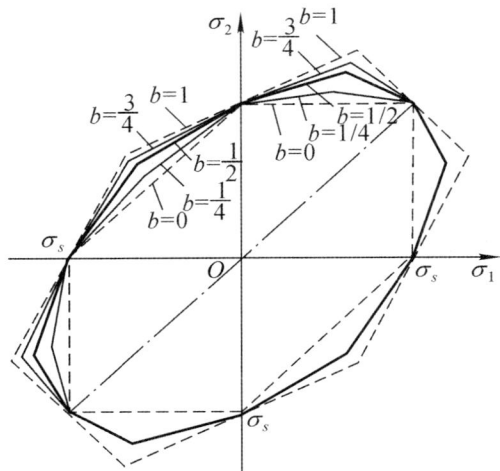

Fig. 3.25. Yield loci of UST in plane stress

Case 1 $\sigma_1 \geqslant \sigma_2 > 0$, $\sigma_3 = 0$
The unified yield criterion with $\alpha = \sigma_t/\sigma_c = 1$ in the plane stress state is

$$f = \sigma_1 - \frac{b}{1+b}\sigma_2 = \sigma_y; \quad \text{if} \quad \sigma_2 \leqslant \frac{1}{2}\sigma_1, \qquad (3.49a)$$

$$f' = \frac{1}{1+b}\sigma_1 + \frac{b}{1+b}\sigma_2 = \sigma_y, \quad \text{if} \quad \sigma_2 \geqslant \frac{1}{2}\sigma_1. \qquad (3.49b)$$

Case 2 $\sigma_1 \geqslant 0$, $\sigma_2 = 0$, $\sigma_3 < 0$
The unified yield criterion for $\alpha = \sigma_t/\sigma_c = 1$ materials in plane stress state is

$$f = \sigma_1 - \frac{1}{1+b}\sigma_3 = \sigma_{y,}, \quad \text{if} \quad \frac{1}{2}(\sigma_1 + \sigma_3) \geqslant 0, \qquad (3.50a)$$

$$f' = \frac{1}{1+b}\sigma_1 - \sigma_3 = \sigma_y, \quad \text{if} \quad \frac{1}{2}(\sigma_1 + \sigma_3) \geqslant 0. \qquad (3.50b)$$

Case 3 $\sigma_1 = 0$, $\sigma_2 \geqslant \sigma_3 < 0$
The unified yield criterion for $\alpha = \sigma_t/\sigma_c = 1$ materials in plane stress state is

$$f = -\frac{1}{1+b}(b\sigma_2 + \sigma_3) = \sigma_y, \quad \text{if} \quad \sigma_2 \leqslant \frac{1}{2}\sigma_3, \qquad (3.51a)$$

$$f' = \frac{1}{1+b}\sigma_1 - \sigma_3 = \sigma_y, \quad \text{if} \quad \sigma_2 \geqslant \frac{1}{2}\sigma_1. \qquad (3.51b)$$

Draw a yield locus in plane stress state for $b=1$ and $\alpha = \sigma_t/\sigma_c = 1$ material.

Problem 3.30 Draw a yield locus in plane stress state for b=0.6 and $\alpha = \sigma_t/\sigma_c = 1$ material.

Problem 3.31 Draw a yield locus in plane stress state for b=1/2 and $\alpha = \sigma_t/\sigma_c = 1$ material.

Problem 3.32 Draw a yield locus in plane stress state for b=0 and $\alpha = \sigma_t/\sigma_c = 1$ material.

Problem 3.33 Show the cross-sectional shapes of the unified strength theory when b=1 and α=1/3 (twin-shear strength theory) on the deviatoric planes, on the meridian planes, and on the σ_1-σ_2 plane with $\sigma_3 = 0$.

Problem 3.34 Draw a yield locus of a new failure criterion from the unified strength theory taking any value of b.

References

Altenbach H, Kolupaev VA (2008) Remarks on Model of Mao-Hong Yu. The Eighth International Conference on Fundamentals of Fracture (ICFF VIII), Zhang TY, Wang B, Feng XQ (eds.), 270-271

Besseling JF, van der Giessen E (1994) Mathematical modelling of inelastic deformation. Chapman & Hall, London

Chen WF (1982) Plasticity in reinforced concrete. McGraw-Hill, New York

Chen WF (1998) Concrete plasticity: past, present and future. In: Yu MH, Fan SC (eds.) Strength Theory: Applications, Developments and Prospects for the 21st Century. Science Press, Beijing, New York, 7-48

Chen WF, et al. (1994) Constitutive equations for engineering materials. Vol. 2: Plasticity and modeling, Elsevier, Amsterdam

Chen WF, Saleeb AF (1994) Constitutive equations for engineering materials. Vol. 1: Elasticity and modeling (revised ed.) Elsevier, Amsterdam. 259-304, 462-489

Drucker DC, Prager W (1952) Soil mechanics and plastic analysis for limit design. Quart. Appl. Math., 10:157-165

Fan SC, Qiang HF (2001) Normal high-velocity impact concrete slabs—a simulation using the meshless SPH procedures. In: Valliappan S, Khalili N (eds.) Computational Mechanics—New Frontiers for New Millennium. Elsevier, Amsterdam, 1457-1462

Fan SC, Yu MH, Yang SY (2001) On the unification of yield criteria. J. Appl. Mech. ASME, 68:341-343

Haythornthwaite RM (1961) Range of yield condition in ideal plasticity. J. Eng. Mech. ASCE, 87(6):117-133

Hencky H (1925) Uber das Wesen der plastischen Verformung (The nature of plastic deformation). Zeitschrift des Vereins Deutscher Ingenieure, 69(20): May 16-Sept 26 (in German)

Huber MT (1904) Przyczynek do podstaw wytorymalosci. Czasopismo Technizne, 22(81) (Lwow, 1904; Pisma, 2, PWN, Warsaw, 1956)

Meyer WJ (1985) Concepts of mathematical modelling. McGraw-Hill, New York

von Mises R (1913) Mechanick der festen Körper im plastisch deformablen Zustand. Nachrichten von der Königlichen Gesellschaft der wissenschaften zu Göettinger. Mathematisch-Physikalische Klasse, 582-592

Mohr O (1900) Welche Umstande bedingen die Elastizitatsgrenze und den Bruch eines Materials. Zeitschrift des Vereins Deutscher Ingenieure, 44:1524-1530; 1572-1577

Mohr O (1905) Abhandlungen aus den Gebiete der Technischen Mechanik. Verlag von Wilhelm Ernst and Sohn, 1905, 1913, 1928

Nadai A (1931) Plasticity. McGraw-Hill, New York

Paul B (1961) A modification of the Coulomb-Mohr theory of fracture. J. Appl. Mech. 28:259-268

Pisarenko GS, Lebedev AA (1976) Deformation and strength of material under complex stressed state. Naukova Dumka, Kiev (in Russian)

Shen ZJ (2004) Reviews to "Unified strength theory and its applications". Adv. Mech., 34(4):562-563 (in Chinese)

Tayler AB (1986) Mathematical models in applied mechanics. Clarendon Press, Oxford

Teodorescu PP (2006) Review of "Unified strength theory and its applications. Springer, Berlin, 2004". Zentralblatt MATH 2006, Cited in Zbl. Reviews, 1059.74002 (02115115)

Tresca H (1864) Sur l'e coulement des corps solids soumis a de fortes pression. Comptes Rendus hebdomadaires des Seances de l' Academie des Sciences, Rend, 59:754-758

Wu KKS, Lahav O, Rees MJ (1999) The large-scale smoothness of the Universe. Nature, 397:225-230

Yu MH (1961a) General behavior of isotropic yield function. Res. Report of Xi'an Jiaotong University. Xi'an, China (in Chinese)

Yu MH (1961b) Plastic potential and flow rules associated singular yield criterion. Res. Report of Xi'an Jiaotong University. Xi'an, China (in Chinese)

Yu MH (1983) Twin shear stress yield criterion. Int. J. Mech. Sci., 25(1):71-74

Yu MH (1988) Three main series of yield and failure functions in plasticity, rock,soil and concrete mechanics. Researches on the Twin Shear Strength Theory. Xi'an Jiaotong University Press, Xi'an, China, 1-34 (in Chinese)

Yu MH (1992) A new system of strength theory. Xi'an Jiaotong University Press, Xi'an (in Chinese)

Yu MH (1994) Unified strength theory for geomaterials and its application. Chinese J. Geotech. Eng., 16(2):1-10 (in Chinese, English abstract)

Yu MH (2002a) Concrete strength theory and its applications. Higher Education Press, Beijing

Yu MH (2002b) Advances in strength theories for materials under complex stress state in the 20th century. Applied Mechanics Reviews ASME, 55(3):169-218

Yu MH (2004) Unified strength theory and its applications. Springer, Berlin

Yu MH, He LN (1983) Non-Schmid effect and twin shear stress criterion of plastic deformation in crystals and polycrystalline metals. Acta Metallurgica Sinica 19(5):190-196 (in Chinese, English abstract)

Yu MH, He LN (1991) A new model and theory on yield and failure of materials under complex stress state. In: Mechanical Behavior of Materials-6, Pergamon Press, Oxford, 3:841-846

Yu MH, He LN, Song LY (1985) Twin shear stress theory and its generalization. Scientia Sinica (Sciences in China), English ed. Series A, 28(11):1174-1183

Zhang LY (2005) The 3D images of geotechnical constitutive models in the stress space. Chinese J. Geotech. Eng., 27(1):64-68

Zhang XS, Guan H, Loo YC (2001) UST failure criterion for punching shear analysis of reinforcement concrete slab-column connections. Computational Mechanics—New Frontiers for New Millennium, Valliappan S, Khalili N (eds.) Elsevier, Amsterdam, 299-304

Zienkiewicz OC, Pande GN (1977) Some useful forms of isotropic yield surfaces for soil and rock mechanics. Finite Elements in Geomechanics. Gudehus G ed. Wiley, London, 179-190

Zyczkowski M (1981) Combined loadings in the theory of plasticity. Polish Scientific Publishers, PWN and Nijhoff

4

Theorems of Limit Analysis

4.1 Introduction

To understand plastic limit analysis it is helpful to review the behavior of an elastic-plastic solid or structure subjected to mechanical loading. An inelastic solid will yield at a specific magnitude of the applied load. The corresponding load is called the elastic limit of the structure. If the external load exceeds the elastic limit, a plastic region starts to spread through the structure. With further expansion of the yield area, the displacement of the structure progressively increases. At another critical load, the plastic region becomes so large as not to resist the unconstrained plastic flow in the solid. The load cannot be increased beyond this point. The collapse load is called the plastic limit of the structure. Plastic limit analysis involves an associated flow rule of the adopted yield criterion. The plastic limit load is also registered as the load-bearing capacity of the structure.

Limit analysis and design of steel structures have been well explored (Symonds and Neal, 1951; Neal, 1956; Hodge, 1959; 1963; Baker and Heyman, 1969; Heyman, 1971; Save and Massonnet, 1972; Horne, 1979; Zyczkowski, 1981; Mrazik et al., 1987; Save et al., 1997). Exploitation of the strength reserve of the load-bearing capacity yields a design of structures with increased admissible loads or decreased cross-sections, which results in a reduction in the amounts of materials and costs.

To save material, one of the choices is to transfer part of the load from the most highly stressed cross-sections to those that are understressed in the elastic state. The number of fully exploited cross-sections can be increased by the redistribution of the internal forces. The load-bearing capacity of structures may be more accurately estimated by choosing an appropriate strength theory or yield criterion.

In plastic limit analysis, direct integration of the equilibrium equations governed by certain yield condition leads to the load-bearing capacity for specific boundary conditions. The associated flow rule is often used to determine the velocity field. Only in some special cases is it possible to derive closed form solutions. But in general it is always feasible to derive approximations through numerical integration of the basic equations.

Determination of the load-bearing capacity of a structure is the simplest when the yield curve is polygonal in shape, as is the case for the Tresca yield criterion and the twin-shear yield criterion. The reason is that only linear equations need to be solved when these kinds of criteria are applied. For other criteria nonlinear equations are involved. Numerical techniques are more appropriate. Thus, the replacement of the Tresca yield conditions by the Huber-von Mises criterion usually renders the analytical solution impractical. However, the unified yield criterion and the unified strength criterion have the advantages of the piecewise linear form, and uniform solutions of the load bearing capacity with respect to different yield conditions can be derived for some simple structures, such as the axial-symmetrical plates, cylinders, tubes and thick-wall vessels.

Ma and He (1994), Ma et al. (1993; 1994; 1995a; 1995b; 1995c) gave a unified plastic limit solution to circular plates under uniform loads and partially uniform loads. Ma and Hao (1998) derived a unified solution to simply supported and clamped circular plates with the Yu's unified yield criterion. Further applications of the unified yield criterion to plastic limit analysis of circular plates under arbitrary loads were reported by Ma et al. (1999). The unified solutions of the limit speed of the rotating disc and cylinder using the unified yield criterion were given by Ma et al. (2001).

The unified plastic limit solution to circular plates under uniform loads and partially uniform loads using the Yu unified strength theory for SD materials was presented by Wei and Yu (2001; 2002), Wang and Yu (2002).

A general formulation of limit design theorems for perfectly plastic materials was given by Gvozdev (1938; 1960). However his work was not known in the Western world until the late 1950s, and before that a very similar theory had been developed by Prager at Brown University (Drucker et al., 1952; Prager, 1947).

One of the most important developments in plastic theory is the upper and lower bound theorems. The contents of these theorems were known by intuition long before Gvozdev's and Prager's school works. However, a complete and precise formulation was given by Gvozdev, Drucker and Greenberg. And Prager's formulation has been proved very valuable. These important principles were also stated by Prager (1947), Hill (1950), Mendelson (1968), Kachanov (1971), Save and Masonnet (1972), Martin (1975), Chen (1975), Zyczkowski (1981), and Nielsen (1999).

The early applications of plasticity to structural concrete were mainly for those reinforced concrete structures whose strength was governed by reinforcement. For such structures a plastic limit design has been standardized. Examples are the yield hinge method for beams and frames (Baker and Heyman, 1969) and the yield line theory for slabs.

The theorems of limit analysis were first presented by Gvozdev in 1938 and independently proved by Hill in 1950 for rigid-perfectly-plastic materials and by Drucker et al. in 1952 for elastic-perfectly-plastic materials. The general forms of the theorems of limit analysis are described in the following sections.

4.2 Perfectly Plastic Solid

A perfectly-plastic solid refers to the material undergoing unlimited plastic deformation under a constant yield stress σ_Y. Fig.4.1 schematically shows the difference among elastic, perfectly-plastic (ideal plastic), hard and soft behavior of material. The value of σ_Y is different for different materials, and even for the same material in different environmental conditions. In the following context, strains and strain rates refer to the plastic quantities unless it is explicitly stated otherwise.

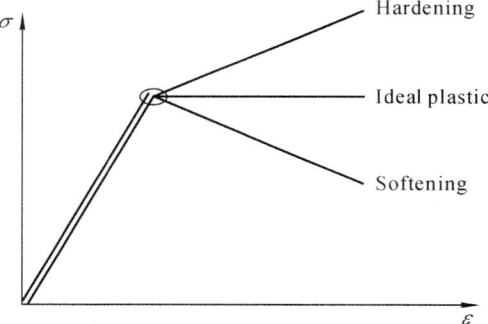

Fig. 4.1. Behavior of elastic, strain hardening, and perfect-plastic materials

At the incipience of plastic flow, it is assumed that strains are very small. Hence strains and displacement are related through Eqs.(2.19) and (2.20), whereas strain rates are derived from displacement rates (or velocities) through Eqs.(2.19) and (2.20).

4.3 Power of Dissipation

At the incipience of plastic flow for a specific point in the stress space, where the stress state is described by $(\sigma_x, \sigma_y, \sigma_z, \tau_{xy}, \tau_{yz}, \tau_{xz})$, and the strain rate by $(\dot{\varepsilon}_x, \dot{\varepsilon}_y, \dot{\varepsilon}_z, \dot{\gamma}_{xy}, \dot{\gamma}_{yz}, \dot{\gamma}_{xy})$, the power of the stress per unit volume of material is

$$d = \sigma_x\dot{\varepsilon}_x + \sigma_y\dot{\varepsilon}_y + \sigma_z\dot{\varepsilon}_z + \tau_{xy}\dot{\gamma}_{xy} + \tau_{yz}\dot{\gamma}_{yz} + \tau_{xz}\dot{\gamma}_{xz}. \tag{4.1}$$

For purely plastic strains this power is dissipated as heat during plastic flow. Therefore, it is called "power of dissipation", which is essentially positive.

Eq.(4.1) can be put into vector form as

$$d = \{\sigma\}^{\mathrm{T}}\{\dot{\varepsilon}\}. \tag{4.2}$$

If elastic strain rates are neglected so that $\{\dot{\varepsilon}\}$ represents the plastic strain rate, the scalar product in Eq.(4.2) is the specific rate of energy dissipation.

The yield surface has the expression

$$\sigma_R(\sigma_x, \cdots, \tau_{xy}, \cdots) - \sigma_y = 0. \tag{4.3}$$

Any stress state at the yield limit is represented by a stress point on this surface. For perfectly plastic materials, σ_R depends on only the stress state instead of the strain state because these materials do not exhibit workhardening. The yield surface is therefore a fixed surface in the six-dimensional space.

The yield surface can be represented by the equations of $\sigma_R(\sigma_x, \sigma_y, \sigma_z, \tau_{xy}, \tau_{yz}, \tau_{yz}) = \sigma_Y$, where $\sigma_R(\sigma_x, \sigma_y, \sigma_z, \tau_{xy}, \tau_{yz}, \tau_{yz})$ is a potential function for the strain rates because normality of $\{\dot{\varepsilon}\}$ to the surface at the stress point P gives

$$\dot{\varepsilon}_x = \lambda\frac{\partial\sigma_R}{\partial\sigma_x}, \cdots, \gamma_{xy} = \lambda\frac{\partial\sigma_R}{\partial\sigma_{xy}}, \cdots, \qquad (x, y, z), \tag{4.4}$$

where λ is a positive scalar factor. When generalized to vertices and flats, Eq.(4.4) is also called the plastic potential flow law.

4.4 Lower-bound Theorem

If a stress distribution balances the applied load, and is below yield or at yield throughout the structure, the structure will not collapse or will just be at the point of collapse. This gives a lower bound of the limit load and is called the lower bound theorem. The maximum lower bound is the limit load.

We can define a statically admissible stress field as being in internal equilibrium when in balance with the external load λ_p, and not exceeding the yield limit anywhere. A multiplier λ is used to define the load magnitude acting on the structure. The multiplier λ corresponding to a statically admissible stress field is called a statically admissible multiplier. The lower bound theorem can be stated by saying that the limit load factor λ^0 is the largest statically admissible multiplier λ^-, i.e.,

$$\lambda^- \leqslant \lambda^0. \tag{4.5}$$

4.5 Upper-bound Theorem

The structure will collapse if there is any compatible pattern of plastic deformation for which the rate of the external forces at work is equal to or exceeds the rate of internal dissipation. It gives the upper bound of the limit or collapse load, and thus is called the upper bound theorem. The minimum upper bound is the limit load.

The upper bound theorem can be stated in view of the admissible multiplier as follows: the limit load factor λ^0 is the smallest kinematically admissible multiplier λ^+, i.e.,

$$\lambda^+ \geqslant \lambda^0. \tag{4.6}$$

The above theorems define the upper and lower bounds for the limit load. They can be summarized as

$$\lambda^- \leqslant \lambda^0 \leqslant \lambda^+ . \tag{4.7}$$

4.6 Fundamental Limit Theorems

When considering a structure which is subjected to a system of loads that start from zero and increase quasi-statically and proportionally, the term "quasi-static" indicates that the loading process is sufficiently slow for all dynamic effects to be disregarded. The term "proportionally" implies that the ratios of the stresses at the same locations of any two different loads are constant throughout the structure. A specific type of loading is determined by the loading location, the direction, and the ratios of stresses at different locations. Choosing one of the loads, we use its magnitude P as a measure for the loading intensity. The variable P is then called the loading parameter.

For beams and frames the transition from a purely elastic region through restricted plastic deformation to unrestricted plastic flow has been extensively studied. For complex structures however it is not straightforward. Emphasis has been put on the direct determination of the limit state in which the plastic deformation in the plastic zones is no longer restricted by the adjacent non-plastic zones and the structure begins to flow under constant loads. The intensity of loading for this limit state is called the limit load, which is usually denoted as P_l.

Limit analysis of a structure is concerned with the limit states of structures under loads. The incipience of the limit state of unrestrained plastic flow is characterized by two phenomena:

a) The stresses are in equilibrium state with the applied loads P, and satisfy the yield condition $\sigma_R = \sigma_Y$ all over the domain. Such a stress field is called statically admissible.

b) The flow mechanism satisfies the kinematical boundary conditions, and for energy balance the power of the applied loads P is equal to the power dissipated in the plastic flow. Such a flow mechanism is called kinematically admissible.

For a given type of loading there may be numerous statically admissible stress fields. Each of the fields corresponds to a certain intensity of loading, which is denoted as P_-. Similarly, for a given kinematically admissible mechanism and a given type of loading, an intensity of loading P_+ can be defined in such a manner that the power of the loads at this intensity of loading is equal to the power of dissipation in the yield mechanism.

The fundamental theorems for limit analysis can then be stated as follows:

a) Static (or lower bound) theorem: the limit load P_l is the largest of all loads P_- corresponding to statically admissible stress fields.

b) Kinematical (or upper bound) theorem: the limit load P_l is the smallest of all loads P_+ corresponding to kinematically admissible mechanisms.

4.7 Important Remarks

4.7.1 Exact Value of the Limit Load (Complete Solution)

Assuming that a statically admissible stress field and a kinematically admissible mechanism that correspond to the same load P have been identified, there are $P \leqslant P_l$ and $P \geqslant P_l$ according to the aforementioned two fundamental theorems. Hence, $P = P_l$ is the exact limit load. It very often happens that it is possible to associate a statically admissible stress field and a kinematically admissible mechanism by the plastic potential flow law. The work equation defining P_+ can then be regarded as a virtual work equation expressing the equilibrium of the associated stress field. Consequently $P_+ = P_-$, and denoting this common value as P, we have $P = P_l$.

A combined theorem is thus derived when it is possible to associate a statically admissible stress field and a kinematically admissible mechanism by the plastic potential flow law and the load P corresponding simultaneously to both fields is the exact limit load P_l.

The two fields of above form are called a complete solution of the limit analysis of a structure. For practical application, one usually starts from a mechanism or from a statically admissible stress field and then searches for the other field.

4.7.2 Elastic-plastic and Rigid-plastic Bodies

With the elastic-plastic idealization, the limit state matches the incipient unrestrained plastic flow. For an example of a relevant displacement δ versus the applied load P (Fig.4.2 (a)), the first part is a ray OA (elastic range), the second part a curve AB (elastic-plastic range, restricted plastic flow) followed by the part parallel to the axis, which indicates unrestrained plastic flow.

With the rigid-plastic idealization, all deformations up to the onset of unrestrained plastic flow at the limit load P_l, sometimes also called the yield-point load, vanish (Fig.4.2 (b)).

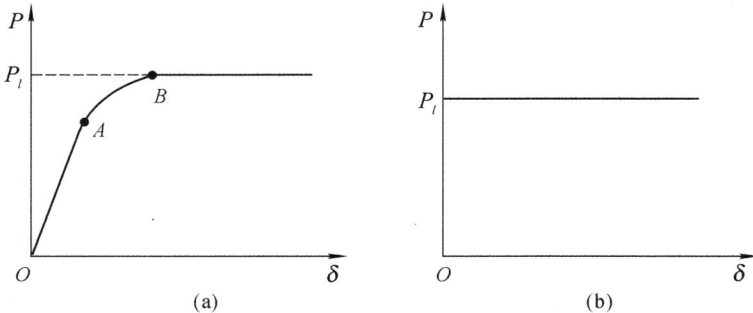

Fig. 4.2. Elasto-plastic (a) and rigid-plastic (b) idealization cases

On the other hand, the fundamental theorems of limit analysis are identical for both idealizations. They are based exclusively on the concepts of statically admissible stress fields and kinematically admissible plastic strain rate fields, irrespective of the elastic or rigid nature of the material before yielding. Thus the lower bound P_-, the upper bound P_+, and the complete solutions are valid for both idealizations.

4.7.3 Load-bearing Capacity

To derive the load-bearing capacity of a structure it requires:
   a) an equilibrium stress field satisfying $\sigma_R \leqslant \sigma_Y$;
   b) a field of plastic strains at impending unrestrained plastic flow.
From the limit analysis point of view, the stress field is statically admissible. The strain field specifies a kinematically admissible mechanism that corresponds to the stress field by the plastic potential flow law. Consequently, a "load-bearing capacity" determined by the deformation theory is the exact limit load for limit analysis.

4.7.4 Uniqueness

The limit load for proportional loading P_l is unique, since it simultaneously matches a statically admissible stress field ($\{\sigma\}$) and a kinematically admissible strain rate field ($\{\dot{\varepsilon}\}$). If there exist several limit loads, the fundamental theorems indicate that P^* must be equal to any one of them. Hence P_l is unique and coincides with P^*.

References

American Society of Mechanical Engineers (1995) Cases of ASME boiler and pressure vessel code, Case N-47. ASME, New York

Baker JF, Heyman J (1969) Plastic design of frames: fundamentals. Cambridge University Press, London

Chakrabarty J (1987) Theory of plasticity, McGraw-Hill, New York

Chen WF (1975) Limit analysis and soil plasticity. Elsevier, Amsterdam

Chen WF (1998) Concrete plasticity: past, Present and future. In: Yu MH, Fan SC (eds.) Strength Theory: Applications, Developments and Prospects for the 21st Century, Science Press, Beijing, New York, 7-48

Chen WF, Saleeb AF (1981) Constitutive equations for engineering materials. Vol.1, Elasticity and modeling; Vol.2, plasticity and modeling, Wiley, New York

Drucker DC, Prager W (1952) Soil mechanics and plastic analysis for limit design. Quart. Appl. Math., 10:157-165

Drucker DC, Prager W, Greenberg HJ (1952) Extended limit design theorems for continuous media. Quart. Appl. Math., 9:381-389

Gvozdev AA (1938) The determination of the value of the collapse load for statically indeterminate systems undergoing plastic deformation. Proceedings of the Conference on Plastic Deformations, Akademiia Nauk SSSR, Moscow, 19-33. (Translated into English by Haythornthwaite RM (1960). Int. J. Mech. Sci., 1:322-355)

Heyman J (1971) Plastic design of frames: applications. Cambridge University Press, London

Hill R (1950) The mathematical theory of plasticity. Clarendon, Oxford

Hodge PG Jr (1959) Plastic analysis of structures. McGraw-Hill, New York

Hodge PG Jr (1963) Limit analysis of rotationally symmetric plates and shells. Prentice-Hall, New Jersey

Horne MR (1979) Plastic theory of structures. Pergamon Press, Oxford

Huang WB, Zeng GP (1989) Solving some plastic problems by using the twin shear stress criterion. Acta Mechanica Sinica, 21(2):249-256 (in Chinese, English abstract)

Johnson W, Mellor PB (1973) Engineering plasticity. Van Nostrand Reinhold, London

Kachanov LM (1971) Foundations of the theory of plasticity. North-Holland Publ. Co., Amsterdam

Li YM (1988) Elasto-plastic limit analysis with a new yield criterion (twin-shear yield criterion). J. Mech. Strength, 10(3):70-74 (in Chinese, English abstract)

Liu XQ, Ni XH, Yan S, et al (1998) Application of the twin shear strength theory in strength calculation of gun barrels. In: Yu MH, Fan SC (eds.) Strength Theory: Applications, Developments and Prospects for the 21st Century. Science Press, New York, Beijing, 1039-1042

Ma GW, Hao H (1999) Investigation on limit angular velocity of annular disc. In: Wang CM, Lee KH, Ang KK (eds.) Computational Mechanics for the Next Millennium, Elsevier, Amsterdam, 1:399-404

Ma GW, Hao H, Iwasaki S (1999a) Plastic limit analysis of a clamped circular plates with the unified yield criterion. Structural Engineering and Mechanics, 7(5):513-525

Ma GW, Hao H, Iwasaki S (1999b) Unified plastic limit analysis of circular plates under arbitrary load. J. Appl. Mech., ASME, 66(2):568-578

Ma GW, Hao H, Miyamoto Y (2001) Limit angular velocity of rotating disc with unified yield criterion. Int. J. Mech. Sci., 43:1137-1153

Ma GW, He LN (1994) Unified solution to plastic limit of simply supported circular plate. Mechanics and Practice, 16(6):46-48 (in Chinese)

Ma GW, He LN, Yu MH (1993) Unified solution of plastic limit of circular plate under portion uniform loading. The Scientific and Technical Report of Xi'an Jiaotong University, 1993-181 (in Chinese)

Ma GW, Iwasaki S, Miyamoto Y, Deto H (1999c) Dynamic plastic behavior of circular plates using the unified yield criterion, Int. J. Solids Struct., 36(3):3257-3275

Ma GW, Yu MH, Iwasaki S, Miyamoto Y (1994) Plastic analysis of circular plate on the basis of twin shear stress criterion. In Proceedings of International Conferences on Computational Method on Structural & Geotechnical Engineering. III, Hong Kong, 930-935

Ma GW, Yu MH, Iwasaki S, Miyamoto Y (1995a) Elasto-plastic response of axisymmetrical solids under impulsive loading. Acta Mechanica Solida Sinica, 8(Special Issue):641-645

Ma GW, Yu MH, Iwasaki S, Miyamoto Y, Deto H (1995b) Unified plastic limit solution to circular plate under portion uniform loading. Journal of Structural Engineering, JSCE, 41A:385-392

Ma GW, Yu MH, Iwasaki S, Miyamoto Y, Deto H (1995c) Unified elasto-plastic solution to rotating disc and cylinder. Journal of Structural Engineering, JSCE, 41A:79-85

Martin JB (1975) Plasticity: Fundamentals and general results. The MIT Press

Mendelson A (1968) Plasticity: theory and application. Macmillan, New York

Mrazik A, Skaloud M, Tochacek M (1987) Plastic design of steel structures. Ellis Horwood, New York

Neal BG (1956) The Plastic methods of structural analysis. Wiley, New York (2nd ed. 1963)

Ni XH, Liu XQ, Liu YT, Wang XS (1998) Calculation of stable loads of strength-differential thick cylinders and spheres by the twin shear strength theory. In: Yu MH, Fan SC (eds.) Strength Theory: Applications, Developments and Prospects for the 21st Century. Science Press, New York, Beijing, 1043-1046

Nielsen MP (1999) Limit analysis and concrete plasticity (2nd ed.). CRC Press, Boca Raton, London

Prager W (1947) An introduction to the mathematical theory of plasticity. J. Appl. Phys., 18:375-383

Save MA, Massonnet CE (1972) Plastic analysis and design of plates, shells and disks. North-Holland, Amsterdam

Save MA, Massonnet CE, Saxce G de (1997) Plastic limit analysis of plates, shells and disks. Elsevier, Amsterdam

Symonds PS, Neal BG (1951) Recent progress in the plastic methods of structural analysis. J. Franklin Inst., 252:383-407, 469-492

Wang F, Fan SC (1998) Limit pressures of thick-walled tubes using different yield criteria. In: Yu MH, Fan SC (eds.) Strength Theory: Applications, Developments and Prospects for the 21st Century. Science Press, New York, Beijing, 1047-1052

Wang YB, Yu MH, Li JC, et al. (2002) Unified plastic limit analysis of metal circular plates subjected to border uniformly distributed loading. J. Mechanical Strength, 24: 305-307 (in Chinese)

Wang YB, Yu MH, Wei XY, et al. (2002) Unified plastic limit analyses of circular plates. Engineering Mechanics, 19:84-88 (in Chinese)

Wei XY (2001) Investigation of long rod penetrating target. PhD. thesis, Xi'an Jiaotong Uni., Xi'an, China (in Chinese)

Wei XY, Yu MH (2001) Unified plastic limit of clamped circular plate with strength differential effect in tension and compression. Chinese Quart. Mechanics, 22:78-83 (in Chinese)

Wei XY, Yu MH (2002) Unified solutions for plastic limit of annular plate. J. Mech. Strength, 24:140-143 (in Chinese)

Yu MH (2002) Advances in strength theories for materials under complex stress state in the 20th century. Applied Mechanics Reviews, ASME, 55(3):169-218

Yu MH (2004) Unified strength theory and its applications. Springer, Berlin

Yu MH, He LN (1991) A new model and theory on yield and failure of materials under complex stress state. Proceedings, Mechanical Behavior of Materials-VI, Vol.3:851-856

Zyczkowski M (1981) Combined loadings in the theory of plasticity. Polish Scientific Publishers, PWN and Nijhoff

5

Plastic Limit Analysis for Simply Supported Circular Plates

5.1 Introduction

The circular plate has been used widely as an important structural element in many branches of engineering. Reliable prediction of the load-bearing capacity of circular plates is crucial for optimum structural design. The load-bearing capacity of circular plates by using the Tresca yield criterion and Huber-von Mises criterion has been given by Hopkins and Wang (1954), Hopkins and Prager (1954), and Ghorashi (1994), et al. The design of circular plates based on the plastic limit load was discussed by Hu (1960). Nine cases including a simply supported circular plate, clamped circular plate, annular plate, a built-in at inner edge and simply supported along the outer edge plate, shearing force along the outer edge and built-in at the inner edge, etc. were studied (Hu, 1960). A systematical summary was given by Mroz and Sawczuk (1960), Hodge (1959; 1963), Save and Massonnet (1972), Zyczkowski (1981), Save (1985) and Save et al. (1997).

Huang et al. (1989) applied the twin-shear stress criterion to derive the plastic limit transverse pressure for the simply supported circular plate. Previous studies showed that the limit analysis method is effective for the analysis of circular plates in the plastic limit state. However, the Tresca criterion, Huber-von Mises criterion, and twin-shear stress criterion are only applicable for certain materials. For instance, the Tresca criterion requires the shear strength and tensile strength of the material to satisfy the relation $\tau_s = 0.5\sigma_s$; the Huber-von Mises criterion is suitable for materials with $\tau_s = 0.577\sigma_s$, and the twin-shear stress criterion is valid for the materials with $\tau_s = 0.677\sigma_s$. All one of the solutions mentioned above is single solution adopted for only one kind of material.

A new unified solution to the plastic limit of a simply supported circular plate by using of the unified yield criterion was presented by Ma and He (1994), Ma et al. (1993; 1994; 1995a; 1995b; 1999). Unified plastic limit analysis of metal circular plates subjected to border uniformly distributed loading was derived by Wang et al. (2002). The unified solution can be adapted for more kinds of non-SD materials. The unified solution for simply supported circular plates using SD materials was derived by Wang and Yu (2002; 2003). The unified plastic limit of the plate for non-SD materials is a special case of the unified solution of the plate for SD materials, such as rock and concrete materials (Chen, 1975; 1981; 1988).

In this chapter, plastic limit analyses of simply supported circular plates with non-SD materials and SD materials under various transverse loading using the unified yield criterion and the unified strength theory are presented. Exact and unified solutions of the plastic limit load, moment field and velocity field in the plastic limit state are derived. The moment field and velocity field with respect to the Tresca criterion, the Huber-von Mises criterion (closed-form solution), and the twin-shear stress criterion are compared. This chapter presents an effective analytical method to compute the exact plastic limit load for circular plates using a piecewise linear yield criterion.

5.2 Basic Equations of Circular Plate

When considering a circular plate of radius a and thickness h subjected to axisymmetric transverse loading $P(r)$, a stress element of the circular plate is considered (Fig.5.1). Because of the axisymmetry of the structure and the loading, the non-zero stresses are the radial stress σ_r, the circumferential stress σ_θ, and the shear stress $\tau_{rz} = \tau_{rz}$. In the plastic limit state, the generalized stresses can be expressed as (Hodge, 1963; Chakrabarty, 1987)

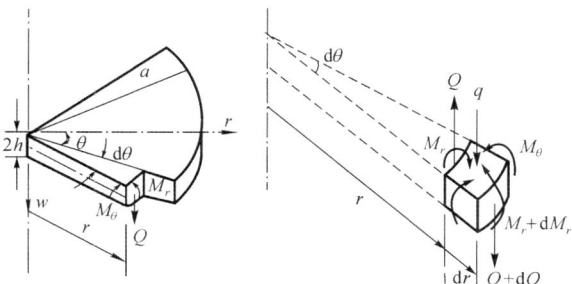

Fig. 5.1. Internal forces in a circular plate element

$$M_r = \int_{-h/2}^{h/2} \sigma_r z \mathrm{d}z, \quad M_\theta = \int_{-h/2}^{h/2} \sigma_\theta z \mathrm{d}z, \tag{5.1a}$$

$$Q_{rz} = \int_{-h/2}^{h/2} \tau_{rz} \mathrm{d}z, \quad M_0 = \int_{-h/2}^{h/2} \sigma_y z \mathrm{d}z = \sigma_y h^2/4, \tag{5.1b}$$

where M_r, M_θ and M_0 are the radial, circumferential, and ultimate (fully plastic) bending moments, respectively; Q_{rz} is the transverse shear force which is generally not assumed to influence the plastic yielding. Defining dimensionless variables of $r = R/a$, $m_r = M_r/M_0$, $m_\theta = M_\theta/M_0$, $p(r) = P(r)a^2/M_0$, the equilibrium equation of a circular plate subjected to a constant uniform load can be written with reference to the axisymmetric condition as

$$d(rm_r)/dr - m_\theta = -\int_0^r p(r)rdr, \qquad (5.2)$$

where $p(r)$ is the transversely distributed loading per unit area.

The equilibrium equation for a uniformly-loaded circular plate can be simplified as

$$d(rm_r)/dr - m_\theta = -\frac{p}{2}r^2. \qquad (5.3)$$

The relations between the curvature rate and the rate of deflection are

$$\dot{k}_r = -d^2\dot{w}/dr^2 \quad \text{and} \quad \dot{k}_\theta = -d\dot{w}/(rdr), \qquad (5.4)$$

where $\dot{w}$, $\dot{k}_r$ and $\dot{k}_\theta$ are non-dimensional deflection rate, non-dimensional curvature rates in radial, and circumferential directions, respectively. The dimensionless deflection is defined as $w = W/a$, where W is the actual deflection, and a is the radius of circular plate. According to the associated flow rule,

$$\dot{k}_r = \dot{\lambda}\partial F/\partial m_r, \quad \dot{k}_\theta = \dot{\lambda}\partial F/\partial m_\theta, \qquad (5.5)$$

where $\dot{\lambda}$ is a plastic flow factor, F is plastic potential which is the same as the yield function according to the associated flow rule.

5.3 Unified Solutions of Simply Supported Circular Plate for Non-SD Materials

The plate is assumed to be made of a rigid-perfectly-plastic material, which satisfies the unified yield criterion. Fig.5.2 shows the generalized unified yield criterion in terms of m_r and m_θ. Fig.5.3 illustrates the flow vector of the curvature velocity at the corners when the unified yield criterion parameter b=0.5. The unified yield criterion is a piecewise linear function, and it has the form of

$$m_\theta = a_i m_r + b_i, \quad (i = 1, ..., 12). \tag{5.6}$$

Table 5.1 lists the respective constants a_i and b_i for the five lines L_i $(i = 1, ..., 5)$ of AB, BC, CD, DE and EF in Fig.5.2.

Substituting the yield criterion into Eq.(5.3) and then integrating Eq.(5.3), the radial moment m_r located on the segments L_i is obtained as

$$m_r = \frac{b_i}{1 - a_i} - \frac{pr^2}{2(3 - a_i)} + c_i r^{-1 + a_i}, \quad (i = 1, ..., 5), \tag{5.7}$$

where c_i $(i=1, ..., 5)$ are integral constants to be determined by boundary and continuity conditions. The field of velocity corresponding to each side L_i is obtained by equating Eq.(5.4) and Eq.(5.5). Considering the yield condition Eq.(5.6), the velocity field is integrated as

$$\dot{w} = \dot{w}_0(c_{1i} r^{1-a_i} + c_{2i}), \quad (i = 1, ..., 5), \tag{5.8}$$

where c_{1i}, c_{2i} $(i=1, ..., 5)$ are the integral constants, $\dot{w}_0$ is the velocity at the plate center.

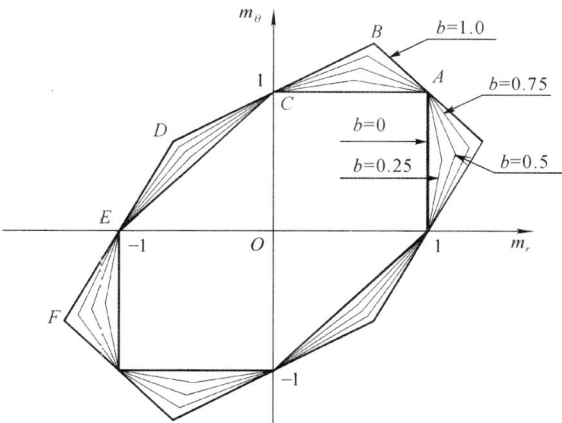

Fig. 5.2. Internal forces in a circular plate element

Table 5.1. Constants a_i and b_i in the unified yield criterion

	AB $(i = 1)$	BC $(i = 2)$	CD $(i = 3)$	DE $(i = 4)$	EF $(i = 5)$
a_i	$-b$	$b/(1 + b)$	$1/(1 + b)$	$1 + b$	$(1 + b)/b$
b_i	$1 + b$	1	1	$1 + b$	$(1 + b)/b$

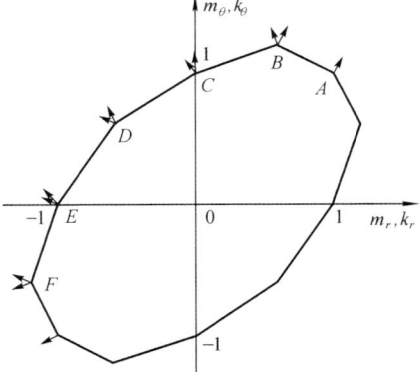

Fig. 5.3. Internal forces in a circular plate element

5.3.1 Uniformly Distributed Load

For a simply supported circular plate under uniformly distributed pressure (Fig.5.4), in the plastic limit state, moments at the center ($r=0$) of the simply supported circular plate satisfy $m_r = m_\theta = 1$ (point A at the yield curve in Fig.5.2). According to the boundary condition of the plate and the requirement of stable flow of the plastic strains (Hodge, 1963), moments at the simply supported edge ($r = 1$) satisfy $m_r = 0$ and $m_\theta = 1$ (point C at the yield curves in Fig.5.2). Bending moments at each point in the plate are located on the sides AB and BC in view of the normality requirement of plasticity.

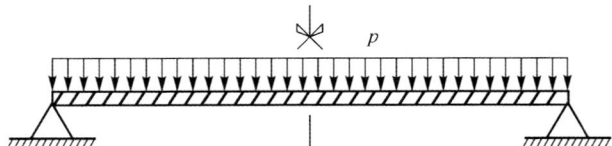

Fig. 5.4. Simply supported circular plate under uniformly distributed load

Assuming r_1 is a non-dimensional radius of a ring where the moments correspond to point B in Fig.5.2, the boundary conditions and continuity conditions can be put as: (1) $m_r(r = 0)=1$; (2) $m_r(r = r_1)$ is continuous and equals to $(1+b)/(2+b)$; (3) $m_r(r = 1)=1$; (4)$\dot{w}(r = 0) = \dot{w}_0$; (5)$\dot{w}(r = r_1)$ and $d\dot{w}/dr(r = r_1)$ are continuous; (6)$\dot{w}(r = 1) = 0$. The integral coefficients c_1, c_2, c_{11}, c_{12}, and c_{22} in Eq.(5.7) and Eq.(5.8) are then derived as

$$\begin{cases} c_1 = 0, \quad c_2 = -(1+b) + \dfrac{1+b}{2+b}\dfrac{3+b}{3+2b}r_1^{-2}, \\[2mm] c_{11} = -\dfrac{r_1^{-b(2+b)/(1+b)}}{(1+b)^2 - (2b+b^2)r_1^{1/(1+b)}}, \\[2mm] c_{21} = 1, \\[2mm] c_{12} = -c_{22} = -\dfrac{(1+b)^2}{(1+b)^2 - (2b+b^2)r_1^{1/(1+b)}}. \end{cases} \tag{5.9}$$

The plastic limit load p is derived as

$$p = \frac{6+2b}{2+b}\frac{1}{r_1^2}, \tag{5.10}$$

where r_1 satisfies the equation of

$$-(3+2b)(2+b) + (3+b)r_1^{-2} + 2b(2+b)r_1^{1/(1+b)} = 0. \tag{5.11}$$

Eq.(5.11) can be solved by half-interval search of r_1 in the interval of $(0, 1)$ for a given value of b in the range of 0 to 1. The convergence with sufficient accuracy of Eq.(5.11) gives the approximation of r_1.

For a special case when $b=0$, the plastic solution becomes

$$\begin{cases} m_r = 1 - r^2, \; m_\theta = 1, \\[1mm] \dot{w} = \dot{w}_0(1-r), \\[1mm] p = 6, \; r_1 = 1\big/\sqrt{2}, \end{cases} \tag{5.12}$$

which is the same as those given by other researchers using the maximum principal stress criterion and the Tresca criterion (Hopkins and Prager, 1953; Hodge, 1963). Figs.5.5 and 5.6 show the moment fields and velocity fields of a simply supported circular plate with respect to three different criteria, namely, the Tresca criterion ($b=0$), the Huber-von Mises criterion ($b=0.5$), and the maximum deviatoric stress criterion or the twin-shear stress criterion ($b=1$). The plastic limit load p corresponding to the three criteria are 6.000, 6.489, and 6.839, respectively, when b equals 0, 0.5 or 1. The plastic limit load versus unified yield criterion parameter b of a simply supported circular plate is shown in Fig.5.6. The plastic limit load with respect to the Huber-von Mises criterion obtained by Hopkins and Wang (1954) is 6.51, which is very close to the present result with $b=0.5$. The upper bound limit load derived from the expanded Tresca hexagon which circumscribes the Huber-von Mises ellipse (Hopkins and Wang, 1954) is 6.83. The plastic limit loads obtained with $b=0.5$ and $b=1$ of the unified yield criterion differ considerably from that with $b=0$ by 8.15% and 14.0%, respectively.

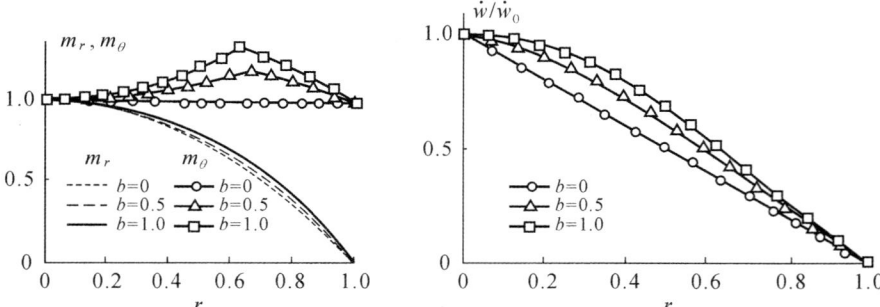

Fig. 5.5. Moment fields and velocity fields of simply supported circular plate

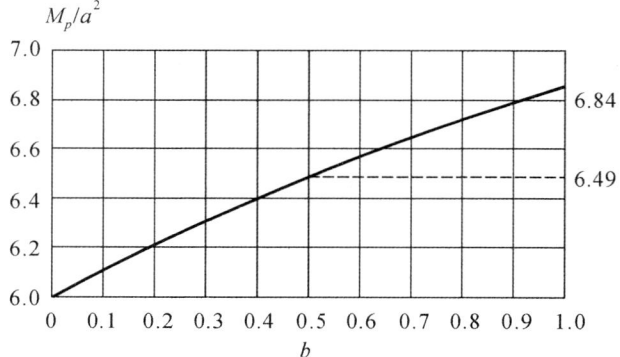

Fig. 5.6. Plastic limit load versus unified yield criterion parameter b of simply supported circular plate

It is seen that the choice of strength theory has a significant influence on the results of elasto-plastic analysis and the load-bearing capacities of a simply supported circular plate for non-SD materials. The unified yield criterion provides us with an effective approach to study these effects.

5.3.2 Arbitrary Axisymmetrical Load

This section presents the exact solution of a circular plate under an arbitrarily distributed axisymmetrical load. The plastic solution of a simply supported circular plate with a varying loading radius of the partial-uniform pressure in Fig.5.7(a) and the arbitrary loading variation in Fig.5.7(b) under different boundary conditions are discussed.

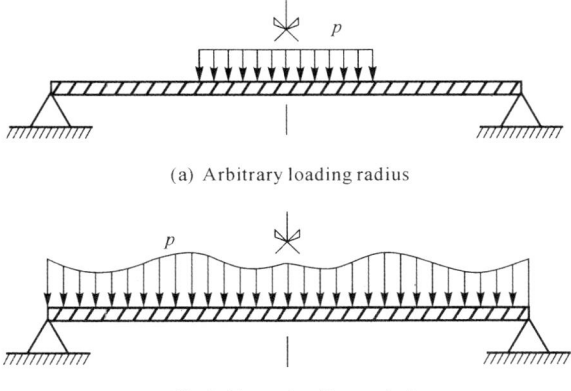

(a) Arbitrary loading radius

(b) Arbitrary loading variation

Fig. 5.7. Plastic limit load versus unified yield criterion parameter b of simply supported circular plate

5.3.2.1 Arbitrary Loading Radius

Defining dimensionless variables, $r = R/a, m_r = M_r/M_0, m = M/M_0$, and $q(r) = p(r)a^2/M_0$, the equilibrium equation of a circular plate subjected to a constant uniform load can be derived with reference to the axisymmetric condition as

$$\mathrm{d}(rm_r)/\mathrm{d}r - m_\theta = -pr^2/2, \quad 0 \leqslant r \leqslant r_p, \qquad (5.13)$$

$$\mathrm{d}(rm_r)/\mathrm{d}r - m_\theta = -pr_p^2/2, \quad r_p \leqslant r \leqslant 1, \qquad (5.14)$$

where $r_p = R_p/a$ is the normalized loading radius of the circular plate; R_p is the loading radius. $r_p = 1$ implies that the entire plate is uniformly loaded, whereas $r_p = 0$ indicates a point loaded at the center. When the unified yield criterion expressed by generalized stresses (Fig.5.2) is used, the expression of the limit condition is the same as that in Eq.(5.6).

For a circular plate under arbitrarily distributed load, the center point of the plate satisfies $m_r = m_\theta = 1$ (point A in Fig.5.2), and the simply supported boundary condition leads to $m_r = 0$ (point C in Fig.5.2). Stress states of all the points in the plate are still on the parts AB and BC. There are two possible cases, i.e., Case (1) $r_p \leqslant r_0$ and Case (2) $r_p > r_0$, where r_0 is the radius of the ring with the moments corresponding to the yield point B in Fig.5.2. These two cases are illustrated in Fig.5.8.

From the boundary conditions and continuity conditions, there are: (1) $m_r = 1$ at $r = 0$; (2) $m_r = 0$ at $r = 1$; (3) m_r is continuous at $r = r_p$; (4) $m_r = d_1 = (1+b)/(2+b)$ and is continuous at $r = r_0$. These conditions will be used to derive the integration coefficients in the moment equations.

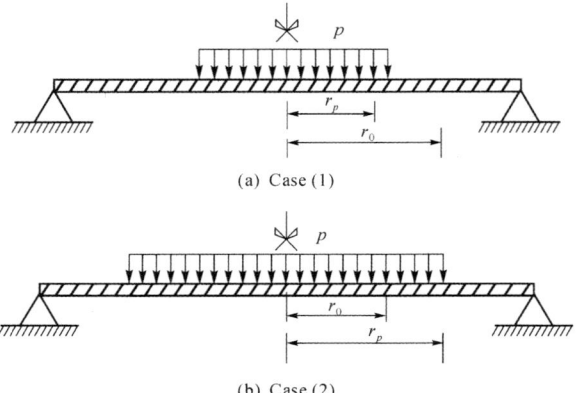

(a) Case (1)

(b) Case (2)

Fig. 5.8. Plastic limit load versus unified yield criterion parameter b of simply supported circular plate

Case (1)

The moment fields of the plate corresponding to the first case can be integrated as

$$m_r = \frac{b_1}{1 - a_1} - \frac{pr^2}{2(3 - a_1)} + c_1 r^{-1+a_1}, \quad 0 \leqslant r \leqslant r_p, \tag{5.15a}$$

$$m_r = \frac{b_1}{1 - a_1} - \frac{pr_p^2}{2(1 - a_1)} + c_2 r^{-1+a_1}, \quad r_p \leqslant r \leqslant r_0, \tag{5.15b}$$

$$m_r = \frac{b_2}{1 - a_2} - \frac{pr_p^2}{2(1 - a_2)} + c_3 r^{-1+a_2}, \quad r_0 \leqslant r \leqslant 1, \tag{5.15c}$$

where c_1, c_2, and c_3 are integration coefficients. The integration coefficients, the plastic limit load p, and the demarcating radius r_0 are derived with application of the boundary and continuity conditions as

$$c_1 = 0, \tag{5.16a}$$

$$\frac{b_1}{1 - a_1} - \frac{pr_p^2}{2(3 - a_1)} = \frac{b_1}{1 - a_1} - \frac{pr_p^2}{2(1 - a_1)} + c_2 r_p^{-1+a_1}, \tag{5.16b}$$

$$\frac{b_1}{1 - a_1} - \frac{pr_p^2}{2(1 - a_1)} + c_2 r_0^{-1+a_1} = d_1, \tag{5.16c}$$

$$\frac{b_2}{1 - a_2} - \frac{pr_p^2}{2(1 - a_2)} + c_3 r_0^{-1+a_2} = d_1, \tag{5.16d}$$

$$\frac{b_2}{1-a_2} - \frac{pr_p^2}{2(1-a_2)} + c_3 = 0. \tag{5.16e}$$

When the loading radius r_p is specified, the unknowns c_2, c_3, p and r_0 in the above simultaneous equations can be derived as

$$c_3 = -\frac{d_1}{1 - r_0^{-1+a_2}}, \tag{5.17}$$

$$p = \frac{2b_2}{r_p^2} - \frac{2(1-a_2)d_1}{(1-r_0^{-1+a_2})r_p^2}, \tag{5.18}$$

$$c_2 = \frac{1+a_1}{(1-a_1)(3-a_1)r_p^{-1+a_1}} \left[b_2 - \frac{(1-a_2)d_1}{1-r_0^{-1+a_2}} \right], \tag{5.19}$$

where the demarcating radius r_0 can be calculated from the equation of

$$\frac{b_1}{1-a_1} - d_1 - \left[-\frac{1-a_2}{1-a_1} \frac{d_1}{1-r_0^{-1+a_2}} + \frac{b_2}{1-a_1} \right]$$
$$+ \frac{1+a_1}{(1-a_1)(3-a_1)r_p^{-1+a_1}} \left[b_2 - \frac{(1-a_2)d_1}{1-r_0^{-1+a_2}} \right] r_0^{-1+a_1} = 0. \tag{5.20}$$

When the derived values are substituted into Eqs.(5.15a)~(5.15c), the moment fields of the plate are calculated.

Case (2)
The corresponding moment fields of Case (2) can be derived as

$$m_r = \frac{b_1}{1-a_1} - \frac{pr^2}{2(3-a_1)} + c_1 r^{-1+a_1}, 0 \leqslant r \leqslant r_0, \tag{5.21a}$$

$$m_r = \frac{b_2}{1-a_2} - \frac{pr^2}{2(3-a_2)} + c_2 r^{-1+a_2}, r_0 \leqslant r \leqslant r_p, \tag{5.21b}$$

$$m_r = \frac{b_2}{1-a_2} - \frac{pr_p^2}{2(1-a_2)} + c_3 r^{-1+a_2}, r_p \leqslant r \leqslant 1. \tag{5.21c}$$

With the same boundary and continuity conditions as for Case (1), the integral coefficients c_1, c_2 and c_3, the plastic limit load p, and the demarcating radius r_0 satisfy

$$c_1 = 0, \tag{5.22a}$$

$$\frac{b_1}{1-a_1} - \frac{pr_0^2}{2(3-a_1)} = d_1, \tag{5.22b}$$

$$\frac{b_2}{1 - a_2} - \frac{pr_0^2}{2(3 - a_2)} + c_2 r_0^{-1 + a_2} = d_1, \tag{5.22c}$$

$$\frac{b_2}{1 - a_2} - \frac{pr_p^2}{2(3 - a_2)} + c_2 r_p^{-1 + a_2} = \frac{b_2}{1 - a_2} - \frac{pr_p^2}{2(1 - a_2)} + c_3 r_p^{-1 + a_2}, \tag{5.22d}$$

$$\frac{b_2}{1 - a_2} - \frac{pr_p^2}{2(1 - a_2)} + c_3 = 0. \tag{5.22e}$$

The integral coefficients, the plastic limit load and the demarcating radius can be derived as

$$p = \left(\frac{b_1}{1 - a_1} - d_1 \right) \frac{2(3 - a_1)}{r_0^2}, \tag{5.23a}$$

$$c_2 = \left[\left(\frac{b_1}{1 - a_1} - d_1 \right) \frac{3 - a_1}{3 - a_2} - \left(\frac{b_2}{1 - a_2} - d_1 \right) \right] r_0^{1 - a_2}, \tag{5.23b}$$

$$c_3 = -\frac{b_2}{1 - a_2} + \left(\frac{b_1}{1 - a_1} - d_1 \right) \frac{3 - a_1}{1 - a_2} \frac{r_p^2}{r_0^2}, \tag{5.23c}$$

where r_0 can be numerically solved from Eq.(5.22d) by substituting Eqs.(5.23a) $\sim$(5.23c) into Eq.(5.22d).

Assuming that $r_p = r_0 = r_{p0}$, the two cases are identical. The value of r_{p0} in this special case can be derived as $r_{p0} = 1/2^{1+b}$ by solving Eqs.(5.18) and (5.22a) with application of $r_p = r_0 = r_{p0}$. When $r_p \leqslant r_{p0}$, the equations derived in Case (1) are adopted. On the other hand, when $r_p > r_{p0}$, the counterparts in Case (2) are adopted. Moment fields with six different values of loading radius r_p, i.e., 1.0, 0.75, 0.5, 0.25, 0.1 and 0.00001 are shown in Fig. 5.9. It can be seen that the moment field varies with the unified yield criterion parameter b. The plastic limit load increases with the increase of b. Table 5.2 gives the plastic limit load p with respect to different values of r_p and b.

Table 5.2. Plastic limit loads with different values of r_p

Criterion	$r_p = 1$	$r_p = 0.75$	$r_p = 0.5$	$r_p = 0.25$	$r_p = 0.1$	$r_p = 0.00001$
$b = 0.0$ (Tresca)	6.0000	7.1111	12.000	38.400	214.29	2×10^{10}
$b = 0.5$(Mises)	6.4887	7.6666	12.886	40.901	224.70	2×10^{10}
$b = 1.0$ (Twin-shear)	6.8392	8.0638	13.509	42.669	232.69	2×10^{10}

According to the hypothesis of the maximum principal stress condition or the Tresca condition, m_θ is equal to 1 in the whole circular plate regardless of the variation in the loading radius r_p. It is identical to the results of the unified yield criterion with $b = 0$, which is obviously unreasonable. When $0 < b \leqslant 1$, m_θ varies with the radius variable r and the loading radius r_p. Compared with that $b = 0$ or with the Tresca criterion, the varying tendency of m_θ seems more reasonable.

When r_p approaches zero, the problem is approximately the case of the circular plate under concentrated load at the center. The unified yield criterion reflects the moment singularity at the center of the circular plate under a concentrated load.

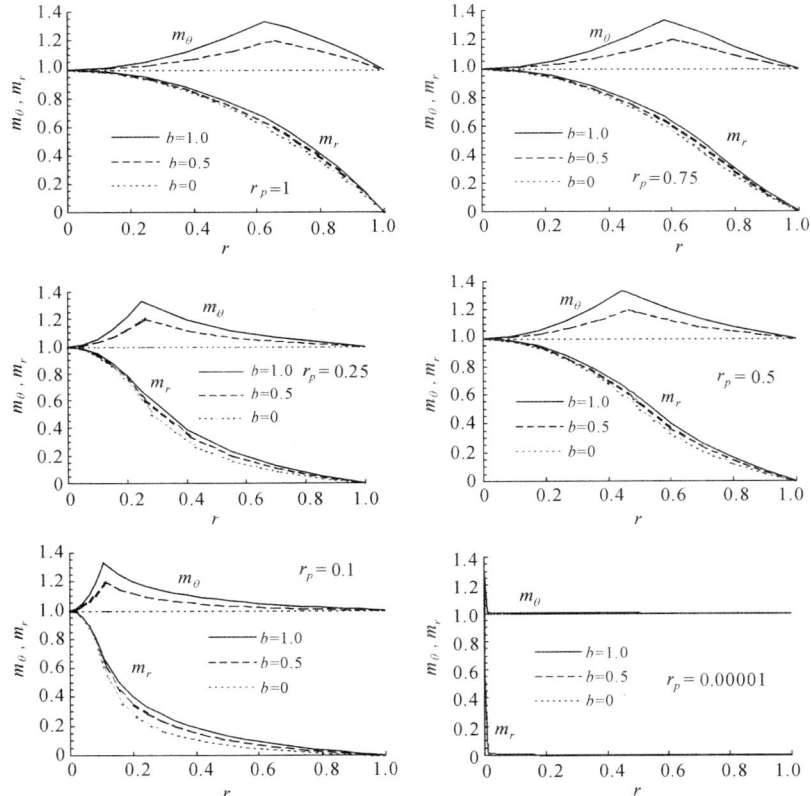

Fig. 5.9. Moment fields with different loading radii

The boundary conditions, continuity conditions, and the velocity field when the plate is subjected to a partial-uniform load are the same as those when the plate is under a uniformly distributed load, except that the demarcating radius r_0 is a function of the loading radius r_p. The plastic limit load derived in the present study satisfies the equilibrium conditions and the yield conditions. The velocity field of deflection which is compatible with the motion mechanism, is obtained. Therefore the solution of a plastic limit load given here is a complete solution. Velocity profiles corresponding to six different values of loading radius r_p, namely $r_p=1$, $r_p=0.75$, $r_p=0.5$, $r_p=0.25$, $r_p=0.1$, and $r_p=0.00001$ are plotted in Fig.5.10.

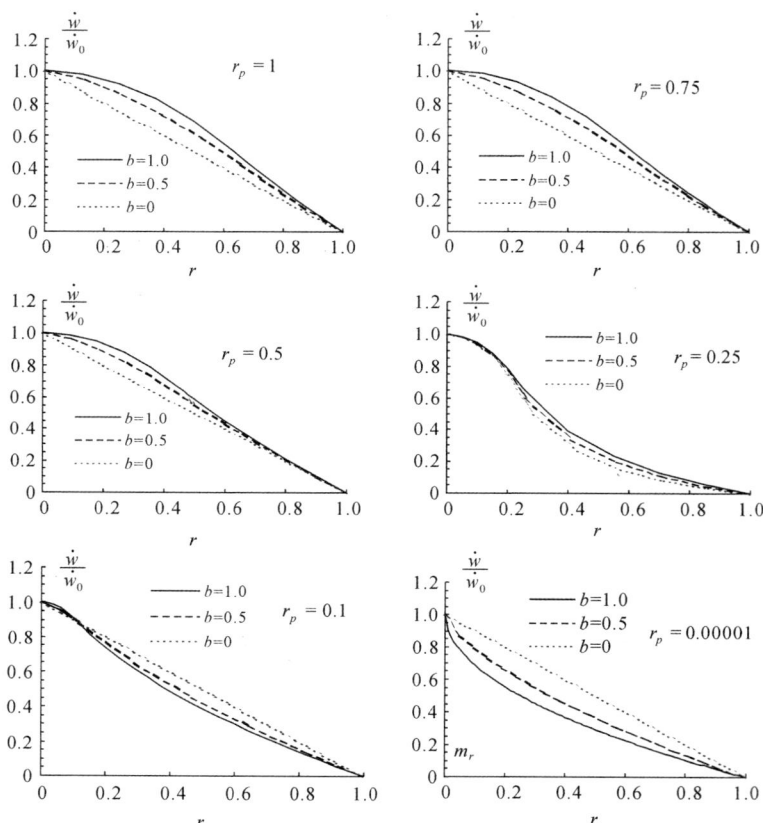

Fig. 5.10. Moment fields with different loading radii

It is seen that the velocity field with $b=0$ is independent of the loading radius r_p which is always a straight line and is not smooth at the plate center. The velocity field derived with the yield conditions with $0 < b \leqslant 1$, however, are functions of r_p and are smoothly connected at the plate center. This indicates that the velocity field in the plastic limit state in terms of the unified yield criterion is again more reasonable than that based on the Tresca criterion.

To verify the current results, the plastic limit solutions with specific parameters in the equations are explicitly given and compared with reported results:

(1) When $b = 0$, the moment field, velocity field, and the plastic limit load are obtained as

$$m_r = 1 - \frac{r^2}{(3 - 2r_p)r_p^2}, \quad m_\theta = 1, \quad \text{when} \quad 0 \leqslant r \leqslant r_p, \tag{5.24a}$$

$$m_r = 1 - \frac{3r - 2r_p}{3 - 2r_p}, \quad m_\theta = 1, \quad \text{when} \quad r_p \leqslant r \leqslant 1, \tag{5.24b}$$

$$\dot{w} = \dot{w}_0(1 - r), \tag{5.24c}$$

$$p = \frac{6}{(3 - 2r_p)r_p^2}. \tag{5.24d}$$

This result is identical to the solution in terms of the Tresca criterion given by Hodge (1963).

(2) When $r_p = 1$, the whole plate is under a uniformly distributed load. The solution is exactly the same as that in Section 5.3.1.

(3) Denoting P_T as the total load on the plate, i.e., $P_T = \pi r_p^2 p$, when r_p approaches zero, it approximates to a concentrated loading case. From the solution of the first case, it can be derived that $\lim_{r_p \to 0} P_T \equiv 2\pi$ which is independent of the variable b. It agrees with the results using the Tresca criterion and the Huber-von Mises criterion (Hodge, 1963).

5.3.2.2 Arbitrary Loading Distribution

Defining dimensionless variable $p(r) = P(r)a^2/M_0$ for a circular plate of radius a and thickness h subjected to an arbitrarily distributed axisymmetrical transverse pressure $\mu P(r)$, where μ is a plastic limit load factor, and $P(r)$ is a load distribution function, the equilibrium equation of the circular plate can be written with application of the axisymmetric condition as

$$\mathrm{d}(rm_r)/\mathrm{d}r - m_\theta = -\int \mu p(r)r\mathrm{d}r. \tag{5.25}$$

Substituting the yield criterion into Eq.(5.25) and then integrating Eq. (5.25), m_r located on the segments L_i are obtained as follows:

$$m_r = \frac{b_i}{1 - a_i} - r^{-1+a_i} \int r^{-a_i} \left[\int \mu p(r)\mathrm{d}r \right] \mathrm{d}r + c_i r^{-1+a_i}, \quad (i = 1, ..., 5), \tag{5.26}$$

where c_i $(i = 1, ..., 5)$ are integration constants and can be determined from the continuity and boundary conditions.

Assuming the load function $p(r) = \sum_{j=1}^{\infty} p_j r^{j-1}$, Eq.(5.26) becomes

$$m_r = \frac{b_i}{1 - a_i} - \mu \sum_{j=1}^{\infty} p_j \frac{r^{j+1}}{(j+1)(j+2-a_i)} + c_i r^{-1+a_i}. \quad (i = 1, ..., 5) \quad (5.27)$$

The field of velocity corresponding to the five sides L_i is obtained as

$$\dot{w} = \dot{w}_0(c_{1i} r^{1-a_i} + c_{2i}) \quad (i = 1, ..., 5), \tag{5.28}$$

where c_{1i} and c_{2i} $(i = 1, ..., 5)$ are integration constants, and $\dot{w}_0$ is the velocity at the plate center.

The plastic limit load of a plate is always taken to be the total limit load on the plate. The dimensionless total limit load of the plate is obtained as

$$P_T = 2\pi \int_0^1 \mu p(r) r \mathrm{d}r \quad \text{or} \quad P_T = 2\pi\mu \sum_{j=1}^{\infty} \frac{p_j}{j+1}. \tag{5.29}$$

In the plastic limit state, moments at the center $(r = 0)$ of a simply supported circular plate satisfy $m_r = m_\theta = 1$ (point A on the yield curves in Fig. 5.2), and moments at the simply supported edge $(r = 1)$ satisfy $m_r = 0$ and $m_\theta = 1$ (point C on the yield curves in Fig. 5.2). Bending moments of other points in the plate are on the lines AB and BC according to the normality requirement of plasticity. Thus, index "i" in Eqs.(5.27) and (5.28) for a simply supported circular plate takes values of 1 or 2 only corresponding to the line AB or the line BC in Fig.5.2, respectively. Assuming r_1 is a non-dimensional radius of a ring where the moments exactly correspond to point B in Fig.5.2, the boundary and continuity conditions can then be described as: (1) $m_r(r = 0) = 1$; (2) $m_r(r = r_1)$ is continuous and equal to d_1; (3) $m_r(r = 1) = 1$; (4) $\dot{w}(r = 0) = \dot{w}_0$; (5) $\dot{w}(r = r_1)$ and $\mathrm{d}\dot{w}/\mathrm{d}r(r = r_1)$ are continuous, (6) $\dot{w}(r = 1) = 0$. Accordingly, the integral coefficients c_1, c_2, c_{11}, c_{12}, c_{21} and c_{22} in Eqs.(5.27) and (5.28) can be determined as

$$c_1 = 0, \quad c_2 = -\frac{b_2}{1 - a_2} + \mu \sum_{j=1}^{\infty} \frac{p_j}{(j+1)(j+2-a_2)}, \tag{5.30}$$

$$c_{11} = -\frac{r_1^{-b(2+b)/(1+b)}}{(1+b)^2 - (2b+b^2)r_1^{1/(1+b)}}, \quad c_{21} = 1, \tag{5.31}$$

$$c_{12} = -c_{22} = -\frac{(1+b)^2}{(1+b)^2 - (2b+b^2)r_1^{1/(1+b)}}. \tag{5.32}$$

The loading factor μ is derived as

$$\mu = \frac{-d_1 + \frac{b_1}{1-a_1}}{\sum_{j=1}^{\infty} \frac{p_j r_1^{j+1}}{(j+1)(j+2-a_1)}}, \tag{5.33}$$

where r_1 satisfies the equation

$$d_1 = \frac{b_2}{1 - a_2} - \mu \sum_{j=1}^{\infty} \frac{p_j r_1^{j+1}}{(j+1)(j+2-a_2)} + c_2 r_1^{-1+a_2}. \tag{5.34}$$

The above equation is solved by a half-interval search method for r_1 in the interval $(0, 1)$ for a given value of b between 0 and 1. Substituting the value of r_1 into Eqs.(5.30)~(5.33), the moments and velocity distributions in Eqs.(5.27) and (5.28) can then be derived.

For a special case of $b=0$, the plastic solution becomes

$$m_r = 1 - \mu \sum_{j=1}^{\infty} \frac{p_j}{(j+1)(j+2)} r^{j+1}, \ m_\theta = 1,$$
$$\dot{w} = \dot{w}_0(1 - r) \text{ and } \mu = \frac{1}{\sum\limits_{j=1}^{\infty} \frac{p_j}{(j+1)(j+2)}}, \tag{5.35}$$

which are the same as those given by Ghorashi (1994) using the maximum principal stress criterion and the Tresca criterion. If a uniformly distributed load is applied, it becomes

$$m_r = 1 - r^2, \ m_\theta = 1, \ \dot{w} = \dot{w}_0(1 - r) \text{ and } \mu = 6, \tag{5.36}$$

which are identical to the results given by Hodge (1963).

Table 5.3 lists the plastic limit load factors and the total limit loads of the simply supported circular plate for five linearly distributed load functions in terms of the three particular yield criteria, namely, the Tresca criterion, the Huber-von Mises criterion (approximated by the unified yield criterion with $b=0.5$), and the twin shear stress criterion.

Table 5.3. Plastic limit loads for linearly distributed load

$p(r)$	r		$1 + r$		1		$2 - r$		$1 - r$	
		P_T		P_T		P_T		P_T		P_T
$b = 0$	12.00	25.13	4.00	20.94	6.00	18.85	4.00	16.76	12.01	12.57
$b = 1/2$	12.78	26.78	4.31	22.56	6.49	20.38	4.34	18.17	13.02	13.64
$b = 1$	13.36	27.98	4.53	23.73	6.84	21.49	4.58	19.17	13.75	14.40

From Table 5.3, different yield criteria make differences in the plastic limit load factor, the total limit load, and the load distribution function. The values in terms of the Huber-von Mises criterion are about 6.5% to 8.5% higher, while the values with the twin shear stress criterion are about 10.5% to 14.5% higher than those with the Tresca criterion. The increasing load distribution along the plate radius leads to minimal changes among different yield criteria. On the other hand, it makes significant differences to the total limit load, implying the increasing load distribution and improves the load-bearing capacity of a plate. For a uniformly distributed load, the load factor corresponding to the three criteria (unified yield criterion with $b=0$, $b=0.5$, and $b=1$) are 6.00, 6.49 and 6.84, respectively. The load factor with respect to the Huber-von Mises criterion reported by Hopkins and Wang (1954) is 6.51, which approximates closely the result using the unified yield criterion with $b=0.5$. Fig.5.11 and Fig.5.12 illustrate schematically the moment fields and velocity fields corresponding to these criteria for a simply supported circular plate subjected to the two types of linearly distributed load, i.e., $p = r$ and $p = 1 - r$. It is seen that the radial moment does not change much, while the circumferential moment varies significantly with respect to different yield criteria. The circumferential moment is not constant if the unified yield criterion with non-zero parameter b is applied. Locations of the maximum circumferential moment shift with the loading condition and the yield criterion. The larger r_1, the larger the total limit loads. The velocity fields with respect to the unified yield criterion (UYC) with non-zero parameter b ($0 < b \leqslant 1$) distribute nonlinearly along the plate, while the velocity field with respect to the Tresca criterion (or UYC with $b=0$) varies linearly and is singular at the plate center. The distribution of velocity also depends on the loading function as illustrated in Figs.5.11(b) and 5.12(b).

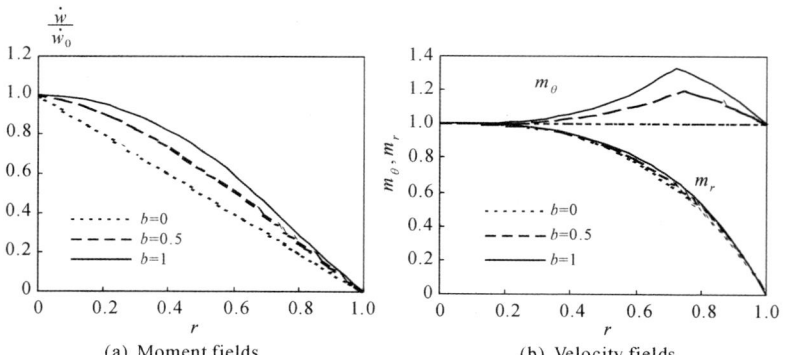

(a) Moment fields   (b) Velocity fields

Fig. 5.11. Moment and velocity fields of simply supported circular plate ($p = r$)

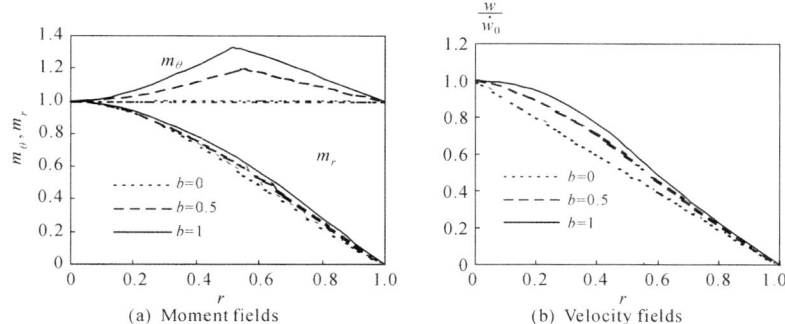

Fig. 5.12. Moments and velocity fields of simply supported circular plate ($p = 1-r$)

5.3.2.3 Edge Moment and Partial-uniform Load

For a plate loaded by edge moment and partial-uniform load as shown in Fig.5.13, there are two possible cases, i.e., Case (1) $d \leqslant r_0$ and Case (2) $d > r_0$, where d is the loading radius and r_0 is the dividing radius at which the moments m_r and m_θ correspond to the yield point B in Fig.5.2.

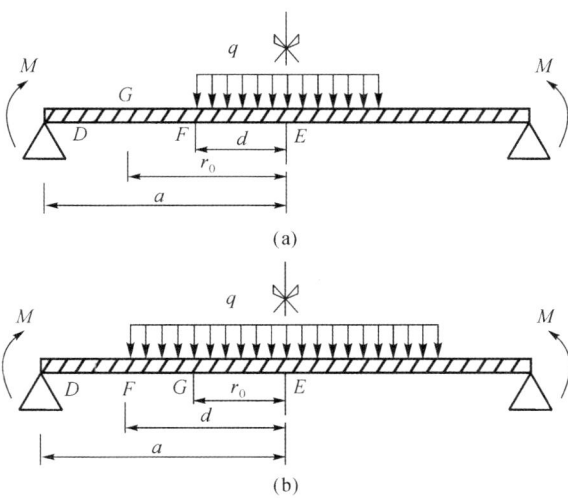

Fig. 5.13. Simply supported circular plate under partial-uniform load and edge moment

Case (1)

When point G lies on the line segment DF (Fig.5.13(a)), the equilibrium equations for EF, FG and GD are

$$\begin{cases} r\dfrac{dm_r}{dr} = (1+b)(m_0 - m_r) - \dfrac{q}{2}r^2 \\ m_\theta = (1+b)m_0 - bm_r \end{cases} \quad \text{for } EF, \qquad (5.37a)$$

$$\begin{cases} r\dfrac{dm_r}{dr} = (1+b)(m_0 - m_r) - \dfrac{q}{2}d^2 \\ m_\theta = (1+b)m_0 - bm_r \end{cases} \quad \text{for } FG, \qquad (5.37b)$$

$$\begin{cases} r\dfrac{dm_r}{dr} = m_0 - \dfrac{1}{1+b}m_r - \dfrac{q}{2}d^2 \\ m_\theta = m_0 - \dfrac{b}{1+b}m_r \end{cases} \quad \text{for } GD, \qquad (5.37c)$$

where q is the plastic limit load, and q satisfies

$$q = \frac{2(1+b)(3+b)m_0}{(2+b)\left[(3+b) - 2\left(\frac{d}{r_0}\right)^{1+b}\right]d^2}, \qquad (5.38)$$

and r_0 satisfies

$$2(1+b)\left(\frac{d}{r_0}\right)^{1+b} + (3+b)\left(\frac{a}{r_0}\right)^{\frac{1}{1+b}} - 2(2+b)\left(\frac{d}{r_0}\right)^{1+b}\left(\frac{a}{r_0}\right)^{\frac{1}{1+b}}$$
$$- \frac{(2+b)m_b}{(1+b)}\left[(3+b-2\left(\frac{d}{r_0}\right)^{1+b}\right]\left(\frac{a}{r_0}\right)^{\frac{1}{1+b}} = 0, \qquad (5.39)$$

where $m_b = M_b/M_0$.

Case (2)
When point G is on line segment EF (Fig.5.13(b)), the equilibrium equations are

$$\begin{cases} r\dfrac{dm_r}{dr} = (1+b)(m_0 - m_r) - \dfrac{q}{2}r^2 \\ M_\theta = (1+b)m_0 - bm_r \end{cases} \quad \text{for } EG, \qquad (5.40)$$

$$\begin{cases} r\dfrac{dm_r}{dr} = m_0 - \dfrac{1}{1+b}m_r - \dfrac{q}{2}r^2 \\ m_\theta = m_0 - \dfrac{b}{1+b}m_r \end{cases} \quad \text{for } GF, \qquad (5.41)$$

$$\begin{cases} r\dfrac{dm_r}{dr} = m_0 - \dfrac{1}{1+b}m_r - \dfrac{q}{2}d^2 \\ m_\theta = m_0 - \dfrac{b}{1+b}m_r \end{cases} \quad \text{for } GD, \qquad (5.42)$$

where q satisfies

$$q = \frac{6+2b}{2+b} \frac{m_0}{r_0^2},$$ (5.43)

and r_0 satisfies

$$\frac{3+b}{3+2b}\left(\frac{d}{r_0}\right)^{\frac{3+2b}{1+b}} 2(2+b)\left(\frac{a}{r_0}\right)^{\frac{1}{1+b}} + (3+b)\left(\frac{d}{r_0}\right)^2\left[\left(\frac{a}{r_0}\right)^{\frac{1}{1+b}} - \left(\frac{d}{r_0}\right)^{\frac{1}{1+b}}\right]$$

$$+ \frac{(2+b)m_b}{(1+b)}\left(\frac{a}{r_0}\right)^{\frac{1}{1+b}} + \frac{2b(2+b)}{3+2b} = 0.$$

(5.44)

When $d = r_0$, point F and G overlap, the moment fields of Case (1) and Case (2) become the same, and r_0 satisfies

$$r_0 = \left(\frac{1}{2} + \frac{(2+b)m_b}{2(1+b)}\right)^{1+b} a.$$ (5.45)

5.3.2.4 Edge Moment and Partial-linear Load

For a circular plate subjected to partial-linear load and edge moment as shown in Fig.5.14, there also are two possible cases, i.e. Case (1) $d \leqslant r_0$ and Case (2) $d > r_0$.

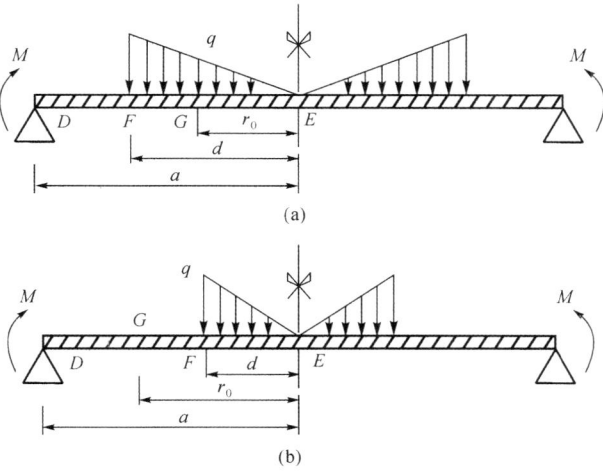

(a)

(b)

Fig. 5.14. Partial-linear load and edge moment

Case (1)

When point G lies on line segment EF (Fig.5.14(a)), the plastic limit loading is

$$q = \frac{3(4+b)}{2+b} \frac{dm_0}{r_0^2}, \qquad (5.46)$$

and r_0 satisfies

$$\left(\frac{(2+b)m_b}{(1+b)} + \frac{(4+b)d^3 - (2+b)r_0^3}{r_0^3} \right) \left(\frac{a}{r_0} \right)^{\frac{1}{1+b}}$$

$$- \frac{3(1+b)(4+b)}{(4+3b)} \left(\frac{d}{r_0} \right)^3 \left(\frac{d}{r_0} \right)^{\frac{1}{1+b}} + \frac{3b(2+b)}{4+3b} = 0. \qquad (5.47)$$

Case (2)

When point G is on line segment FD (Fig.5.14(b)), the plastic limit loading is

$$q = \frac{3(1+b)(4+b)m_0}{(2+b) \left[(4+b) - 3 \left(\frac{d}{r_0} \right)^{1+b} \right] d^2}, \qquad (5.48)$$

and r_0 satisfies

$$3(1+b) \left(\frac{d}{r_0} \right)^{1+b} + (4+b) \left(\frac{a}{r_0} \right)^{\frac{1}{1+b}} - 3(2+b) \left(\frac{d}{r_0} \right)^{1+b} \left(\frac{a}{r_0} \right)^{\frac{1}{1+b}}$$

$$- \frac{(2+b)m_b}{(1+b)} \left[(4+b) - 3 \left(\frac{d}{r_0} \right)^{1+b} \right] \left(\frac{a}{r_0} \right)^{\frac{1}{1+b}} = 0 \qquad (5.49)$$

When $d = r_0$, i.e., points F and G overlap, and r_0 satisfies

$$r_0 = \left(\frac{2}{3} + \frac{(2+b)m_b}{3(1+b)} \right)^{1+b} a. \qquad (5.50)$$

The relationship between the plastic limit load and parameter b is given in Table 5.4 and Fig.5.15 for the loading cases of (1) partial-uniform loading with edge moment, and (2) partial-linear load with edge moment, where $d = 0.6a$ and $M_b = 0.3M_0$. It is seen that the plastic limit load is different with respect to different yield criteria. When $b = 0$ (Tresca criterion), the plastic limit load is the minimum; when $b = 1$ (the twin-shear yield criterion), the plastic limit load is the maximum.

Table 5.4. Relationships of limit load q and parameter b

Loading type	b	0	0.1	0.2	0.3	0.4	0.5
1	r_0	0.7000	0.6797	0.6576	0.6418	0.6284	0.6171
	q	6.4815	6.6926	6.8850	7.0610	7.2226	7.3715
2	r_0	0.7412	0.7189	0.7002	0.6844	0.6709	0.6592
	q	10.6061	10.9061	11.1799	11.4303	11.6603	11.8721
Loading type	b	0.6	0.7	0.8	0.9	1.0	
1	r_0	0.6074	0.5990	0.5915	0.5847	0.5783	
	q	7.5093	7.6374	7.7568	7.8863	7.9729	
2	r_0	0.6491	0.6402	0.6324	0.6254	0.6192	
	q	12.0679	12.2495	12.4186	12.5765	12.7243	

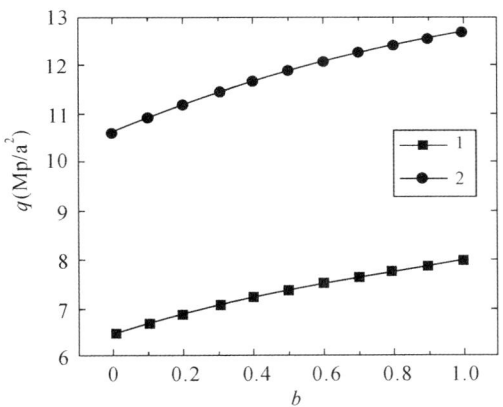

Fig. 5.15. Plastic limit load q versus unified yield criterion parameter b

5.4 Unified Solutions of Simply Supported Circular Plate for SD Materials

The unified solutions of a circular plate for non-SD materials are extended to SD materials. The unified plastic limit of a clamped circular plate with SD materials (strength differential effect in tension and compression) by using the unified strength theory was derived by Wei and Yu (2001). Unified plastic limit analyses of simply supported circular plates with different tensile and compressive strength under uniform annular load were derived by Wang and Yu (2002; 2003a). Plastic limit analysis of simply supported circular plates with different tensile and compressive strength under linear distributed load was given by Wang and Yu (2003). In this section, we will analyze the plastic limit load of simply supported circular plates that are made of SD materials. A parameter α is introduced, which is the ratio of the negative limit bending moment and positive limit bending moment, $\alpha = +m_p/ - m_p$. The yield loci in terms of the generalized stresses m_r and m_θ is shown in Fig.5.16 with respect to different values of b.

The generalized yield criterion for a plate is similar to the yield criterion in the plane stress state. The generalized unified yield criterion for a plate is similar to the unified yield criterion in the plane stress state. However, two cases have to be considered, i.e., (a) $+m_p \neq -m_p$ and (b) $+m_p = -m_p$. They are shown in Fig.5.16.

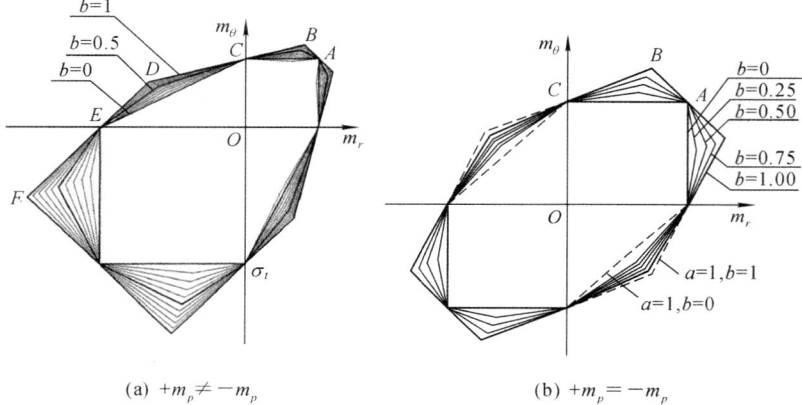

(a) $+m_p \neq -m_p$ (b) $+m_p = -m_p$

Fig. 5.16. Generalized unified yield criterion for plate

The yield criterion of the unified strength theory in terms of generalized stresses is

$$m_r - \frac{\alpha b}{1+b}m_\theta = m_p, \quad m_\theta - \frac{\alpha b}{1+b}m_r = m_p, \tag{5.51a}$$

$$m_r - \frac{\alpha}{1+b}m_\theta = m_p, \quad m_\theta - \frac{\alpha}{1+b}m_r = m_p, \tag{5.51b}$$

$$\frac{\alpha}{1+b}(bm_r + m_\theta) = -m_p, \quad \frac{\alpha}{1+b}(bm_\theta + m_r) = -m_p, \tag{5.51c}$$

$$\frac{1}{1+b}(m_r + bm_\theta) = m_p, \quad \frac{1}{1+b}(m_\theta + bm_r) = m_p, \tag{5.51d}$$

$$\frac{1}{1+b}(m_r + \alpha m_\theta) = m_p, \quad \frac{1}{1+b}(m_\theta + \alpha m_r) = m_p, \tag{5.51e}$$

$$\frac{b}{1+b}(m_r + \alpha m_\theta) = m_p, \quad \frac{b}{1+b}(m_\theta + \alpha m_r) = m_p. \tag{5.51f}$$

The center point of the simply supported plate satisfies $m_r(r = 0) = m_\theta(r = 0)$ (point A in Fig.5.16); and simply supported boundary satisfies $m_r(r = a) = 0$ (point C in Fig. 5.16), $m_\theta = (1 + \alpha)\, m_r$ (point B in Fig.5.16). Stress states of all points in the plate are located on parts AB and BC. The yield conditions of parts AB, BC in Fig.5.16 can be expressed as

$$AB: \quad \frac{b}{1+b} m_r + \frac{1}{1+b} m_\theta = m_p, \qquad (5.52a)$$

$$BC: \quad m_\theta - \frac{\alpha b}{1+b} m_r = m_p. \qquad (5.52b)$$

5.4.1 Partial-uniform Load

There are two possible cases where the plate is subjected to a partial-uniform load as shown in Figs.5.17(a) and 5.17(b), i.e., Case (1) $d \leqslant r_0$ and Case (2) $d > r_0$, respectively, where d is the loading radius. The moments m_r and m_θ at point G with a radius of r_0 are located at point B in Fig.5.16.

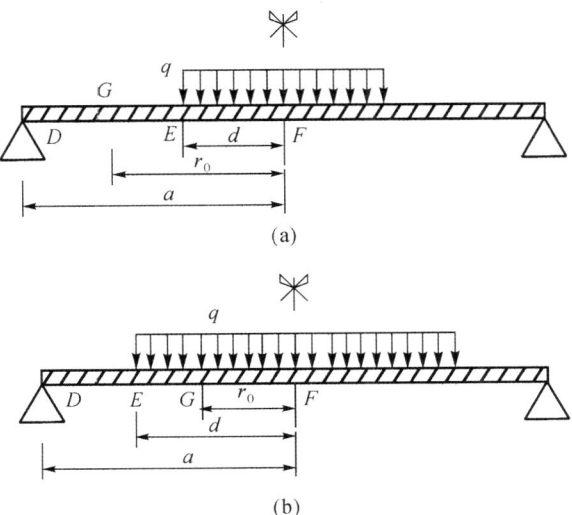

Fig. 5.17. Simply support circular plate subjected to partial-uniform load

Case (1)
When point G lies on segment DE, the equilibrium equations for different parts are given by

$$EF : \begin{cases} r\dfrac{dm_r}{dr} = (1+b)m_p - (1+b)m_r - \dfrac{pr^2}{2}, \\ m_\theta = (1+b)m_p - bm_r, \end{cases} \tag{5.53a}$$

$$GE : \begin{cases} r\dfrac{dm_r}{dr} = (1+b)m_p - (1+b)m_r - \dfrac{pd^2}{2}, \\ m_\theta = (1+b)m_p - bm_r, \end{cases} \tag{5.53b}$$

$$DG : \begin{cases} r\dfrac{dm_r}{dr} = m_p - \dfrac{1+b-\alpha b}{1+b}m_r - \dfrac{pd^2}{2}, \\ m_\theta = m_p + \dfrac{\alpha b}{1+b}m_r. \end{cases} \tag{5.53c}$$

Case (2)
When point G is located on segment EF, the equilibrium equations are

$$GF : \begin{cases} r\dfrac{dm_r}{dr} = (1+b)m_p - (1+b)m_r - \dfrac{pr^2}{2}, \\ m_\theta = (1+b)m_p - bm_r, \end{cases} \tag{5.54a}$$

$$EG : \begin{cases} r\dfrac{dm_r}{dr} = m_p - \dfrac{1+b-\alpha b}{1+b}m_r - \dfrac{pr^2}{2}, \\ m_\theta = m_p + \dfrac{\alpha b}{1+b}m_r, \end{cases} \tag{5.54b}$$

$$DG : \begin{cases} r\dfrac{dm_r}{dr} = m_p - \dfrac{1+b-\alpha b}{1+b}m_r - \dfrac{pd^2}{2}, \\ m_\theta = m_p + \dfrac{\alpha b}{1+b}m_r. \end{cases} \tag{5.54c}$$

The boundary and continuity conditions are: (1) at point F, $m_r(r=0)$ is a finite value; (2) at point D, $m_r(r=0)=0$; (3) at point E, $m_r(r=d)$ is continuous; (4) at point G, $m_r(r=r_0)$ continuous and equal to $(1+b)m_p/(1+b+\alpha)$.

Case (1)
Integrate Eqs.(5.53a) and (5.53c),

$$m_r = m_p - \frac{pr^2}{6+2b} + c_1 r^{-(1+b)} \quad m_\theta = (1+b)m_p - bm_r \quad \text{for} \quad EF, \tag{5.55a}$$

$$m_r = m_p - \frac{pd^2}{2(1+b)} + c_2 r^{-(1+b)} \quad m_\theta = (1+b)m_p - bm_r \quad \text{for} \quad GE, \tag{5.55b}$$

$$m_r = \frac{1+b}{1+b-\alpha b}(m_p - \frac{pd^2}{2}) + c_3 r^{-\frac{1+b-\alpha b}{1+b}} \quad m_\theta = m_p + \frac{\alpha b}{1+b}m_r \quad \text{for} \quad DG. \tag{5.55c}$$

With reference to the boundary and continuity conditions (1) to (4), we have

$$c_1 = 0, \tag{5.56a}$$

$$c_2 = \frac{pd^{3+b}}{(1+b)(3+b)}, \tag{5.56b}$$

$$c_3 = \left[-\frac{1+b}{1+b-\alpha b} \left(m_p - \frac{pd^2}{2} \right) \right] a^{\frac{1+b-\alpha b}{1+b}}. \tag{5.56c}$$

The plastic limit load is

$$p = \frac{2(1+b)(3+b)\alpha m_p}{(1+b+\alpha)d^2 \left[(3+b) - 2\left(\frac{d}{r_0}\right)^{1+b} \right]}, \tag{5.57}$$

where r_0 satisfies

$$2\alpha(1+b)\left(\frac{d}{r_0}\right)^{1+b} + (3+b)(1+b-\alpha b)\left(\frac{a}{r_0}\right)^{\frac{1+b-\alpha b}{1+b}} \\ -2(1+b+\alpha)\left(\frac{d}{r_0}\right)^{1+b}\left(\frac{a}{r_0}\right)^{\frac{1+b-\alpha b}{1+b}} = 0. \tag{5.58}$$

Substituting c_1, c_2 and c_3 into Eqs.(5.55a) and (5.55c), the moment fields for Case (1) can be derived.

Case (2)

When point G located between E and F, from Eq.(5.54),

$$\begin{cases} m_r = m_p - \dfrac{pr^2}{6+2b} + c_4 r^{-(1+b)} \\ m_\theta = (1+b)m_p - bm_r \end{cases} \quad \text{for } GF, \tag{5.59a}$$

$$\begin{cases} m_r = \dfrac{1+b}{1+b-\alpha b}m_p - \dfrac{(1+b)pr^2}{2(3+3b-\alpha b)} + c_5 r^{-\frac{1+b-\alpha b}{1+b}} \\ m_\theta = m_p + \dfrac{\alpha b}{1+b}m_r \end{cases} \quad \text{for } EG, \tag{5.59b}$$

$$\begin{cases} m_r = \dfrac{1+b}{1+b-\alpha b}m_p - \dfrac{1+b}{1+b-\alpha b}\dfrac{pd^2}{2} + c_6 r^{-\frac{1+b-\alpha b}{1+b}} \\ m_\theta = m_p + \dfrac{\alpha b}{1+b}m_r \end{cases} \quad \text{for } DG. \tag{5.59c}$$

Applying the boundary and continuity conditions (1) to (4), we obtain

$$c_4 = 0, \tag{5.60a}$$

$$c_5 = -\frac{1+b}{1+b-\alpha b}m_p\alpha^{\frac{1+b-\alpha b}{1+b}} + \frac{(1+b)pd^2}{2(1+b-\alpha b)}\left(\alpha^{\frac{1+b-\alpha b}{1+b}} - d^{\frac{1+b-\alpha b}{1+b}}\right)$$

$$+ \frac{(1+b)p}{2(3+3b-\alpha b)}d^{\frac{3+3t-\alpha b}{1-b}},$$

(5.60b)

$$c_6 = \left(-\frac{1+b}{1+b-\alpha b}m_p + \frac{1+b}{1+b-\alpha b}\frac{pd^2}{2}\right)\alpha^{\frac{1+b-\alpha b}{1+b}}.$$

(5.60c)

The plastic limit load is derived as

$$p = \frac{(6+2b)\alpha m_p}{(1+b-\alpha)r_0^2},$$

(5.61)

where r_0 satisfies

$$\alpha(1+b) - \frac{\alpha(3+b)(1+b-\alpha b)}{3+3b-\alpha b} - (1+b+\alpha)\left(\frac{a}{r_0}\right)^{\frac{1+b-\alpha b}{1+b}}$$

$$+ \alpha(3+b)\left(\frac{d}{r_0}\right)^2\left[\left(\frac{a}{r_0}\right)^{\frac{1+b-\alpha b}{1+b}} - \left(\frac{d}{r_0}\right)^{\frac{1+b-\alpha b}{1+b}}\right]$$

(5.62)

$$+ \frac{\alpha(3+b)(1+b-\alpha b)}{3+3b-\alpha b}\left(\frac{d}{r_0}\right)^{\frac{3+3b-\alpha b}{1+b}} = 0.$$

Special Case
When $r_0 = d$, i.e., point G overlaps point E, the moment fields of the two cases are the same as

$$\begin{cases} m_r = m_p - \dfrac{pr^2}{6+2b} + c_7 r^{-(1+b)} \\ m_\theta = (1+b)m_r - bm_r \end{cases} \qquad \text{for } EF,$$

(5.63a)

$$\begin{cases} m_r = \dfrac{1+b}{1+b-\alpha b}\left(m_p - \dfrac{pd^2}{2}\right) + c_8 r^{-\frac{1+b-\alpha b}{1+b}} \\ m_\theta = m_p + \dfrac{\alpha b}{1+b}m_r \end{cases} \qquad \text{for } DE.$$

(5.63b)

From the boundary and continuity conditions (1) to (4),

$$c_7 = 0,$$

(5.64a)

$$c_8 = \left[-\frac{1+b}{1+b-\alpha b}\left(m_p - \frac{pd^2}{2}\right)\right]\alpha^{\frac{1+b-\alpha b}{1+b}},$$

(5.64b)

the plastic limit load is simplified as

$$p = \frac{\alpha(6 + 2b)m_p}{(1 + b + \alpha)d^2}, \tag{5.65}$$

where $d = r_0$ and

$$\left(\frac{\alpha}{d}\right)^{\frac{1+b-\alpha b}{1+b}} = \frac{-2\alpha}{1 + b - \alpha b - 2\alpha}. \tag{5.66}$$

Denoting the critical loading radius as d_0, when $d \leqslant d_0$, i.e. Case (1), Eq.(5.58) gives a unique solution in the region of $d < r \leqslant a$, while Eq.(5.62) has no solution in $0 < r \leqslant d$. Thus the equations for Case (1) are adopted to solve the plastic limit load and moment fields. When, on the other hand, Eq.(5.58) has no solution in $d < r \leqslant a$, while Eq.(5.62) can be used to solve r_0 in the region of $0 < r \leqslant d$. In this case, point G is on the segment EF. The plastic limit load and moment fields can be obtained from Case (2).

Figs.5.18 to 5.21 show the moment fields when $\alpha = 0.1$, $d = a$, $0.5a$, $0.1a$ and $0.00001a$, respectively. The plastic limit loads with respect to different values of unified yield criterion parameter b are plotted in Fig.5.22.

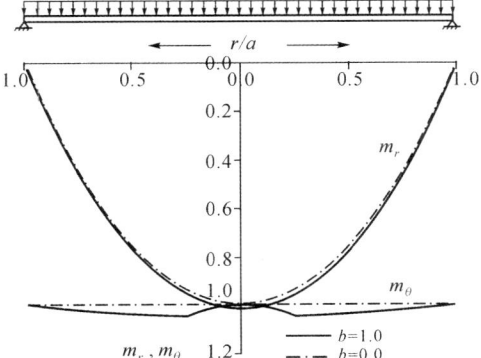

Fig. 5.18. Moment fields when $d = a$

From Fig.5.18 to Fig.5.21, the moment fields depend on the unified yield criterion parameter b. The higher the parameter b, the higher the corresponding moment and the plastic limit load. When $b = 0$, the moment m_θ is independent of the loading radius, and $m_\theta = m_p$. When $b \neq 0$, m_θ varies along the radial direction, and the variation of m_θ gives a more reasonable representation than that with $b = 0$. When d approaches zero, which corresponds to a concentrated loading case, the solution of m_θ has no singularity at $r = 0$ when the parameter b is equal to 0. When $b \neq 0$, the solutions of both m_θ and m_r have singularity at $r = 0$. It can be said that when using the unified strength theory, the singularity of the moment fields at the plate center for a concentrated loading case can be properly represented.

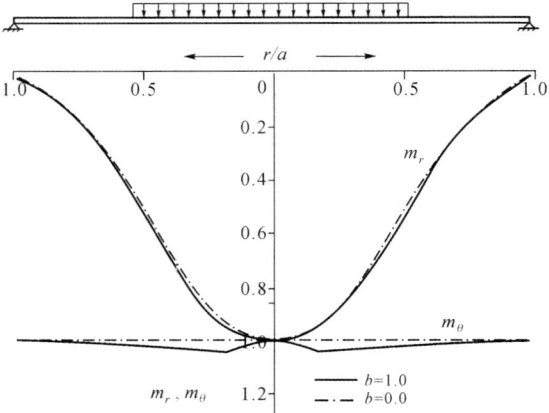

Fig. 5.19. Moment fields when $d = 0.5a$

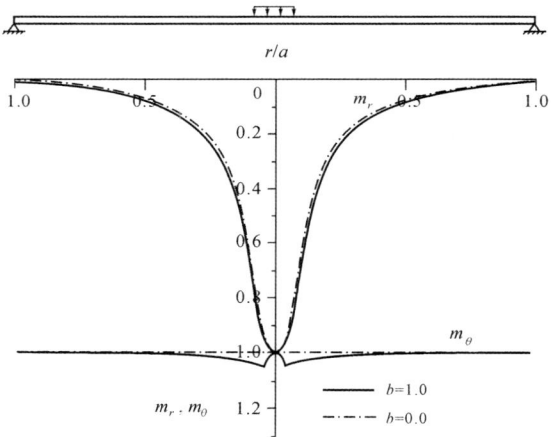

Fig. 5.20. Moment fields when $d = 0.1a$

5.4.2 Linearly Distributed Load

For a simply supported circular plate under linearly distributed load, two different loading cases are discussed in the following context.

Case (1)
For a linear pressure load as shown in Fig.5.23, when the moment of point F falls on point B in Fig.5.16, the moment fields satisfy the boundary and continuity conditions of (1) $m_r = m_\theta$ at $r = 0$; (2) m_r and m_θ are continuous at point F or $r = r_0$; (3) $m_r = 0$ at $r = a$ of the outer edge.

The equilibrium equations of lines EF and FD are

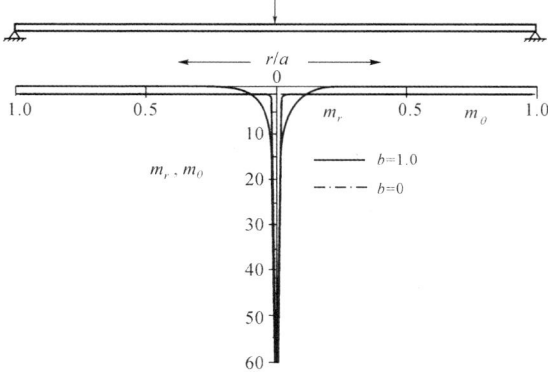

Fig. 5.21. Moment fields when $d = 0.00001a$

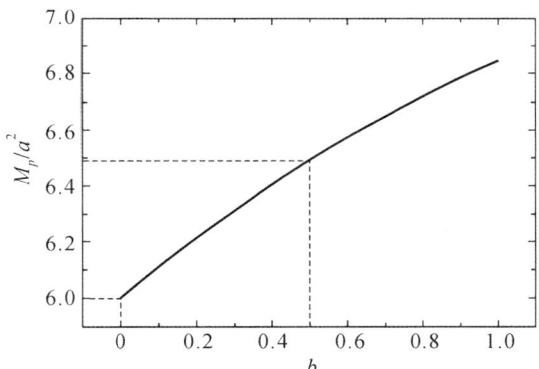

Fig. 5.22. Plastic limit loads with respect to different values of unified yield criterion parameter b

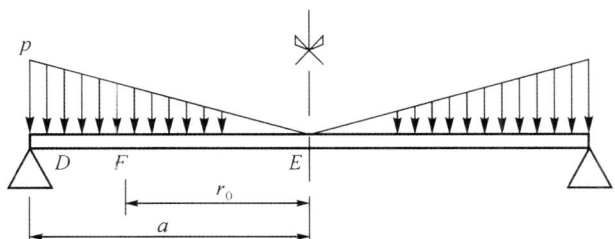

Fig. 5.23. Linearly distributed loading (Case (1))

$$\begin{cases} r\dfrac{\mathrm{d}m_r}{\mathrm{d}r} = (1+b)m_p - (1+b)m_r - \dfrac{p}{3a}r^3 \\ m_\theta = (1+b)m_p - bm_r \end{cases} \quad \text{for } EF, \qquad (5.67a)$$

$$\begin{cases} r\dfrac{\mathrm{d}m_r}{\mathrm{d}r} = m_p - \dfrac{1+(1-\alpha)b}{1+b}m_r - \dfrac{p}{3a}r^3 \\ m_\theta = (1+b)m_p - \dfrac{\alpha b}{1+b}m_r \end{cases} \quad \text{for } FD. \qquad (5.67b)$$

Solving Eq.(5.67) with references to the boundary and continuity conditions yields

$$
\begin{cases}
m_r = c_1 r^{-(1+b)} + m_p - \dfrac{pr^3}{3a(4+b)} \\[2mm]
m_\theta = -c_1 b r^{-(1+b)} + m_p + \dfrac{bpr^3}{3a(4+b)}
\end{cases}
\qquad \text{for } EF, \qquad (5.68a)
$$

$$
\begin{cases}
m_r = c_2 r^{-\frac{1+(1-\alpha)b}{1+b}} + \dfrac{1+b}{1+(1-\alpha)b} m_p - \dfrac{(1+b)pr^3}{3a(4+4b-b\alpha)} \\[2mm]
m_\theta = c_2 \dfrac{\alpha b}{1+b} r^{-\frac{1+(1-\alpha)b}{1+b}} + \dfrac{1+b}{1+(1-\alpha)b} m_p - \dfrac{\alpha b pr^3}{3a(4+3b)}
\end{cases}
\text{for } FG.
$$

$$(5.68b)$$

The two coefficients c_1 and c_2 are derived as

$$ c_1 = 0, \qquad (5.69a) $$

$$ c_2 = \left[-\dfrac{(1+b)m_p}{1+(1-\alpha)b} + \dfrac{(1+b)pa^2}{3(4+4b-b\alpha)} \right] \alpha^{\frac{1+(1-\alpha)b}{1+b}}. \qquad (5.69b) $$

The limit loading is

$$ p = \dfrac{3\alpha(4+b)\alpha m_p}{(1+\alpha+b)r_0^3}, \qquad (5.70) $$

and r_0 satisfies

$$
\alpha(4+b)[1+(1-\alpha)b]\left(\dfrac{a}{r_0}\right)^3 \left(\dfrac{a}{r_0}\right)^{\frac{1+(1-\alpha)b}{1+b}} - (4+4b-b\alpha)(1+\alpha
$$
$$
+b)\left(\dfrac{a}{r_0}\right)^{\frac{1+(1-\alpha)b}{1+b}} + 3\alpha b(1+\alpha+b) = 0 \qquad (5.71)
$$

Case (2)

For a linear pressure load as shown in Fig.5.24, when the moment of point F is at point B in Fig.5.16, the equilibrium equations of lines EF and FD are

$$
\begin{cases}
r\dfrac{dm_r}{dr} = (1+b)m_p - (1+b)m_r - \dfrac{pr^2}{2} + \dfrac{p}{3a}r^3 \\[2mm]
m_\theta = (1+b)m_p - bm_r
\end{cases}
\text{for } EF, \quad (5.72a)
$$

$$
\begin{cases}
r\dfrac{dm_r}{dr} = m_p - \dfrac{1+(1-\alpha)b}{1+b}m_r - \dfrac{pr^2}{2} + \dfrac{p}{3a}r^3 \\[2mm]
m_\theta = (1+b)m_p - \dfrac{\alpha b}{1+b}m_r
\end{cases}
\text{for } FD. \quad (5.72b)
$$

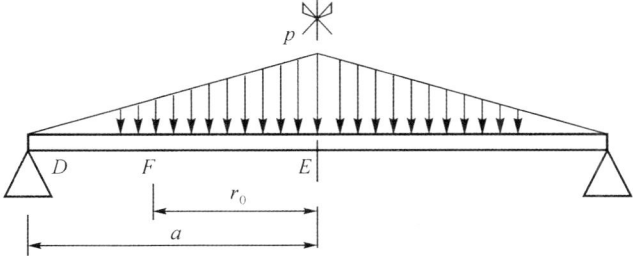

Fig. 5.24. Linearly distributed loading (Case (2))

Eq.(5.72) can be solved with application of the boundary and continuity conditions as

$$
\begin{cases}
m_r = c_3 r^{-(1+b)} + m_p - \dfrac{pr^2}{2(3+b)} + \dfrac{pr^3}{3a(4+b)} \\[2mm]
m_\theta = -c_1 b r^{-(1+b)} + m_p + \dfrac{bpr^2}{2(3+b)} - \dfrac{bpr^3}{3a(4+b)}
\end{cases}
\quad \text{for } EF, \qquad (5.73a)
$$

$$
\begin{cases}
m_r = c_4 r^{-\frac{1+(1-\alpha)b}{1+b}} + \dfrac{1+b}{1+(1-\alpha)b} m_p \\[2mm]
\quad - \dfrac{(1+b)pr^2}{2(3+3b-b\alpha)} + \dfrac{(1+b)pr^3}{3\alpha(4+4b-b\alpha)} \\[4mm]
m_\theta = c_4 \dfrac{\alpha b}{1+b} r^{-\frac{1+(1-\alpha)b}{1+b}} + \dfrac{1+b}{1+(1-\alpha)b} m_p \\[2mm]
\quad + \dfrac{\alpha b pr^2}{2(3+3b-b\alpha)} - \dfrac{\alpha b pr^3}{3\alpha(4+3b)}
\end{cases}
\quad \text{for } FD. \qquad (5.73b)
$$

The two coefficients c_3 and c_4 are derived as

$$
c_3 = 0, \qquad (5.74a)
$$

$$
c_4 = \left[\frac{(1+b)(6+6b-b\alpha)pa^2}{6(3+3b-b\alpha)(4+4b-b\alpha)} - \frac{(1+b)}{1+(1-\alpha)b} m_p \right] \alpha^{\frac{1}{1+b}}. \qquad (5.74b)
$$

The limit loading is obtained as

$$
p = \frac{6\alpha a(3+b)(4+b)m_p}{(1+\alpha+b)[3a(4+b)r_0^2 - 2(3+b)r_0^3]}, \qquad (5.75)
$$

and r_0 satisfies

$$6\alpha ab(1 + \alpha + b)(4 + b)(4 + b - b\alpha)r_0^2 - (1 + \alpha + b)(3 + 3b$$

$$- b\alpha)(4 + 4b - b\alpha)(3a^2(4 + b)r_0^2 - 2(3 + b)r_0^3) \left(\frac{a}{r_0} \right)^{\frac{(1+(1-\alpha)b)}{1+b}} \qquad (5.76)$$

$$- 6\alpha b(1 + \alpha + b)(3 + 3b - b\alpha)(3 + b)r_0^3 = 0.$$

From Eqs.(5.70) and (5.75), the relationship of the plastic limit load and the unified strength theory parameters b and α can be determined as shown in Figs.5.25 and 5.26. It is seen that the plastic limit load is significantly affected by the unified yield criterion parameters b and α. When α is given, the plastic limit load increases with the increase in b. When $b = 0$, which corresponds to the Mohr-Coulomb criterion, the plastic limit load is the minimum, and when $b = 1$ corresponding to the twin-shear strength criterion, the plastic limit load gives the maximum value. For any specific value of parameter b, the plastic limit load increases with the increase in α. When $\alpha = 1$, it gives the same results as those based on the unified yield criterion.

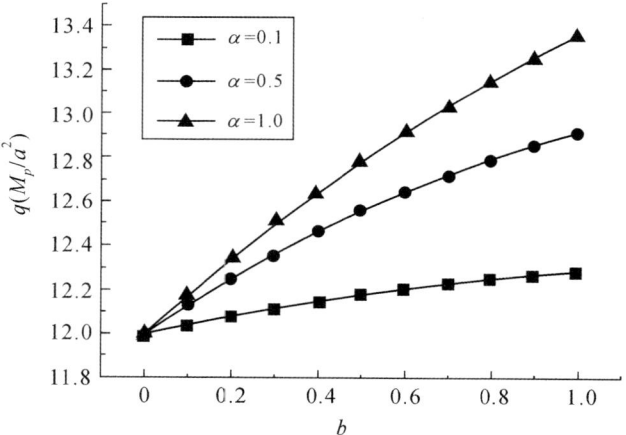

Fig. 5.25. Relation curves of limit loading and unified strength theory parameters b, α (Case (1))

It is seen from Fig.5.25 and Fig.5.26 that the unified strength theory parameters b and the ratio of material strength in tension and in compression α have a significant influence on the limit bearing capacity of a simply supported circular plate. The unified strength theory provides us with an effective approach for studying these effects and for raising the bearing capacities of engineering structures more than the Tresca criterion and the Mohr-Coulomb criterion ($b = 0$). So, there is a considerable economical benefit in using the

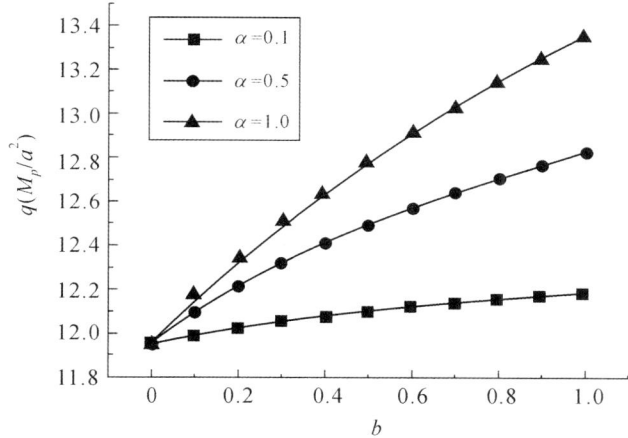

Fig. 5.26. Relation curves of limit loading and unified strength theory parameters b, α (Case (2))

new results if the strength of material is adapted to the new yield criterion ($b > 0$). This brings a tremendous energy saving and reduction in pollution.

5.5 Summary

The unified solution to the plastic limit load for simply supported circular plates made of either non-SD materials or SD materials under various loading conditions are derived. They are obtained by applying the unified strength theory in the plane stress state. The unified solution gives a series of new results, and establishes a relationship between various results and encompasses solutions using the Tresca criterion, the maximum stress criterion, the Huber-von Mises criterion, the Mohr-Coulomb criterion and the twin-shear criterion as special cases. The plastic limit load of a simply supported circular plate under a uniformly distributed load is $p = (6 + 2b)/(2 + b)r_0^2$ for non-SD materials, $p = \alpha(6 + 2b)/((1 + b + \alpha)r_0^2)$ for SD materials, where p is the normalized plastic limit load, and $p = Pa^2/M_p$. A series of solutions for various materials can be deduced from the unified solution. For an easier understanding of the current results, some specific solutions are given below:

Non-SD materials:

$p = 6.0$, it follows the Tresca criterion (or the unified yield criterion with b=0);

$p = 6.51$, it follows the Huber-von Mises criterion (Hopkins-Wang 1954, numerical integrated method);

$p = 6.46$, it follows the Huber-von Mises criterion (Sokolovsky's solution, 1955);

$p = 6.49$, it follows the unified yield criterion with $b = 1/2$ (Ma and He, 1994; Ma et al., 1995);

$p = 6.84$, it satisfies the twin-shear yield criterion (Li, 1988; Huang and Zeng, 1989) or the unified yield criterion with $b = 1$ (Ma and He, 1994; Ma et al, 1995).

SD materials:

$p = 6\alpha/((1+\alpha)r_0^2)$ for SD materials satisfying the Mohr-Coulomb criterion ($b = 0$);

$p = 14\alpha/((3+2\alpha)r_0^2)$ for SD materials satisfying a new criterion ($b = 1/2$);

$p = 9\alpha/((2 + \alpha)r_0^2)$ for SD materials satisfying the twin-shear strength criterion ($b = 1$).

A series of research exercises were carried out to show the effects of strength theory on the results of elasto-plastic analysis and the load-bearing capacities of a simply supported circular plate for non-SD materials and SD materials. The choice of strength theory has a significant influence on these results. The unified yield criterion and unified strength theory provide us with an effective approach for studying these effects. The unified plastic limit of a clamped circular plate for non-SD materials and SD materials will be described in Chapter 6.

5.6 Problems

Problem 5.1 Determine the limit bearing capacity of a simply supported circular plate by using of the Tresca criterion.

Problem 5.2 Determine the limit bearing capacity of a simply supported circular plate by using of the unified yield criterion ($b = 0$).

Problem 5.3 Determine the limit bearing capacity of a simply supported circular plate by using of the unified yield criterion ($b = 0.5$).

Problem 5.4 Determine the limit bearing capacity of a simply supported circular plate by using of the unified yield criterion ($b = 0.8$).

Problem 5.5 Determine the limit bearing capacity of a simply supported circular plate by using of the unified yield criterion ($b = 1.0$).

Problem 5.6 Determine the limit bearing capacity of a simply supported circular plate by using of the unified strength theory ($b = 0$).

Problem 5.7 Determine the limit bearing capacity of a simply supported circular plate by using of the unified strength theory ($b = 0.5$).

Problem 5.8 Determine the limit bearing capacity of a simply supported circular plate by using of the unified strength theory ($b = 0.8$).

Problem 5.9 Determine the limit bearing capacity of a simply supported circular plate by using of the unified strength theory ($b = 1.0$).

Problem 5.10 A simply supported circular plate under uniform annular load is shown in Fig.5.27. The relationship of limit load q and b for a special case is shown in Fig.5.28. Please derive the unified solution for the plate. The referenced figure similar to Fig.5.8 is shown in Fig.5.29.

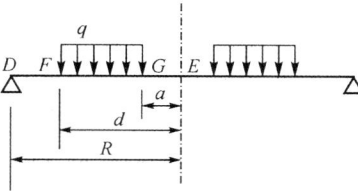

Fig. 5.27.

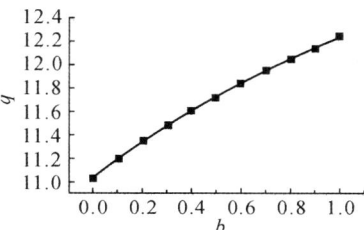

Fig. 5.28.

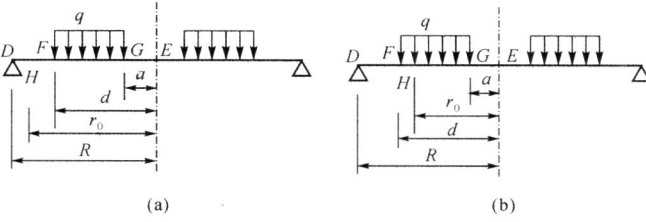

(a)   (b)

Fig. 5.29.

Problem 5.11 A simply supported circular plate is under linear and uniform load, as shown in Fig.5.30. Please derive the unified solution for the plate.

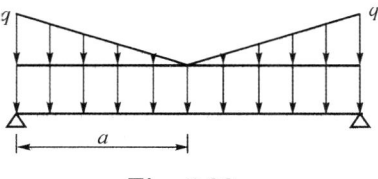

Fig. 5.30.

Problem 5.12 A simply supported circular plate is under linear and uniform load as shown in Fig.5.31. Please derive the unified solution for the plate.

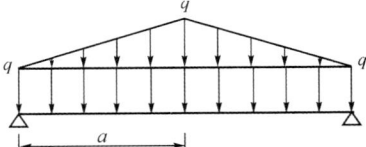

Fig. 5.31.

References

Chakrabarty J (1987) Theory of plasticity. McGraw-Hill, New York

Chen WF (1975) Limit analysis and soil plasticity. Elsevier, Amsterdam

Chen WF (1988) Concrete Plasticity: Past, present and future. In: Yu MH, Fan SC (eds.) Strength Theory: Applications, Developments and Prospects for the 21st Century, Science Press, Beijing, New York, 7-48

Chen WF, Saleeb AF (1981) Constitutive equations for engineering materials. Vol.1, Elasticity and Modelling; Vol.2, Plasticity and Modelling, Wiley, New York

Ghorashi M (1994) Limit analysis of circular plates subjected to arbitrary rotational symmetric loading. Int. J. Mech. Sci., 36(2):87-94

Hodge PG Jr (1959) Plastic analysis of structures. McGraw-Hill, New York

Hodge PG Jr (1963) Limit analysis of rotationally symmetric plates and shells. Prentice-Hall, New Jersey

Hopkins HG, Prager W (1953) The load carrying capacities of circular plates. J. Mech. Phys. Solids, 2:1-13

Hopkins HG, Prager W (1954) On the dynamic of plastic circular plates. Z. Angew. Math. Phys., 5: 317-330

Hopkins HG, Wang AJ (1954) Load carrying capacities for circular plates of perfectly-plastic material with arbitrary yield condition. J. Mech. Phys. Solids, 3:117-129

Hu LW (1960) Design of circular plates based on plastic limit load. J. Eng. Mech., ASCE, 86(1):91-115

Ma GW, Hao H, Iwasaki S (1999) Plastic limit analysis of a clamped circular plate with unified yield criterion. Structural Engineering and Mechanics, 7(5):513-525

Ma GW, Hao H, Iwasaki S (1999) Unified plastic limit analysis of circular plates under arbitrary load. J. Appl. Mech., ASME, 66(6):568-570

Ma GW, He LN (1994) Unified solution to plastic limit of simply supported circular plate. Mechanics and Practice, 16(6):46-48 (in Chinese, English abstract)

Ma GW, He LN, Yu MH (1993) Unified solution of plastic limit of circular plate under portion uniform loading. The Scientific and Technical Report of Xi'an Jiaotong University, 1993-181 (in Chinese)

Ma GW, Iwasaki S, Miyamoto Y, Deto H (1999) Dynamic plastic behavior of circular plate using unified yield criterion. Int. J. Solids Struct. 36(3):3257-3275

Ma GW, Yu MH, Iwasaki S, et al. (1994) Plastic analysis of circular plate on the basis of the Twin shear unified yield criterion. In: Lee PKK, Tham LG, Cheung YK (eds.) Proceedings of International Conference on Computational Methods in Structural and Geotechnical Engineering. China Translation and Printing Ltd., Hong Kong, 3:930-935.

Ma GW, Yu MH, Iwasaki S, Miyamoto Y (1995) Unified elasto-plastic solution to rotating disc and cylinder. J. Structural Engineering (Japan), 41A:79-85

Ma GW, Yu MH, Miyamoto Y, et al. (1995) Unified plastic limit solution to circular plate under portion uniform load. J. Structural Engineering (in English, JSCE), 41A:385-392

Mroz Z, Sawczuk A (1960) The load carrying capacity of annular plates. Bulletin of Science Akademia of SSSR, Moscow, 3:72-78 (in Russian)

Save MA (1985) Limit analysis of plates and shells: Research over two decades. J. Struct. Mech., 13:343-370

Save MA, Massonnet CE (1972) Plastic analysis and design of plates, shells and disks. North-Holland Publ. Co., Amsterdam

Save MA, Massonnet CE, Saxce G de (1997) Plastic limit analysis of plates, shells and disks. Elsevier, Amsterdam

Wang AJ, Hopkins HG (1954) On the plastic deformation of built-in circular plates under impulsive load. J. Mech. Phys. Solids, 3:22-37

Wang F (1988) Nonlinear finite analysis of RC plate and shell using unified strength theory, Ph. D. Thesis, Nanyang Technological University, Singapore

Wang YB, Yu MH (2003a) Unified plastic limit analysis of simply supported circular plates with strength differential effect in tension and compression. Chinese J. Applied Mechanics, 20(1):144-148

Wang YB, Yu MH (2003b) Plastic limit analysis of simply supported circular plates with different tensile and compressive strength under linear distributed load. China Civil Engineering Journal, 36(8):31-36 (in Chinese, English abstract)

Wang YB, Yu MH, Li JC, et al. (2002) Unified plastic limit analysis of metal circular plates subjected to border uniformly distributed loading. J. Mech. Strength, 24:305-307 (in Chinese, English abstract)

Wang YB, Yu MH, Wei XY (2002) Unified plastic limit analyses of circular plates under uniform annular load. Engineering Mechanics, 19(1):84-88 (in Chinese, English abstract)

Wei XY, Yu MH (2001) Unified plastic limit of clamped circular plate with strength differential effect in tension and compression. Chinese Quart. Mechanics, 22(1):78-83 (in Chinese)

Yu MH (1992) A new system of strength theory. Xi'an Jiaotong University Press, Xi'an (in Chinese)

Yu MH (2002) Advance in strength theory of materials under complex stress state in 20th century. Applied Mechanics Reviews, ASME, 55(3):169-218

Yu MH (2004) Unified Strength Theory and Its Applications. Springer, Berlin

Yu MH, He LN (1991) A new model and theory on yield and failure of materials under complex stress state. In: Mechanical Behavior of Materials-6, Pergamon Press, Oxford, 3:841-846

Zyczkowski M (1981) Combined Loadings in the Theory of Plasticity. Polish Scientific Publishers, PWN and Nijhoff

6

Plastic Limit Analysis of Clamped Circular Plates

6.1 Introduction

The unified solutions of plastic limit analyses for a simply supported circular plate have been described in last chapter. Clamped circular plates are one of the typical structural elements in many branches of engineering. The limit analyses for a clamped circular plate were studied by using the Tresca, Huber-von Mises and Mohr-Coulomb criteria (Wang and Hopkins, 1954; Hu, 1960; Hodge, 1963; Zyczkowski, 1981; Nielsen, 1999).

In this chapter the plastic limit analyses of clamped circular plates are carried out based on the unified strength theory. The unified yield criterion, which is suitable for materials without SD effect, is first employed to derive the moment and velocity fields and the plastic limit load for various transverse loads. A unified solution including a series of solutions for a clamped circular plate was first derived by Ma et al. (1994; 1999). The SD effect was subsequently considered by applying the unified strength theory. The unified solution to a clamped circular plate for SD materials was derived by Wei and Yu (2001). A series of solutions for a clamped circular plate is given. These results are described in this chapter and compared with reported results using the Tresca, Huber-von Mises and Mohr-Coulomb criteria.

6.2 Unified Solutions of Clamped Circular Plate for Non-SD Materials

6.2.1 Uniformly Distributed Load

Considering a clamped circular plate under a uniformly distributed load in Fig.6.1, the normalized moment variables m_r and m_θ are similar to those in Chapter 5. The moment and velocity equations of the plate remain the same as given in Eq.(5.2) to Eq.(5.5).

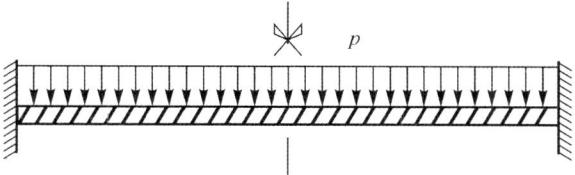

Fig. 6.1. Clamped circular plate

When the SD effect for a material is negligible, the unified strength theory can be simplified to the unified yield criterion which has the piecewise linear form of

$$m_\theta = a_i m_r + b_i \quad (i = 1, ..., 12). \tag{6.1}$$

The clamped circular plate is assumed to be made of a rigid-perfectly-plastic material that satisfies the unified yield criterion as shown in Fig.5.2 (see Chapter 5). The constants a_i and b_i for the five lines L_i $(i = 1, ..., 5)$ of AB, BC, CD, DE, and EF in Fig.5.2 are listed in Table 5.1.

Substituting the yield criterion in Eq.(6.1) into Eq.(5.3) and integrating Eq.(5.3), the radial moment m_r falling on the segments L_i is derived as

$$m_r = \frac{b_i}{1 - a_i} - \frac{p r^2}{2(3 - a_i)} + c_i r^{-1 + a_i} \quad (i = 1, ..., 5), \tag{6.2}$$

where c_i $(i = 1, ..., 5)$ are integration constants for the five lines L_i $(i = 1, ..., 5)$ to be determined by boundary and continuity conditions. The field of velocity corresponding to the lines L_i is derived by equating Eq.(5.4) and Eq.(5.5). Considering the yield condition Eq.(6.1), the velocity field is integrated as

$$\dot{w} = \dot{w}_0 (c_{1i} r^{1 - a_i} + c_{2i}) \quad (i = 1, ..., 5), \tag{6.3}$$

where c_{1i} and c_{2i} $(i = 1, ..., 5)$ are the integration constants, $\dot{w}_0$ is the velocity at the plate center. Eqs.(6.2) and (6.3) are the same as Eqs.(5.7) and (5.8) given in Chapter 5.

In the elastic state the moment fields of a clamped circular plate (Fig.6.1) satisfy $m_\theta = m_r$ at the plate center $(r = 0)$, $m_\theta = \nu m_r$ at the fixed edge, and $m_\theta \geqslant m_r$ at other points in the plate, where ν is the elastic Poisson's ratio. In the plastic limit state, it remains $m_\theta = m_r$ at the plate center due to the symmetry of the plate.

According to the kinematically admissible requirement, $\dot{w}$ is a decreasing function of r which implies $\dot{k}_\theta \geqslant 0$ in Eq.(5.4). Solutions with respect to the Tresca criterion and the Huber-von Mises criterion assume $m_\theta = 0$ and $m_\theta = 0.5$ at the fixed edge, respectively (Hodge, 1963; Hopkins and Wang, 1954).

Providing $m_\theta = \nu_p m_r$ at the outer edge, where ν_p is an edge effect coefficient in the plastic limit state and $0 \leqslant \nu_p \leqslant 0.5$, the moment fields of the entire plate lie on the five sides of AB, BC, CD, DE and EF as shown in Fig.5.2 of the unified yield criterion. Assuming that the points A, B, C, D, and E in Fig.5.2 correspond to dimensionless radii r_0, r_1, r_2, r_3 and r_4 in the plate respectively, there is $0 = r_0 \leqslant r_1 \leqslant r_2 \leqslant r_3 \leqslant r_4 \leqslant 1$. They divide the plate into five parts L_i $(i = 1, ..., 5)$. On the outer edge where $r = 1$, the moments m_r and m_θ lie on the line EF. They are at point F when ν_p is 0.5, and at point E when ν_p is zero. The continuity and boundary conditions of the clamped circular plate give (1) $m_r(r = 0)=1$; (2) $m_r(r = r_i,\ i = 1, ..., 4)$ are continuous; and (3) $m_r(r = 1)= m_\theta / \nu_p = -(1+b)/(1+b-\nu_p b)$ corresponding to the yield line EF in Fig.5.2. Defining $\alpha_1 = r_1/r_2$, $\alpha_2 = r_2/r_3$, $\alpha_3 = r_3/r_4$, and $\alpha_4 = r_4$, there are $r_1 = \alpha_1 \alpha_2 \alpha_3 \alpha_4$, $r_2 = \alpha_2 \alpha_3 \alpha_4$, $r_3 = \alpha_3 \alpha_4$, and $r_4 = \alpha_4$.

The continuity conditions of the radial moment m_r with respect to the yield points A, B, C, D, E leads to

$$m_r(r = r_i) = d_i \quad (i = 0, ..., 5). \tag{6.4}$$

The values of d_i $(i = 0, ..., 5)$ are listed in Table 6.1.

Table 6.1. Constants d_i

Points	A $(i = 0))$	B $(i = 1)$	C $(i = 2)$
d_i	1	$(1 + b)/(2 + b)$	0
Points	D $(i = 3)$	E $(i = 4)$	line EF $(i = 5)$
d_i	$-(1 + b)/(2 + b)$	-1	$-(1 + b)/(1 + b - \nu_p b)$

With the application of the above continuity and boundary conditions, c_i $(i = 1, ..., 5)$ in Eq.(6.2) are determined as

$$c_1 = 0, \tag{6.5a}$$

$$c_2 = -\frac{2b(1 + b)}{3 + 2b} r_1^{1/(1+b)}, \tag{6.5b}$$

$$c_3 = \left[\frac{(1 + b)(3 + b)}{(2 + 3b)(2 + b)} \alpha_1^{-2} - \frac{1 + b}{b} \right] r_2^{b/(1+b)}, \tag{6.5c}$$

$$c_4 = \left[\frac{2(1 + b)}{b(2 + b)} + \frac{(3 + b)}{(2 - b)(2 + b)} \alpha_1^{-2} \alpha_2^{-2} \right] r_3^{-b}, \tag{6.5d}$$

$$c_5 = \left[b + \frac{b(3 + b)}{(2b - 1)(2 + b)} \alpha_1^{-2} \alpha_2^{-2} \alpha_3^{-2} \right] r_4^{-1/b}, \tag{6.5e}$$

where the coefficients α_1, α_2, α_3 and α_4 satisfy the equations of

$$-(3+2b)(2+b) + (3+b)\alpha_1^{-2} + 2b(2+b)\alpha_1^{1/(1+b)} = 0, \tag{6.6a}$$

$$\begin{aligned} &2(1+b)(2+3b) - b(3+b)\alpha_1^{-2}\alpha_2^{-2} \\ &+ \left[b(3+b)\alpha_1^{-2} - (2+3b)(2+b)\right]\alpha_2^{b/(1+b)} = 0, \end{aligned} \tag{6.6b}$$

$$\begin{aligned} &-(2-b)(2+b) - b(3+b)\alpha_1^{-2}\alpha_2^{-2}\alpha_3^{-2} \\ &+ \left[2(1-b)(2-b) + b(3+b)\alpha_1^{-2}\alpha_2^{-2}\right]\alpha_3^{-b} = 0, \end{aligned} \tag{6.6c}$$

$$\begin{aligned} &-\frac{b(1+b)(1-\nu_p)}{1+b-\nu_p b} - \frac{b(3+b)}{(2b-1)(2+b)}\alpha_1^{-2}\alpha_2^{-2}\alpha_3^{-2}\alpha_4^{-2} \\ &+ \left[b + \frac{b(3+b)}{(2b-1)(2+b)}\alpha_1^{-2}\alpha_2^{-2}\alpha_3^{-2}\right]\alpha_4^{-1/b} = 0. \end{aligned} \tag{6.6d}$$

The load-bearing capacity of the clamped circular plate is then derived as

$$p = \frac{2(3+b)}{2+b}r_1^{-2}. \tag{6.7}$$

α_i ($0 \leqslant \alpha_i \leqslant 1$, i=1, ...,4) can be calculated from Eqs.(6.6a)~(6.6d) with half interval search method. Substituting the values of c_i and r_i into Eqs.(6.2), (6.5) and (6.7), the moment fields and the plastic limit load of the plate are then determined.

The continuous and boundary conditions of the velocity field can be expressed as (1) $\dot{w}(r = 0) = \dot{w}_0$; (2) $\dot{w}$ and $d\dot{w}/dr(r = r_i$, $i = 1, ..., 4)$ are continuous; and (3) $\dot{w}(r = 1) = 0$. According to these conditions, the constants c_{1i} and c_{2i} in Eq.(6.3) can be derived as

$$c_{11} = -\frac{\dot{w}_0}{(d_{14} + d_{24})d_{13}d_{12}d_{11} + d_{23}d_{12}d_{11} + d_{22}d_{11} + d_{21}}, \tag{6.8}$$

and

$$\begin{bmatrix} c_{1(i+1)} \\ c_{2(i+1)} \end{bmatrix} = \begin{bmatrix} d_{1i} & 0 \\ d_{2i} & 1 \end{bmatrix} \begin{bmatrix} c_{1i} \\ c_{2i} \end{bmatrix} \quad (i = 1, ..., 4), \tag{6.9}$$

where d_{1i} and d_{2i} ($i = 1, ..., 5$) are constants determined by the continuous conditions, and they have the form of

$$d_{1i} = \frac{1-a_i}{1-a_{i+1}}r_i^{a_{i+1}-a_i}, \quad d_{2i} = -\frac{a_{i+1}-a_i}{1-a_{i+1}}r_i^{1-a_i} \quad (i = 1, ..., 4). \tag{6.10}$$

Substituting these integration constants into Eq.(6.3), the velocity field of the clamped circular plate is then obtained.

Figs.6.2 and 6.3 respectively show the moment fields m_r, m_θ, and velocity field $\dot{w}$ when the coefficient ν_p is 0.5 and $b = 0$, 0.5 and 1. The equilibrium equation, when the Tresca criterion is used, is invalid on the line EF (Hodge, 1963). However, the singularity of the solution can be simply avoided by replacing $b = 0$ by a very small value, e.g., $b = 0.001$, when the unified yield criterion is adopted.

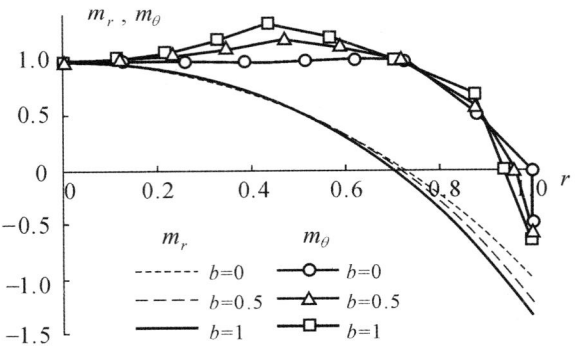

Fig. 6.2. Moment fields of clamped circular plate ($\nu_p = 0.5$)

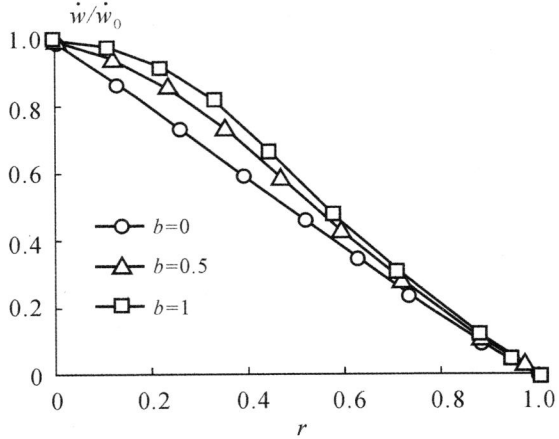

Fig. 6.3. Velocity fields of clamped circular plate ($\nu_p = 0.5$)

The moment and velocity fields when $\nu_p = 0$ and $b = 0$ are the same as those with the Tresca criterion (Hodge, 1963; Save et al., 1972; 1997). The plastic limit load of the clamped circular plate with respect to the Tresca criterion, the Huber-von Mises criterion and the twin-shear stress criterion are 11.265, 11.799 and 12.179 when $\nu_p = 0.0$; 11.277, 12.720, and 13.791 when

ν_p=0.5, respectively. Hopkins and Wang (1954) gave a plastic load of 12.5 based on the Huber-von Mises criterion. They also gave the lower bound and upper bound solutions of the plastic limit load using the Tresca yield hexagons inscribed and circumscribed to the Huber-von Mises ellipse as 11.26 and 13.00 respectively. It is seen that the solution based on the unified yield criterion with ν_p=0.5 and that derived by Hopkins and Wang (1954) are in good agreement with the solution using the Huber-von Mises criterion with the percentage difference of approximately of 1.76%. However, the upper bound plastic limit load (13.00) given by Hodge (1963), Save et al. (1997) is lower than that of the present result (13.791). The plastic limit load of a clamped circular plate obtained with respect to the unified yield criterion of $b = 0.5$ and $b = 1$ differs remarkably from that with $b = 0$ by 12.8% and 22.3%, respectively.

6.2.2 Arbitrary Loading Radius

Considering an arbitrary loading radius as shown in Fig.6.4, the equilibrium equation for the clamped circular plate is the same as Eqs.(5.13) and (5.14) for the simply supported circular plate. The boundary conditions and continuity conditions are the same as those stated in Section 6.2.1. Since that the loading radius is arbitrary, there are a few possible cases with different locations of the loading radius.

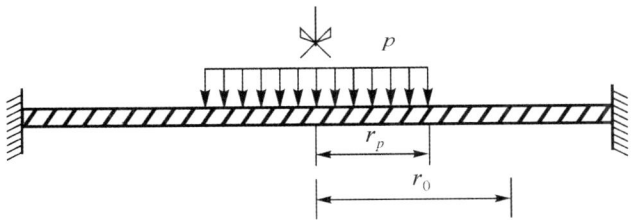

Fig. 6.4. Clamped plate with partial uniform loading

For a specific loading radius r_p or r_j $(j = 1, ..., 4)$ corresponding to the yield points B, C, D, and E in Fig.5.2, the plate is divided into five parts. The loading radius may lie in one of these parts, i.e., $r_{j-1} \leqslant r_p \leqslant r_j$ $(j = 1, ..., 5)$. Considering the particular case of $r_p = r_j$, $j = 1, ..., 5$, and defining the critical radius r_p or r_j as r_{pj} $(j = 1, ..., 5)$, direct integration of the differential Eqs.(5.13) and (5.14) of radial moment m_r with respect to r leads to

$$m_{ri} = \frac{b_i}{1 - a_i} - \frac{pr^2}{2(3 - a_i)} + c_i r^{-1+a_i}, \quad 1 \leqslant i \leqslant j \text{ or } 0 \leqslant r \leqslant r_{pj}, \quad (6.11a)$$

$$m_{ri} = \frac{b_i}{1 - a_i} - \frac{pr_{pj}^2}{2(1 - a_i)} + c_i r^{-1+a_i}, \quad j \leqslant i \leqslant 5 \quad \text{or} \quad r_{pj} \leqslant r \leqslant 1, \quad (6.11b)$$

where c_i $(i = 1, ..., 5)$ are integration constants. Eq.(6.11a) gives the moment distribution in the loading area, while Eq.(6.11b) is the moment distribution in the area outside the loading radius. Applying the boundary and continuity conditions yields

$$\begin{cases} \dfrac{b_i}{1 - a_i} - \dfrac{pr_{i-1}^2}{2(3 - a_i)} + c_i r_{i-1}^{-1+a_i} = d_{i-1} \\[3mm] \dfrac{b_i}{1 - a_i} - \dfrac{pr_i^2}{2(3 - a_i)} + c_i r_i^{-1+a_i} = d_i \end{cases} \quad 1 \leqslant i \leqslant j, (j = 1, ..., 5), \quad (6.12a)$$

and

$$\begin{cases} \dfrac{b_i}{1 - a_i} - \dfrac{pr_{pj}^2}{2(1 - a_i)} + c_i r_{i-1}^{-1+a_i} = d_{i-1} \\[3mm] \dfrac{b_i}{1 - a_i} - \dfrac{pr_{pj}^2}{2(1 - a_i)} + c_i r_i^{-1+a_i} = d_i \end{cases} \quad j + 1 \leqslant i \leqslant 5, \ (j = 1, ..., 4).$$

$$(6.12b)$$

Since the moment at the plate center is a finite value,

$$c_1 = 0. \tag{6.13}$$

The plastic limit load is derived from Eq.(6.12a) as

$$p = 2(3 - a_1)\left(-d_1 + \frac{b_1}{1 - a_1}\right)r_1^{-2} \quad \text{or} \quad p = \frac{2(3 - a_1)}{2 - a_1}r_1^{-2}. \tag{6.14}$$

For the case of $j = 1$, it can be derived from Eq.(6.12b) that

$$c_i = \frac{d_i - d_{i-1}}{r_i^{-1+a_i} - r_{i-1}^{-1+a_i}}, \quad (i = 2, ..., 5), \tag{6.15}$$

$$\eta_{i-1}^{-1+a_i} = 1 - (d_i - d_{i-1})\Big/\left[d_i - \frac{b_i}{1 - a_i} + \frac{3 - a_1}{(1 - a_i)(2 - a_1)}\right], \quad (i = 2, ..., 5), \tag{6.16}$$

where η_i $(i = 1, ..., 4)$ are defined as

$$\eta_{i-1} = r_{i-1}/r_i. \tag{6.17}$$

The values of η_i for this case can be calculated directly from Eq.(6.12b). For the case of $2 \leqslant j \leqslant 5$, it can be derived from Eq.(6.12a) that

$$c_i = \left[d_{i-1} - \frac{b_i}{1 - a_i} + \frac{p r_{i-1}^2}{2(3 - a_i)} \right] r_{i-1}^{1-a_i}, \quad (2 \leqslant i \leqslant j), \tag{6.18}$$

and η_{i-1} $(2 \leqslant i \leqslant j)$ satisfies

$$
\begin{aligned}
&\frac{b_i}{1 - a_i} - \frac{(3 - a_1)\eta_1^{-2} \cdots \eta_{i-2}^{-2}}{(3 - a_i)(2 - a_1)} \eta_{i-1}^{-2} \\
&+ \left[d_{i-1} - \frac{b_i}{1 - a_i} + \frac{(3 - a_1)\eta_1^{-2} \cdots \eta_{i-2}^{-2}}{(3 - a_i)(2 - a_1)} \right] \eta_{i-1}^{1-a_i} = d_i, \quad 2 \leqslant i \leqslant j.
\end{aligned}
\tag{6.19}
$$

The other integration constants can be derived from Eq.(6.12b) as

$$c_i = \frac{d_i - d_{i-1}}{r_i^{-1+a_i} - r_{i-1}^{-1+a_i}}, \quad (j + 1 \leqslant i \leqslant 5), \tag{6.20}$$

and η_{i-1} $(j + 1 \leqslant i \leqslant 5)$ satisfies

$$\eta_{i-1}^{-1+a_i} = 1 - \frac{(d_i - d_{i-1})}{\left[d_i - \frac{b_i}{1-a_i} + \frac{(3-a_1)\eta_1^{-2} \cdots \eta_{j-1}^{-2}}{(1-a_i)(2-a_1)} \right]}, \quad (j + 1 \leqslant i \leqslant 5). \tag{6.21}$$

Thus, for both cases there are

$$r_1 = \eta_1 \eta_2 \eta_3 \eta_4, \; r_2 = \eta_2 \eta_3 \eta_4, \; r_3 = \eta_3 \eta_4, \; r_4 = \eta_4 \text{ and } r_5 = 1. \tag{6.22}$$

Substituting r_i $(i=1, ..., 5)$ and all the integration constants into Eq.(6.11a), Eq.(6.11b) and Eq.(6.14), the moment fields and the plastic limit load for the critical cases are then determined. The critical loading radii r_{jp} $(j = 1, ..., 5)$ can also be calculated from Eq.(6.22).

For an arbitrary loading radius r_p, i.e., $r_{(j-1)p} \leqslant r_p \leqslant r_{jp}$ $(j = 1, ..., 5)$, the moment fields become

$$m_{ri} = \frac{b_i}{1 - a_i} - \frac{p r^2}{2(3 - a_i)} + c_i r^{-1+a_i}, \quad 0 \leqslant r \leqslant r_{j-1}, \tag{6.23a}$$

$$m_{rj1} = \frac{b_j}{1 - a_j} - \frac{p r^2}{2(3 - a_j)} + c_{j1} r^{-1+a_j}, \quad r_{j-1} \leqslant r \leqslant r_p, \tag{6.23b}$$

$$m_{rj2} = \frac{b_j}{1 - a_j} - \frac{p r_p^2}{2(1 - a_j)} + c_{j2} r^{-1+a_j}, \quad r_p \leqslant r \leqslant r_j, \tag{6.23c}$$

$$m_{ri} = \frac{b_i}{1 - a_i} - \frac{p r_p^2}{2(1 - a_i)} + c_i r^{-1+a_i}, \quad r_j \leqslant r \leqslant 1. \tag{6.23d}$$

In comparison with the uniformly distributed load case discussed in Section 6.2.1, Eq.(6.23) introduces one more integration constant with the continuity condition of $m_{rj1}(r = r_p) = m_{rj2}(r = r_p)$. Therefore the integration constants and the dividing radius can be calculated with reference to the boundary and the continuity conditions in a similar manner as the cases discussed above.

The equations of the velocity fields are the same as Eq.(6.3). The continuity and boundary conditions of velocity can be described with (1) $\dot{w}_0(r = 0) = \dot{w}_0$; (2) $\dot{w}_0$ and $d\dot{w}/dr$ ($r = r_i, i = 1, ..., 4$) are continuous; and (3) $\dot{w}(r = 1) = 0$. c_{1i} and c_{2i} in Eq.(6.3) can then be determined following the derivation procedure as derived in Section 6.2.1.

Table 6.2 lists the plastic limit loads p for the Tresca criterion, the Huber-von Mises criterion and the twin-shear stress criterion, with the load uniformly distributed over the entire plate ($r_p = 1$). The results given by Hopkins and Wang (1954) based on the Tresca criterion are also given for comparison. From Table 6.2, the results based on the unified yield criterion can approximate all the previous solutions with the Huber-von Mises criterion, the twin-shear stress criterion and the Tresca criterion. The edge effect, which is a function of the ratio ν_p, also affects the plastic limit load.

Table 6.2. Plastic limit loads for three common yield criteria

Yield criteria	Tresca (ν_p=0)			Mises (ν_p=0.5)			Twin-shear (Yu, 1961)	
	Hopkins (1954)	$b =$ 0.0001	$b =$ 0.001	Hopkins (1954)	$b =$ 0.5	$b = 1$ $\nu_p = 0$	$b = 1$ $\nu_p = 0.5$	
p	11.26	11.259	11.260	12.5	12.720	12.176	13.708	

Again the equilibrium conditions in Eqs.(5.13) and (5.14) on the line EF are invalid if the Tresca criterion is used. They are valid only when the point E corresponds to the fixed edge ($r = 1$), which leads to $\nu_p = 0$, $m_\theta = 0$, and $m_r = -1$ at the plate edge. On the other hand, $m_\theta = 0.5m_r$ or $\nu_p = 0.5$ at the plate edge has to be assumed when the Huber-von Mises criterion is applied according to the plastic flow requirement (Hodge, 1963). Thus the Tresca and the Huber-von Mises criteria adopt two different coefficients for ν_p. When the unified yield criterion is adopted, it is straightforward to extend the yield trajectory to line EF in Fig.5.2 in the plastic limit analysis because ν_p can cover the range from 0 to 0.5 to reflect the edge effect on the plastic limit solutions. It is worth noting that a small value of b, e.g., 0.0001, instead of 0 should be used to avoid the singularity problem as incurred with the Tresca criterion.

Fig.6.5 illustrates the moment fields of the plate in plastic limit state under different load radii for ν_p=0 and ν_p=0.5 respectively, with respect to different criteria. For $\nu_p=0$, it is seen that the influences of yield criteria on the radial moment distributions are insignificant, but they are prominent on the tangential moment distributions. The moment at the plate center is equal to that at the plate edge regardless of the yield criterion. The twin-shear yield criterion ($b = 1$) results in the largest values for both m_r and m_θ, while the Tresca criterion ($b = 0$) leads to the smallest estimations. The maximum tangential moment m_θ based on the unified yield criterion with non-zero parameter b is not equal to 1 as predicted by the Tresca criterion, indicating the maximum m_θ does not occur at the plate center. The highest m_θ is equal to $2(1+b)(2+b)$ and occurs at $r = r_1$. For $\nu_p=0.5$, different criteria affect the moment distributions near the edge significantly. The twin-shear stress criterion results in the largest positive and negative values of m_r and m_θ, while the Tresca criterion results in the smallest.

The velocity fields with $\nu_p=0.5$ are compared in Fig.6.6(a) against the different loading radii and the different criteria. It is seen that the influence of the criteria on the velocity fields is relatively large when the transverse load is uniformly distributed over the whole plate or concentrated at the plate center. The velocity distribution is more concentrated at the center area of the plate as the loading radius reduces. It should be mentioned that the velocity field is not smooth at the plate center for the Tresca criterion. Fig.6.6(b) and Fig.6.6(c) illustrate the edge effects for the unified yield criterion with $b = 1$ and $b = 0.5$, respectively. The influence of ν_p on the velocity distribution is in-significant when the loading radius is large, and it increases with the decrease of r_p. However, the effects become less prominent with the decrease of the parameter b. The velocity distribution with $\nu_p=0$ is always higher than that with $\nu_p=0.5$.

Besides the moment and velocity distributions, the plastic limit load of the circular plate with various loading radii is also an important factor for structural design. Fig.6.7 illustrates the influences of yield criteria, edge effect, and loading radius on P_T, which is defined as the total plastic limit load, i.e., $P_T = \pi r_p^2 p$. It can be seen that the total plastic load P_T increases as any of b, r_p and ν_p increases. The twin-shear yield criterion results in the largest plastic limit loads, while the Tresca criterion gives the smallest estimations. For a demonstration of the effect of different criteria, a difference ratio of plastic limit loads is defined as follows:

$$D_r = \frac{P_T - P_T(\text{Tresca})}{P_T(\text{Tresca})} \times 100\%, \tag{6.24}$$

where P_T (Tresca) is the plastic limit load based on the Tresca criterion.

Fig.6.8 illustrates the percentage differences corresponding to the Huber-von Mises criterion and the twin-shear stress criterion by specifying the unified strength theory parameter b as 0.5 and 1 respectively for the unified yield criterion. It is seen that the percentage difference varies from 10.6% to 13.0% and from 17.8% to 21.7% with respect to the two criteria of different loading radii. The maximum percentage difference between the total plastic limit loads between the Tresca criterion and the twin-shear stress criterion is 21.7% when $r_p = 1$ and $\nu_p = 0.5$. These results imply that the Tresca criterion may significantly underestimate the plastic limit load or bearing capacities of a clamped circular plate. The Tresca criterion (or the unified yield criterion with $b = 0$) cannot reflect properly the edge effects (ν_p) on the plastic limit load either.

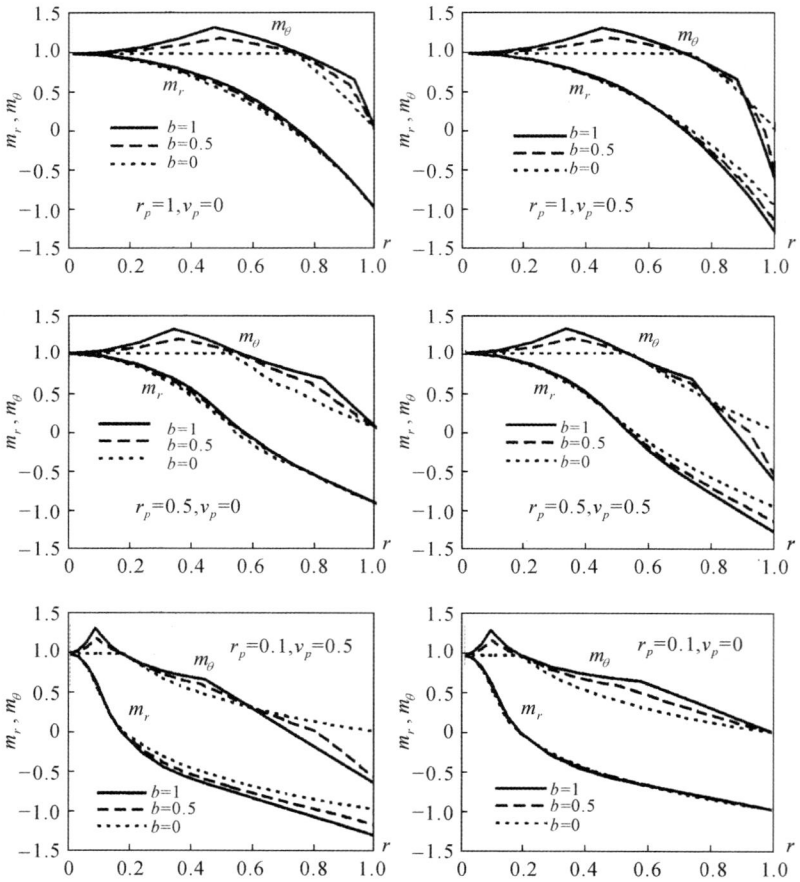

Fig. 6.5. Velocity fields for clamped circular plate

6.2.3 Arbitrary Loading Distribution

Fig.6.9 shows a clamped circular plate under arbitrarily distributed axisymmetrical pressure.

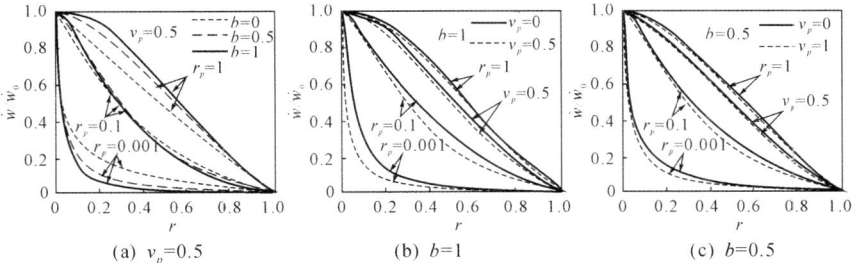

Fig. 6.6. Velocity fields with different loading radii and edge conditions

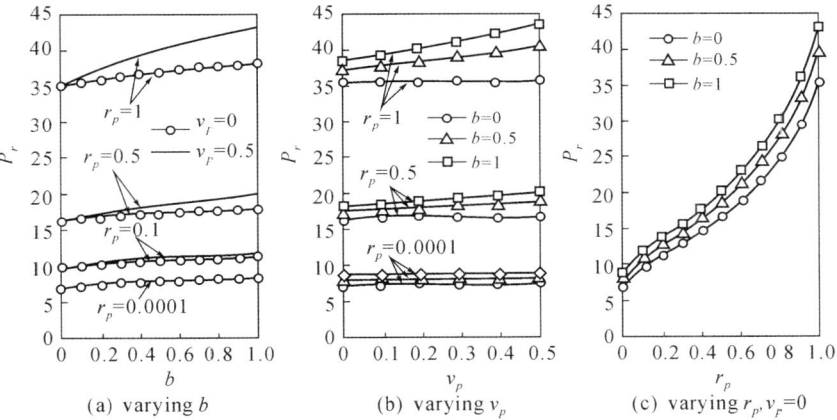

(a) varying b

(b) varying v_p

(c) varying r_p, $v_r=0$

Fig. 6.7. Effects of yield criteria, edge condition and loading radii

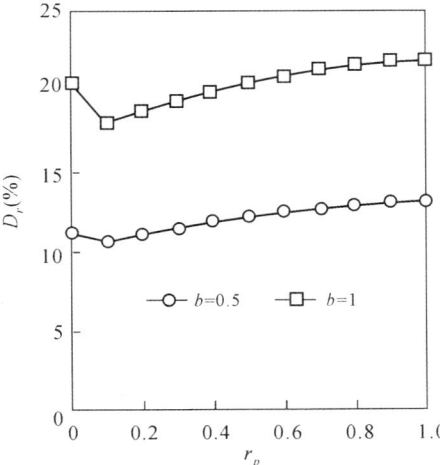

Fig. 6.8. Percentage differences of the plastic limit loads

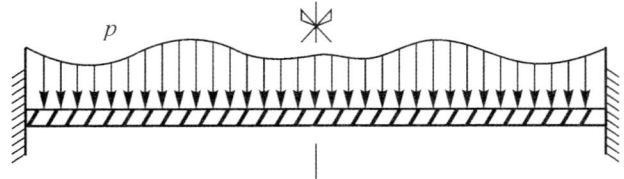

Fig. 6.9. Clamped plate under arbitrary loading distribution

Defining dimensionless variable $p(r) = P(r)a^2/M_0$ for a circular plate of radius a and thickness h subjected to an arbitrarily distributed axisymmetrical transverse pressure $\mu p(r)$, where μ is a plastic limit load factor, and $p(r)$ is a load distribution function, the equilibrium equation can be written with application of the axial-symmetrical condition as

$$\mathrm{d}(rm_r)/\mathrm{d}r - m_\theta = -\int \mu p(r) r \mathrm{d}r. \tag{6.25}$$

Substituting the unified yield criterion of Eq.(6.1) into Eq.(6.25) and integrating Eq.(6.25), m_r corresponding to segments L_i is obtained as

$$m_r = \frac{b_i}{1 - a_i} - r^{-1+a_i} \int r^{-a_i} \left[\int \mu p(r) \mathrm{d}r \right] \mathrm{d}r + c_i r^{-1+a_i}, \quad (i = 1, ..., 5), \tag{6.26}$$

where c_i $(i = 1, ..., 5)$ are integration constants and they can be determined from the continuity and boundary conditions.

Assuming the load function can be expanded as $p(r) = \sum\limits_{j=1}^{\infty} p_j r^{j-1}$, Eq. (6.26) becomes

$$m_r = \frac{b_i}{1 - a_i} - \mu \sum_{j=1}^{\infty} p_j \frac{r^{j+1}}{(j+1)(j+2-a_i)} + c_i r^{-1+a_i}, \quad (i = 1, ..., 5). \tag{6.27}$$

The field of velocity corresponding to the five sides L_i is obtained as

$$\dot{w} = \dot{w}_0(c_{1i} r^{1-a_i} + c_{2i}), \quad (i = 1, ..., 5), \tag{6.28}$$

where c_{1i} and c_{2i} $(i = 1, ..., 5)$ are integration constants, and $\dot{w}_0$ is the velocity at the plate center.

The load-bearing capacity of a plate is always taken as the total limit load on the plate. The dimensionless total limit load of the plate can be derived as

$$P_T = 2\pi \int_0^1 \mu p(r) r dr \quad \text{or} \quad P_T = 2\pi \mu \sum_{j=1}^{\infty} \frac{p_j}{j+1}. \tag{6.29}$$

In the plastic limit state there is $m_\theta = m_r = 1$ at the center of a clamped circular plate. According to the kinematically admissible requirement, $\dot{w}$ is a decreasing function of r, so that $\dot{k}_\theta \geqslant 0$. Hence the moment fields of the entire clamped circular plate lie on the five lines of AB, BC, CD, DE, and EF in Fig.5.2 based on the unified yield criterion. The points A, B, C, D, and E in Fig.5.2 correspond to five dividing dimensionless radii on the plate, which divide the plate into five parts. The five radii are denoted as r_0, r_1, r_2, r_3, and r_4, respectively and there is $r_0 \leqslant r_1 \leqslant r_2 \leqslant r_3 \leqslant r_4 \leqslant 1$. On the outer edge ($r = r_5 = 1$), the moments m_r and m_θ are assumed to correspond to the yield point F. The continuous and the boundary conditions of the clamped circular plate can be expressed as (1) $m_r(r = 0) = 1$; (2) $m_r(r = r_i) = d_i$ ($i = 1, ..., 4$) are continuous, (3) $m_r(r = 1) = d_5$. c_i ($i = 1, ..., 5$) in Eq.(6.27) are determined with application of these conditions,

$$c_1 = 0, c_i = \left[d_{i-1} - \frac{b_i}{1 - a_i} + \mu \sum_{j=1}^{\infty} \frac{p_j r_{i-1}^{j+1}}{(j+1)(j+2-a_i)} \right] r_{i-1}^{1-a_i}, \quad (i = 2, ..., 5). \tag{6.30}$$

The load factor is derived as

$$\mu = \frac{-d_1 + \frac{b_1}{1-a_1}}{\sum_{j=1}^{\infty} \frac{p_j r_1^{j+1}}{(j+1)(j+2-a_1)}}, \tag{6.31}$$

and the dividing radii r_i ($i = 2, ..., 5$) are calculated by the following simultaneous equations:

$$\frac{b_i}{1 - a_i} + \left[d_{i-1} - \frac{b_i}{1 - a_i} + \mu \sum_{j=1}^{\infty} \frac{p_j r_{i-1}^{j+1}}{(j+1)(j+2-a_i)} \right] r_{i-1}^{1-a_i} r_i^{-1+a_i}$$

$$- \mu \sum_{j=1}^{\infty} \frac{p_j r_i^{j+1}}{(j+1)(j+2-a_i)} = d_i, \quad (i = 2, ..., 5). \tag{6.32}$$

Eq.(6.32) can be solved by iterative method. Giving an arbitrary initial value of r_1, r_i ($i = 2, ..., 5$) in Eq.(6.32) can be solved by trial and error. It should be noted that r_5 is equal to 1, which can be used as the convergent criterion. For any load distribution $p(r)$, r_1 should be reduced if the calculated r_5 is less than 1.0 or increased if the calculated r_5 is larger than 1.0.

Substituting the values of c_i and r_i into Eqs.(6.27) and (6.29), the moment fields and the plastic limit load factor of the clamped circular plate are determined. The expression of the velocity fields is the same as that for the uniformly distributed load case.

The plastic limit load factors and the total limit loads of the clamped circular plate under the five different linear load distributions in terms of different criteria (unified yield criterion with different parameter b) are listed in Table 6.3.

Table 6.3. Plastic limit loads of clamped circular plate

$p(r)$	r		$1+r$		1		$2-r$		$1-r$	
	μ	P_T	μ	P_T	μ	P_T	μ	P_T	μ	P_T
$b=0$	22.98	48.13	7.57	39.66	11.28	35.43	7.45	31.26	21.87	22.90
$b=1/2$	25.83	54.09	8.54	44.72	12.72	39.96	8.40	35.19	24.60	25.76
$b=1$	27.78	58.19	9.20	48.19	13.71	43.06	9.05	37.91	26.46	27.71

From Table 6.3, the loading conditions have a significant effect on the plastic limit load. Increasing load distribution leads to a larger plastic limit load. The plastic limit load of a clamped circular plate is about twice that of a simply supported circular plate with similar load distribution. The plastic limit loads with respect to the Huber-von Mises criterion ($b = 0.5$) are 12.4% to 12.8% higher than those with respect to the Tresca criterion ($b = 0$) for the five linear loading conditions. They are 20.9% to 21.5% higher if the twin-shear stress criterion ($b = 1$) is adopted. The differences are more pronounced than for the simply supported circular plate, which indicates the influence of different yield criteria on the load factor and the total limit load varies with different end conditions. For a uniformly distributed load, Hopkins and Wang (1954) presented a numerical solution of the load factor of 12.5 based on the Huber-von Mises criterion. The present result of 12.72 corresponding to $b = 0.5$ is in good agreement with their result since the percentage difference is only 1.76%.

Fig.6.10 and Fig.6.11 respectively show the moment fields m_r, m and velocity field for the two load distribution conditions of $p = r$ and $p = 1 - r$. A solution based on the Tresca criterion can be approximated with a very small value of b, e.g., $b = 0.001$. The moment and velocity fields with $b = 0.001$ of the unified yield criterion are the same as those with the Tresca criterion (Hodge, 1963). Comparing Figs.6.10 and 6.11, the moment profiles vary with the different loading conditions as well as the yield criterion. r_1 is higher in Fig.6.10(a) than that in Fig.6.11(a), implying a higher total limit load when $p = r$. The profiles of velocity fields also depend on the loading condition and the yield criterion. There is no singularity at the central point if the unified yield criterion with non-zero parameter b is adopted.

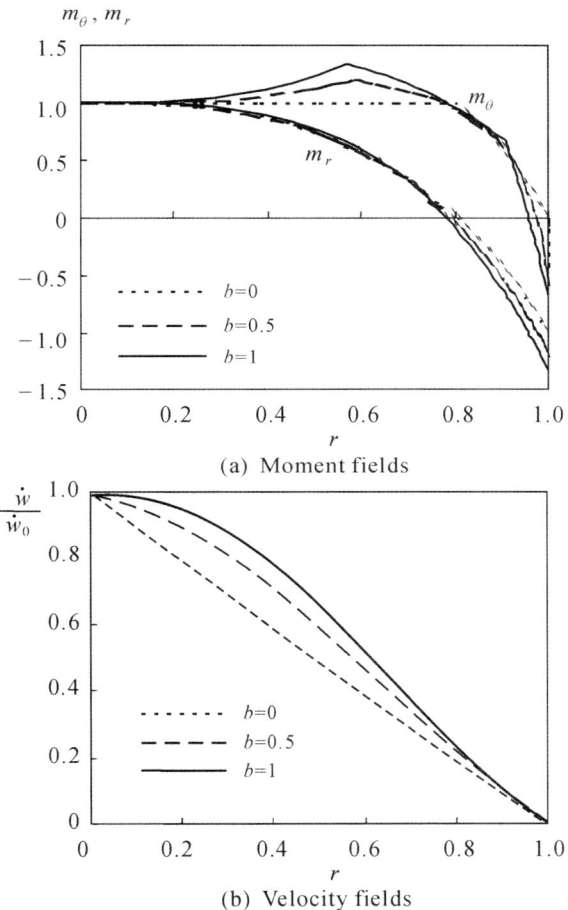

Fig. 6.10. Moment and velocity fields of clamped circular plate $(p = r)$

6.3 Unified Solutions of Clamped Circular Plate for SD Materials

For SD materials, such as rock, concrete, soils, etc. (Chen, 1981; 1998; Nielsen, 1999), the strength function is the same as Eq.(6.1), while the parameters a_i and b_i are related to the strength difference ratio α. The generalized stress yield loci are shown in Fig.6.12(a). Fig.6.12(b) shows the yield loci of the

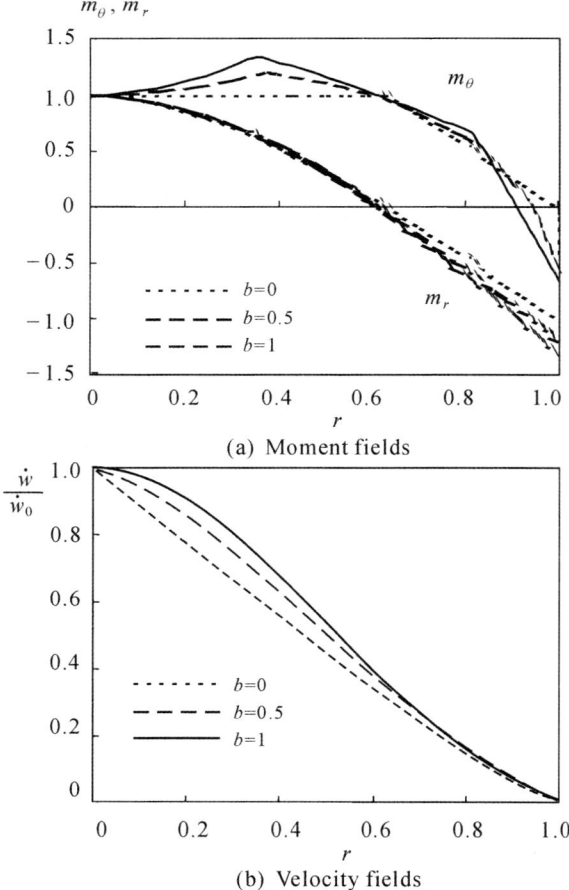

Fig. 6.11. Moment and velocity fields of clamped circular plate, $p = 1 - r$

generalized stress for reinforced concrete plates when the parameter $\alpha = +m_p/-m_p = 0.5$. Table 6.4 gives the values of a_i and b_i in L_i $(i = 1, ..., 6)$, i.e., AB, BC, CD, DE, EF, and FG lines.

Table 6.4. a_i and b_i of unified strength theory

	AB $(i = 1)$	BC $(i = 2)$	CD $(i = 3)$	DE $(i = 4)$	EF $(i = 5)$	FG $(i = 6)$
a_i	$-b$	$\alpha(1+b)/b$	$\alpha(1+b)$	$\alpha(1+b)$	$\alpha(1+b)/b$	$-1/b$
b_i	$1+b$	1	1	$1+b$	$(1+b)/b$	$-(1+b)/\alpha b$

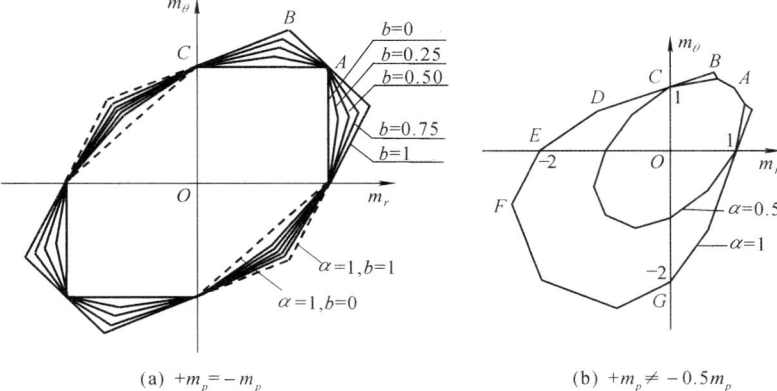

(a) $+m_p = -m_p$ (b) $+m_p \neq -0.5m_p$

Fig. 6.12. Unified strength theory expressed by generalized stresses

In the elastic stage the center of the plate satisfies the condition of $m_r = m_\theta$ ($r=0$). On the fixed boundaries the relation $m_\theta = \mu m_r$ is valid; and m_r and m_θ are both negative. When the plates are in the elasto-plastic stage, at the central point, the condition $m_r = m_\theta$ ($r = 0$) is still valid due to the symmetry of the structure. On the fixed boundaries, m_r and m_θ are both negative while $|m_r| > |m_\theta|$. m_r is $-1/\alpha$ as the line EF reaches point F. The circumferential bending moment gives the maximum negative moment on the fixed boundary. The plastic hinge appears on the fixed boundary. Substituting $m_r = -1/\alpha$ into the equilibrium equation,

$$m_\theta = -\left(\frac{1}{a} - \frac{1}{2}pr^2\right). \tag{6.33}$$

With increasing p, $|m_\theta|$ decreases and subsequently m_r and m_θ approach point E, where $m_\theta = 0$ and $m_r = -1/\alpha$. At the center of the plate, on the other hand, there is $m_r = m_\theta = 1$ when the plastic limit load is reached. Around the plate center, m_r and m_θ are determined by the yield functions of AB and BC. Therefore the internal moment corresponds to lines AB, BC, CD, and DE when the clamped plate falls into the plastic limit state. We assume the non-dimensional radii r_0, r_1, r_2 and r_3 to correspond to points A, B, C, and D respectively. The plates will be divided into four annular areas, where $0 = r_0 \leqslant r_1 \leqslant r_2 \leqslant r_3 \leqslant r_4 = 1$. Defining $\alpha_1 = r_1/r_2$, $\alpha_2 = r_2/r_3$ and $\alpha_3 = r_3/r_4$, there are $r_1 = \alpha_1\alpha_2\alpha_3$, $r_2 = \alpha_2\alpha_3$, and $r_3 = \alpha_3$. The boundary conditions and continuity conditions can be written as (1) $m_r(r = 0) = 1$; (2) $m_r(r = r_i) = 1$ ($i = 1, ..., 3$) continuous; (3) $m_r(r = 1) = -1/\alpha$; (4) $c\dot{w}(r = 0) = \dot{w}_0$; (5) $d\dot{w}/dr(r = r_i)$ and, $i = 1, ..., 3$ continuous; (6) $\dot{w}(r = 1) = 0$. We can calculate the integration constants with application of these boundary and continuity conditions as

$$c_1 = 0, c_2 = \frac{-2\alpha b(1+b)}{(3+3b-\alpha b)(1+b-\alpha b)} r_1^{\frac{1+b-\alpha b}{1+b}}, \tag{6.34a}$$

$$c_3 = \left\{ \frac{\alpha(3+b)(1+b)}{(\alpha+b+1)(3+3b-\alpha)} \alpha_1^{-2} - \frac{1+b}{1+b-\alpha} \right\} r_2^{(1+b-\alpha)/(1+b)}, \tag{6.34b}$$

$$c_4 = \left\{ \frac{(1+b)(1+\alpha)}{\alpha(b+2)(\alpha-1+\alpha b)} + \frac{\alpha(3+b)}{(3-\alpha-\alpha b)(\alpha+b+1)} \alpha_1^{-2} \alpha_2^{-2} \right\} r_3^{(1-\alpha-\alpha b)}, \tag{6.34c}$$

where α_i $(i = 1, ..., 3)$ satisfies

$$(\alpha+b+1)(3+3b-\alpha b) - \alpha(3+b)(1+b-\alpha b)\alpha_1^{-2} - 2\alpha b(\alpha+b+1)\alpha_1^{\frac{1+b-\alpha b}{1+b}} = 0, \tag{6.35a}$$

$$\frac{1}{1-b-\alpha} + \frac{1}{\alpha(b+2)} - \frac{\alpha(3+b)}{(\alpha+b+1)(3+3b-\alpha)} \alpha_1^{-2} \alpha_2^{-2}$$
$$+ \left\{ \frac{\alpha(3+b)}{(\alpha+b+1)(3+3b-\alpha)} \alpha_1^{-2} - \frac{1}{1+b-\alpha} \right\} \alpha_2^{(1+b-\alpha)/(1+b)} = 0, \tag{6.35b}$$

$$\frac{1}{\alpha(1-\alpha-\alpha b)} - \frac{\alpha(3+b)}{(\alpha+b+1)(3-\alpha-\alpha b)} \alpha_1^{-2} \alpha_2^{-2} \alpha_3^{-2}$$
$$+ \left\{ \frac{(1+b)(1+\alpha)}{\alpha(b+2)(\alpha-1+\alpha b)} + \frac{\alpha(3+b)}{(3-\alpha-\alpha b)(\alpha+b+1)} \alpha_1^{-2} \alpha_2^{-2} \right\} \alpha_3^{(1-\alpha-\alpha b)} = 0. \tag{6.35c}$$

α_i $(i = 1, ..., 3)$ can be solved from Eqs.(6.35a)$\sim$(6.35c), and subsequently c_i and r_i can be obtained.

c_{1i} and c_{2i} in Eq.(6.3) can be derived as

$$c_{11} = -\frac{\dot{w}_0}{(d_{14}+d_{24})d_{13}d_{12}d_{11} + d_{23}d_{12}d_{11} + d_{22}d_{11} + d_{21}}, \tag{6.36a}$$

$$c_{21} = 1, \tag{6.36b}$$

and

$$\begin{bmatrix} c_{1(i+1)} \\ c_{2(i+1)} \end{bmatrix} = \begin{bmatrix} d_{1i} & 0 \\ d_{2i} & 1 \end{bmatrix} \begin{bmatrix} c_{1i} \\ c_{2i} \end{bmatrix}, \quad (i = 1, ..., 4), \tag{6.36c}$$

where d_{1i} and d_{2i} $(i = 1, ..., 4)$, are constants that can be expressed as

$$d_{1i} = \frac{1-a_i}{1-a_{i+1}} r_i^{a_{i+1}-a_i}, \quad (i = 1, ..., 3), \tag{6.37a}$$

$$d_{2i} = -\frac{a_{i+1}-a_i}{1-a_{i+1}} r_i^{1-a_i}, \quad (i = 1, ..., 3). \tag{6.37b}$$

The plastic limit load is

$$p = \frac{\alpha(6 + 2b)}{\alpha + b + 1} r_1^{-2}. \tag{6.38}$$

Substituting the coefficients c_i, c_{1i}, and c_{2i} into Eqs.(6.2) and (6.3), and r_i into Eq.(6.38), the internal moment, velocity fields and plastic limit load can then be derived.

When $\alpha = 1$ and $b = 0$, the unified strength theory is simplified to be the Tresca criterion, and the plastic limit load is equal to 11.26, which agrees with the results with respect to the Tresca criterion given by Hopkins and Wang (1954). When $\alpha = 1$ and $b = 0.5$, the unified strength theory gives the linear approximation of the Huber-von Mises criterion and the corresponding plastic limit load derived from Eq.(6.38) is 11.799. When $\alpha = 1$ and $b = 1$, the unified strength theory is simplified to be the twin-shear yield criterion, and the plastic limit load is derived to be 12.176, which is consistent with the results based on the twin-shear yield criterion given by Ma and Iwasaki (1999).

Fig.6.13 shows the variation of the plastic limit load p with respect to the unified strength theory parameters b and α. It is seen that the plastic limit load p deceases with the increase of α for specific tensile strength.

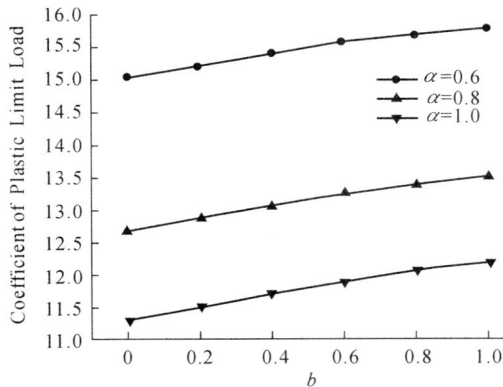

Fig. 6.13. Plastic limit load of clamped circular plate with the unified strength theory parameter b

Figs.6.14 and 6.15 give the fields of internal moments of a clamped circular plate when $\alpha = 1$ and $b = 1$, respectively. It is demonstrated that the effect of different b on the radial bending moment is small while that on the circumferential bending moment is relatively high; and the parameter α has a significant effect on both radial and circumferential bending moments.

Figs.6.16 and 6.17 depict the velocity fields of a clamped circular plate when $\alpha = 1$ and $b = 1$. It is seen that both b and α affect the velocity field significantly.

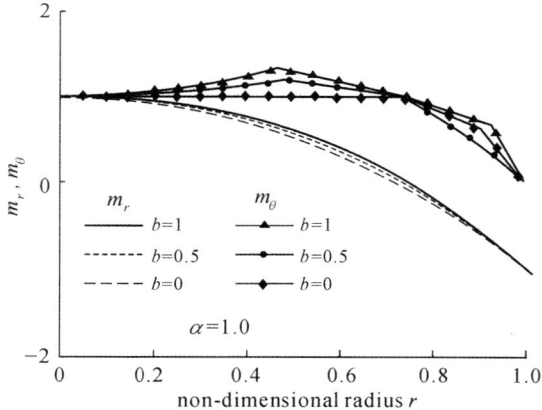

Fig. 6.14. Moment fields of clamped plate when $\alpha = 1$

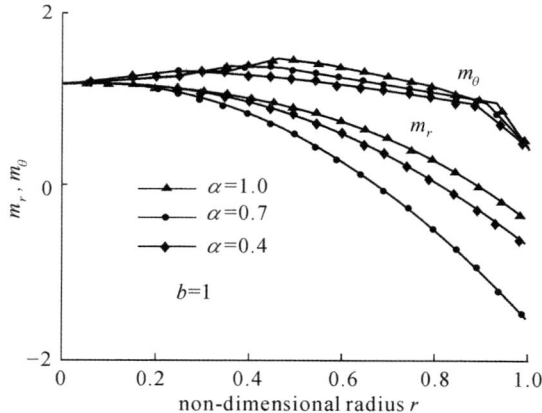

Fig. 6.15. Moment fields of clamped plate when $b = 1$

For different materials, the parameter α is the ratio of the tensile strength σ_t to the compressive strength σ_c, i.e., $\alpha = \sigma_t/\sigma_c$; unified strength theory parameter b may be determined by the shear strength limit of materials τ_0, tensile and compressive strengths, i.e.,

$$b = \frac{(1 + \alpha)\tau_0 - \sigma_t}{\sigma_c - \tau_0}. \tag{6.39}$$

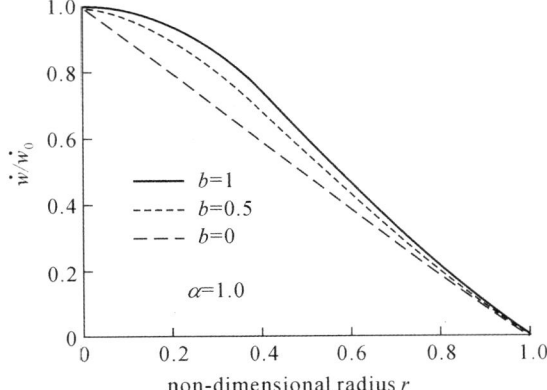

Fig. 6.16. Velocity fields of clamped plate when $\alpha = 1$

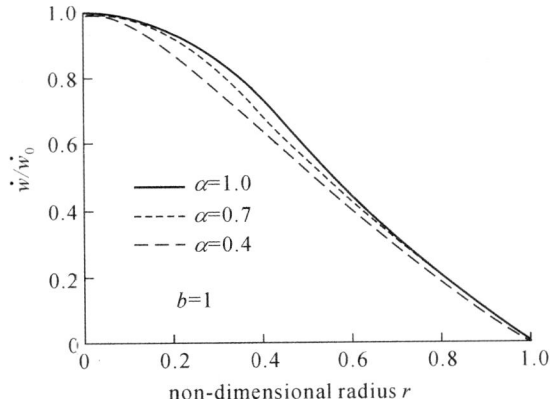

Fig. 6.17. Velocity fields of clamped plate when $b = 1$

6.4 Summary

The plastic limit load, internal moment field and velocity field for a clamped circular plate are derived based on the unified strength theory. With different values of parameters b and α, a series of plastic limit loads are obtained. The results when $b = 0$ and $b = 1$ represent, respectively, the lower bound and upper bound solutions of the plastic limit load. These solutions are closed-form solutions as they satisfy all static and kinematic admissible conditions, i.e., the equilibrium equation, stress boundary conditions, yield condition, flow rule, and velocity boundary conditions. The solutions of non-SD materials with identical tensile and compressive strengths are special cases of the solutions using the unified strength theory for SD materials. It is found that the ratio of tensile and compressive strengths has a significant effect on the plastic limit load. The unified strength theory parameter b also affects the plastic limit load of the plate significantly. The solutions based on the unified strength theory are the systematic solutions for more materials. They may be more accurate for different materials. It considers that different materials behave in different ways with respect to the intermediate principal stress effect compared to those with the Tresca criterion and the Mohr-Coulomb criterion, which ignore the effect of the intermediate principal stress.

6.5 Problems

Problem 6.1 Determine the limit bearing capacity of clamped circular plate by using the Tresca criterion.

Problem 6.2 Determine the limit bearing capacity of clamped circular plate by using the unified yield criterion ($b = 0$).

Problem 6.3 Determine the limit bearing capacity of clamped circular plate by using the unified yield criterion ($b = 0.5$).

Problem 6.4 Determine the limit bearing capacity of clamped circular plate by using the unified yield criterion ($b = 0.8$).

Problem 6.5 Determine the limit bearing capacity of clamped circular plate by using the unified yield criterion ($b = 1.0$).

Problem 6.6 Determine the limit bearing capacity of clamped circular plate by using the unified strength theory ($b = 0$).

Problem 6.7 Determine the limit bearing capacity of clamped circular plate by using the unified strength theory ($b = 0.5$).

Problem 6.8 Determine the limit bearing capacity of clamped circular plate by using the unified strength theory ($b = 0.8$).

Problem 6.9 Determine the limit bearing capacity of clamped circular plate by using the unified strength theory ($b = 1.0$).

Problem 6.10 Write a paper concerning the unified solution of plastic limit analysis for a clamped circular plate under another load.

References

Chen WF (1998) Concrete plasticity: past, present and future. In: Yu MH, Fan SC (eds.) Strength Theory: Applications, Developments and Prospects for the 21st Century, Science Press, Beijing, New York, 7-48

Chen WF, Saleeb AF (1981) Constitutive equations for engineering materials. Vol.1, Elasticity and Modelling; Vol.2, Plasticity and Modeling, Wiley, New York

Hodge PG Jr (1963) Limit analysis of rotationally symmetric plates and shells. Prentice-Hall, New Jersey

Hopkins HG, Prager W (1953) The load carrying capacities of circular plates. J. Mech. Phys. Solids, 2:1-13

Hopkins HG, Wang AJ (1954) Load carrying capacities for circular plates of perfectly-plastic material with arbitrary yield condition. J. Mech. Phys. Solids, 3:117-129

Hu LW (1960) Design of circular plates based on plastic limit load. J. Eng. Mech., ASCE, 86(1):91-115

Ma GW, Hao H, Iwasaki S (1999) Plastic limit analysis of clamped circular plates with unified yield criterion. Struct. Engineering and Mechanics, 7(5):513-525

Ma GW, Hao H, Iwasaki S (1999) Unified plastic limit analysis of circular plates under arbitrary load. J. Appl. Mech., ASME, 66(6):568-570

Ma GW, He LN (1994) Unified solution to plastic limit of simply supported circular plate. Mechanics and Practice, 16(6):46-48 (in Chinese, English abstract)

Ma GW, Iwasaki S, Miyamoto Y, Deto H (1999) Dynamic plastic behavior of circular plate using unified yield criterion. Int. J. Solids Struct., 36(22):3257-3275

Ma GW, Yu MH, Iwasaki S, et al. (1994) Plastic analysis of circular plate on the basis of the twin-shear unified yield criterion. In: Lee PKK, Tham LG, Cheung YK (eds.) Proceedings of International Conference on Computational Methods in Structural and Geotechnical Engineering. China Translation and Printing Ltd., Hongkong, 3:930-935,

Ma GW, Yu MH, Iwasaki S, Miyamoto Y (1995) Unified elasto-plastic solution to rotating disc and cylinder. J. Struct. Eng. (Japan), 41A:79-85

Ma GW, Yu MH, Miyamoto Y, et al. (1995) Unified plastic limit solution to circular plate under portion uniform load. J. Struct. Eng. (in English, J SCE), Vol.41A:385-392

Nielsen MP (1999) Limit analysis and concrete plasticity. (2nd ed.) CRC Press, Boca Raton, London

Save MA, Massonnet CE (1972) Plastic analysis and design of plates, shells and disks. North-Holland, Amsterdam

Save MA, Massonnet CE, Saxce G de (1997) Plastic limit analysis of plates, shells and disks. Elsevier, Amsterdam

Wang AJ, Hopkins HG (1954) On the plastic deformation of built-in circular plates under impulsive load. J. Mech. Phys. Solids, 3:22-37

Wei XY, Yu MH (2001) Unified plastic limit of clamped circular plate with strength differential effect in tension and compression. Chinese Quart. Mechanics, 22(1):78-83

Yu MH (1992) New system of strength theories. Xi'an Jiaotong University Press, Xi'an (in Chinese)

Yu MH (2002) Advance in strength theory of material and complex stress state in the 20th Century. Applied Mechanics Reviews, 55(3):169-218

Yu MH (2004) Unified strength theory and its applications. Springer, Berlin

Yu MH, He LN (1991) A new model and theory on yield and failure of materials under the complex stress state, In: Mechanical Behavior of Materials-6, Pergamon Press, Oxford, 3:841-846

Zyczkowski M (1981) Combined loadings in the theory of plasticity. Polish Scientific Publishers, PWN and Nijhoff

7

Plastic Limit Analysis of Annular Plate

7.1 Introduction

An annular plate is a very common structural component in many branches of engineering, such as mechanical, aeronautic, and civil fields. Limit analysis of the plate is important in revealing its structural behavior and load-bearing capacity. The load-bearing capacity of circular plates has been given in terms of the Huber-von Mises criterion (Hopkins and Prager, 1953) and the Tresca yield criterion (Hopkins and Wang, 1954). The results are applicable for the materials which has $\tau_s = 0.5\sigma_s$ and $\tau_s = 0.577\sigma_s$, respectively. The load-bearing capacity of annular plates was studied by Mroz and Sawczuk (1960), Hodge (1959, 1963), Save and Massonnet (1972), Zyczkowski (1981), Save, Massonnet and Saxce (1997), et al.

Recently, the twin-shear yield criterion and unified strength theory have been applied in many fields (Huang et al., 1989; Li, 1988; Ma et al., 1994; 1995; 1999). The load-bearing capacity of circular and annular plates using an arbitrary yield criterion was given by Aryanpour and Ghorashi (2002). The yield criteria they used are applicable for the materials with identical tensile and compressive strength (non-SD materials). However, many materials have an SD effect (different tensile and compressive strengths), such as concrete, rock, cast iron and polymer. Recent studies have shown that even some high-strength metal materials have an obvious SD effect (Chait, 1972; Rauch et al., 1972; Drucker, 1973; Casey and Sullivan, 1985). Therefore, the plastic limit analysis for the materials having an SD effect is very necessary.

For a simply supported annular plate under uniform load as shown in Fig.7.1, the unified solutions for non-SD materials in terms of the unified yield criterion was given by Ma et al. (1994; 1995). The Tresca-Mohr-Coulomb strength theory has been used for limit analysis. But the effect of intermediate principal stress has not been considered in the Tresca-Mohr-Coulomb strength theory. The unified solution for the plastic limit of an annular plate with the unified strength theory was given by Wei and Yu (2002).

In this chapter, the unified solutions of an annular plate under uniform load for non-SD and SD materials will be presented. The unified solutions of plastic load-bearing capacities, moment fields and velocity fields of an annular plate are obtained. The solutions take into account all the stress components, and can be applied for the non-SD and SD materials with different shear-tension ratios, or different relative effects of the intermediate principal stress.

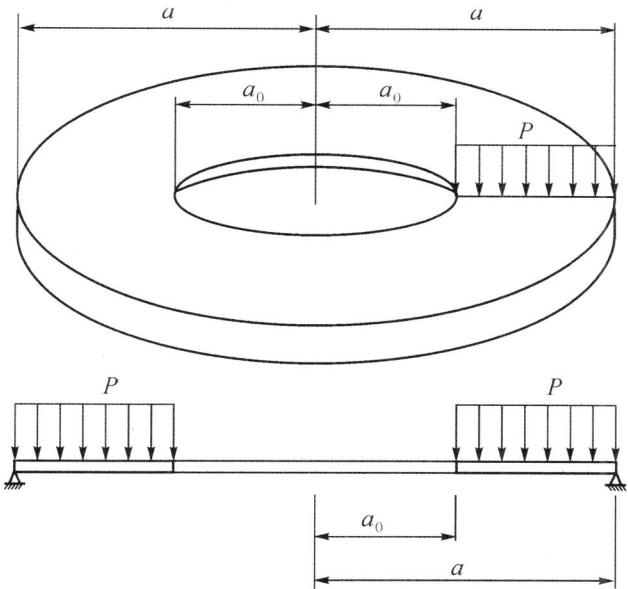

Fig. 7.1. Simply supported annular plate

7.2 Basic Equations for Annular Plate Based on UYC

For an annular plate with a simply supported outer edge and a free inner edge, if the load per unit area is P, the generalized stresses are

$$M_r = \int_{-\frac{h}{2}}^{\frac{h}{2}} \sigma_r z \mathrm{d}z, \quad M_\theta = \int_{-\frac{h}{2}}^{\frac{h}{2}} \sigma_\theta z \mathrm{d}z, \tag{7.1a}$$

$$Q_{rz} = \int_{-\frac{h}{2}}^{\frac{h}{2}} \tau_{rz} z \mathrm{d}z, \quad M_0 = \int_{-\frac{h}{2}}^{\frac{h}{2}} \sigma_t z \mathrm{d}z, \tag{7.1b}$$

where M_r and M_θ are radial and circumferential bending moments, respectively; Q_{rz} is the transverse shear force; M_0 is the ultimate bending moment. Defining

$$r = R/a, \quad m_r = M_r/M_0, \quad m_\theta = M_\theta/M_0,$$
$$p = Pa^2/M_0, \quad \beta = a_0/a,$$

then the moment fields of the entire plate satisfy $m_r \geqslant 0$ and $m_\theta \geqslant 0$.

It implies that the moment components m_r and m_θ lie on the two segments CB and BA in Fig.7.2(b) for the unified yield criterion. Boundary conditions of the annular plate give that $m_r = 0$ and $\dot{w} = \dot{w}_0$ at the inner edge $(r = \beta)$; and $m_r = 0$, $\dot{w} = 0$ at the outer edge $(r = 1)$. The equilibrium equation can be written as

$$\mathrm{d}(rm_r)/\mathrm{d}r - m_\theta = \frac{1}{2}p\beta^2 - \frac{1}{2}pr^2, \tag{7.2}$$

which satisfies the condition of zero shear force at the inner edge.

The varying trajectory of moment components m_r and m_θ along the radial direction can be put into two cases according to the unified yield criterion, namely, (1) $C \to B \to C$ when the inner to outer radius ratio β is larger than the critical value of β_0 (Fig.7.2(a)), and (2) $C \to B \to A \to B \to C$ when β is smaller than β_0 (Fig.7.2(b)).

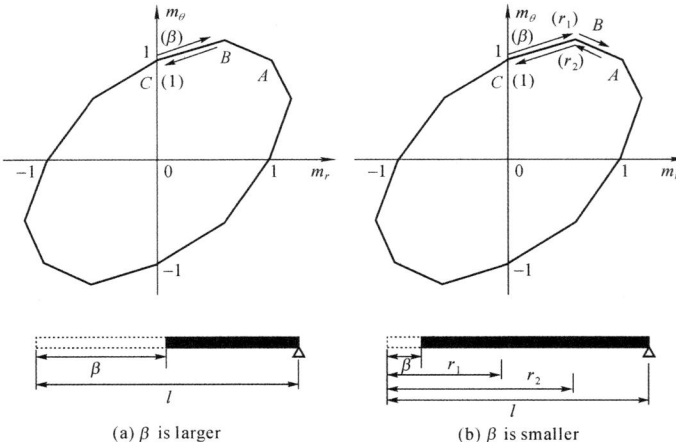

(a) β is larger   (b) β is smaller

Fig. 7.2. Case (1): β is larger, Case (2): β is smaller

The unified yield criterion for the plate can be expressed by generalized stresses m_r and m_θ as follows:

$$\begin{cases} M_r - \dfrac{b}{1+b}M_\theta = \pm M_p, \quad M_r - \dfrac{1}{1+b}M_\theta = \pm M_p, \\[2mm] \dfrac{1}{1+b}(bM_r + M_\theta) = \pm M_p, \quad \dfrac{1}{1+b}M_r - M_\theta = \pm M_p, \\[2mm] \dfrac{1}{1+b}(M_r + bM_\theta) = \pm M_p, \quad \dfrac{b}{1+b}M_r - M_\theta = \pm M_p, \end{cases} \quad (7.3)$$

which can be rewritten as

$$m_\theta = a_i m_r + b_i, \quad (i = 1, ..., 5), \tag{7.4}$$

where a_i and b_i are constants and their values are shown in Table 7.1. According to Eqs.(7.3) and (7.4),

$$m_r = \frac{b_i + \frac{1}{2}p\beta^2}{1 - a_i} - \frac{pr^2}{2(3 - a_i)} + c_i r^{-1+a_i}, \quad (i = 1, ..., 5). \tag{7.5}$$

The values of a_i and b_i are shown in Table 7.1 for L_i ($i = 1, ...,4$) (AB, BC, CD, DE lines).

Table 7.1. Constants a_i and b_i in the unified yield criterion

	AB ($i=1$)	BC ($i=2$)	CD ($i=3$)	DE ($i=4$)	EF ($i=5$)
a_i	$-b$	$b/(1+b)$	$1/(1+b)$	$1+b$	$(1+b)/b$
b_i	$1+b$	1	1	$1+b$	$(1+b)/b$

According to the symmetry of the geometry,

$$\dot{k}_r = -\mathrm{d}^2\dot{w}/\mathrm{d}r^2, \quad \dot{k}_\theta = -(1/r)\mathrm{d}\dot{w}/\mathrm{d}r. \tag{7.6}$$

Based on the associated flow rule,

$$\dot{k}_r = \dot{\lambda}\partial F/\partial m_r, \quad \dot{k}_\theta = \dot{\lambda}\partial F/\partial m_\theta. \tag{7.7a}$$

From Eq.(7.4),

$$\partial F/\partial m_r = -a_i, \quad \partial F/\partial m_\theta = 1.$$

Substituting the above equations into Eq.(7.7a) yields

$$\dot{k}_r = -\dot{\lambda}a_i, \quad \dot{k}_\theta = \dot{\lambda}. \tag{7.7b}$$

The velocity field is derived by substituting Eq.(7.7b) into Eq.(7.6) and integrating Eq.(7.6),

$$w = c_{1i}r^{1-a_i} + c_{2i}, \quad (i = 1, ..., 3). \tag{7.8}$$

7.2.1 Case (1)

The moment at each point in the plate lies on the side CB in Fig.7.2(a) if the ratio of the inner radius to outer radius β is bigger than a critical value β_0. The equilibrium equation can be derived by direct integration of Eq.(7.2) with application of the boundary condition of the outer and inner edges,

$$w = c_{1i}r^{1-a_i} + c_{2i}, \quad (i = 1, ..., 3), \tag{7.9a}$$

$$m_r = (1+b)\left(1 + \frac{1}{2}p\beta^2\right)(1 - r^{-\frac{1}{1+b}}) - \frac{1+b}{6+4b}p(r^2 - r^{-\frac{1}{1+b}}), \tag{7.9b}$$

$$m_\theta = \frac{b}{1+b}m_r + 1. \tag{7.9c}$$

And the plastic limit load is

$$p = \frac{(6+4b)(1 - \beta^{-1/(1+b)})}{(3+2b)\beta^{(1+2b)/(1+b)} - 2(1+b)\beta^2 - \beta^{-1/(1+b)}}. \tag{7.10}$$

The velocity field can be derived from Eq.(7.8) and the boundary conditions as

$$\dot{w} = \dot{w}_0\frac{1 - r^{1/(1+b)}}{1 - \beta^{1/(1+b)}}. \tag{7.11}$$

To satisfy the requirement of statical admissibility, all the moment components (m_θ, m_r) should lie on the segment CB. Thus, the maximum radial moment m_r at $r = r_0$ satisfies

$$m_r(r = r_0) \leqslant (1+b)/(2+b), \tag{7.12}$$

and

$$\partial m_r/\partial r = 0. \tag{7.13}$$

For the critical case of $\beta = \beta_0$, the maximum moment m_r at $r = r_0$ is equal to $(1+b)/(2+b)$, which corresponds to the corner point B in Fig.7.2(a). r_0 can then be deduced from Eqs.(7.9a) and (7.13) as

$$r_0 = \left[\frac{2(1+b)p}{(6+4b)(1 + p\beta^2/2) - p}\right]^{-\frac{1+b}{3+2b}}. \tag{7.14}$$

When Eqs.(7.10) and (7.14) are substituted into Eq.(7.9a), the critical radius ratio β_0 can be calculated by the half interval search method in the interval $(\beta, 1)$ regarding the yield condition $m_r(r = r_0, \beta = \beta_0) = (1+b)/(2+b)$ at point B in Fig.7.2(a).

7.2.2 Case (2)

When the radius ratio $\beta \leqslant \beta_0$, and the moments m_θ and m_r locate on the side BA in Fig.7.2(b), m_r is obtained by integrating Eq.(7.2),

$$m_r = \begin{cases} \dfrac{b_2 + p\beta^2/2}{1 - a_2} - \dfrac{pr^2}{2(3 - a_2)} + c_1 r^{-1+a_2}, & \beta \leqslant r \leqslant r_1, \\[3mm] \dfrac{b_1 + p\beta^2/2}{1 - a_1} - \dfrac{pr^2}{2(3 - a_1)} + c_2 r^{-1+a_1}, & r_1 \leqslant r \leqslant r_2, \\[3mm] \dfrac{b_2 + p\beta^2/2}{1 - a_2} - \dfrac{pr^2}{2(3 - a_2)} + c_3 r^{-1+a_2}, & r_2 \leqslant r \leqslant 1, \end{cases} \qquad (7.15)$$

where a_1, b_1, a_2 and b_2 are defined in Table 7.1; r_1 and r_2 are dividing radii where the moments fall on point B in Fig.7.2(b); c_1, c_2 and c_3 are integration constants. The six unknowns c_1, c_2, c_3, p, r_1 and r_2 in Eq.(7.15) can be numerically calculated from the six equations of boundary and continuous conditions, namely,

    (1) $m_r(r = \beta) = 0$ (point C in Fig.7.2(b));
    (2) $m_r(r = r_1, \beta \leqslant r \leqslant r_1) = (1 + b)/(2 + b)$ (point B in Fig.7.2(b));
    (3) $m_r(r = r_1, r_1 \leqslant r \leqslant r_2) = (1 + b)/(2 + b)$ (point B in Fig.7.2(b));
    (4) $m_r(r = r_2, r_1 \leqslant r \leqslant r_2) = (1 + b)/(2 + b)$ (point B in Fig.7.2(b));
    (5) $m_r(r = r_2, r_2 \leqslant r \leqslant 1) = (1 + b)/(2 + b)$ (point B in Fig.7.2(b));
    (6) $m_r(r = 1) = 0$ (point C in Fig.7.2(b)).

    The procedure to derive c_1, c_2, c_3, p, r_1 and r_2 is similar to that for Case (1).

    The velocity fields are then derived as

$$\dot{w} = \dot{w}_0 \begin{cases} c_{11} r^{1-a_2} + c_{12}, & \beta \leqslant r \leqslant r_1, \\[1mm] c_{21} r^{1-a_1} + c_{22}, & r_1 \leqslant r \leqslant r_2, \\[1mm] c_{31} r^{1-a_2} + c_{32}, & r_2 \leqslant r \leqslant 1, \end{cases} \qquad (7.16)$$

where $\dot{w}_0$ is the velocity at the inner edge; c_{11}, c_{21}, c_{12}, c_{22}, c_{31} and c_{32} are integration constants.

    According to the boundary and continuous conditions of velocity fields: (1) $\dot{w}(r = \beta) = \dot{w}_0$; (2) $\dot{w}$ and $d\dot{w}/dr$ $(r = r_i, \ i = 1, 2)$ are continuous; (3) $\dot{w}(r = 1) = 0$, the integration constants in Eq.(7.16) are determined from the following matrix equations,

$$
\begin{bmatrix}
\beta^{1/(1+b)} & 1 & 0 & 0 & 0 & 0 \\
r_1^{1/(1+b)} & 1 & -r_1^{1/(1+b)} & -1 & 0 & 0 \\
r_1^{-b/(1+b)}\big/(1+b) & 0 & (1+b)r_1^b & 0 & 0 & 0 \\
0 & 0 & r_2^{1+b} & 1 & -r_2^{1+b} & -1 \\
0 & 0 & (1+b)r_2^{-b} & 0 & r_2^{-1/(1+b)}\big/(1+b) & 0 \\
0 & 0 & 0 & 0 & 1 & 1
\end{bmatrix}
\begin{bmatrix}
c_{11} \\ c_{12} \\ c_{21} \\ c_{22} \\ c_{31} \\ c_{32}
\end{bmatrix}
=
\begin{bmatrix}
1 \\ 0 \\ 0 \\ 0 \\ 0 \\ 0
\end{bmatrix}.
$$

$$(7.17)$$

The constants c_{1i} and c_{2i} ($i=1, ..., 3$) can be calculated directly from Eq.(7.17) since the coefficient matrix is constant for the unified strength theory parameter b and ratio β. The velocity field can then be determined by substituting those constants into Eq.(7.16).

7.2.3 Special Case

The above two subsections give the similar solutions when the unified strength theory parameter b is equal to 0 in Eqs.(7.9a)~(7.11), (7.15) and (7.16), which is the exact solution in terms of the Tresca criterion. The moments are

$$
m_r = (1 + \frac{1}{2}p\beta^2)(1 - r^{-1}) - \frac{1}{6}p(r^2 - r^{-1}). \tag{7.18}
$$

The velocity field is obtained from Eq.(7.8) with application of the boundary and continuous conditions as

$$
\dot{w} = \dot{w}_0 \frac{1 - r}{1 - \beta}. \tag{7.19}
$$

And the corresponding plastic limit load is derived from Eq.(7.10) with $b = 0$,

$$
p = \frac{6}{(1 - \beta)(1 + 2\beta)}. \tag{7.20}
$$

7.3 Unified Solutions of Annular Plate for Non-SD Materials

Figs.7.3(a) and 7.3(b) illustrate schematically the radial and circumferential moment fields of an annular plate when the radius ratio $\beta=0.1$ and 0.5, which are representative of the second case and the first case, respectively.

There are two peak values of m_θ (Fig.7.3(a)) for the second case, which occur at the radii $r = r_1$ and $r = r_2$, while there is only one peak value of m_θ (Fig.7.3(b)) for the first case.

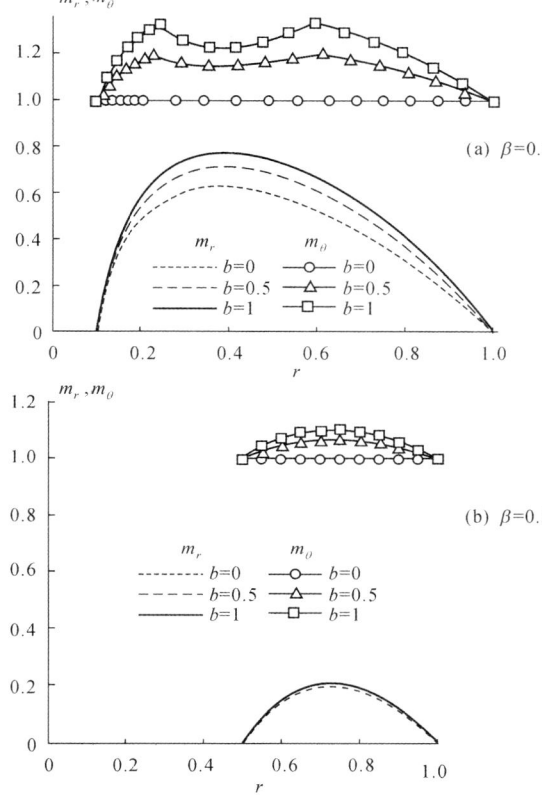

Fig. 7.3. Moment fields of annular plate based on UYC

The influence of yield criteria on the moment distribution is more prominent for the second case than for the first one. The plastic limit loads of an
annular plate in the case of β=0.1 are 5.555, 6.307 and 6.790 respectively for
the three specific cases of the unified yield criterion with $b = 0$, $b = 0.5$ and
$b = 1$.

Figs.7.4(a) and 7.4(b) show the velocity fields of the annular plates with
β=0.1 and β=0.5, respectively. Fig.7.5 illustrates schematically the percentage difference of plastic limit loads in terms of the unified yield criterion with
$b = 0.5$ and $b = 1$ to that in terms of the Tresca criterion when the ratio of
inner radius to outer radius β ranges from 0.01 to 0.99. The maximum percentage differences of plastic limit loads with respect to UYC with $b = 0.5$
and $b = 1$ to that with $b = 0$ are 13.9% and 23.0%, respectively, for the
critical case of $\beta = \beta_0$.

The difference of yield criterion has significant influence on the load-
bearing capacities of circular plates. Fig.7.6 shows the plastic limit loads

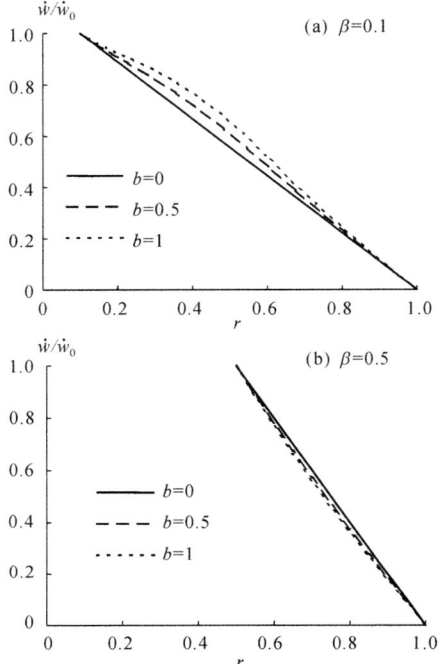

Fig. 7.4. Velocity fields of annular plate based on UYC

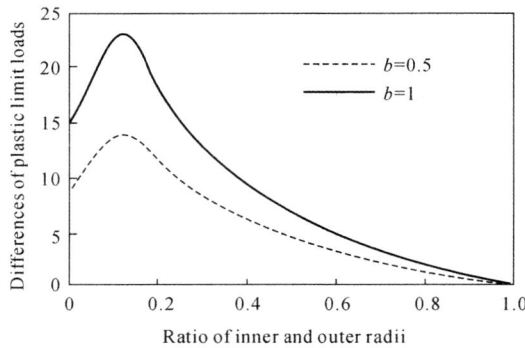

Fig. 7.5. Percentage difference of plastic limit loads by different criteria

with the parameter b varying from 0 to 1 for the three circular plates with different supporting conditions. The plastic limit loads increase with b; the upper bound and lower bound load-bearing capacities are deduced with $b = 1$ and $b = 0$, respectively. Since the solutions are both statically admissible and kinematically admissible, i.e., all the equilibrium equations, stress boundary conditions, yield conditions, flow conditions, and velocity boundary conditions are satisfied, they are the exact solutions to the problems. For fixed boundary condition, a solution based on the Tresca criterion (Hodge, 1963) leads to $m_\theta = 0$ (corresponding to $\nu_p = 0$) according to the moment boundary condition at the edge; the iterative solution based on the Huber-von Mises criterion (Hopkins and Wang, 1954) satisfies $m_\theta = 0.5m_r$ (corresponding to $\nu_p = 0.5$) at the edge according to the plastic flow requirement ($\dot{k}_\theta \geqslant 0$). The present study has derived the unified solutions with the edge effect coefficient ν_p in the range of $0 \leqslant \nu_p \leqslant 0.5$, which satisfies both the moment boundary condition and the plastic flow requirement.

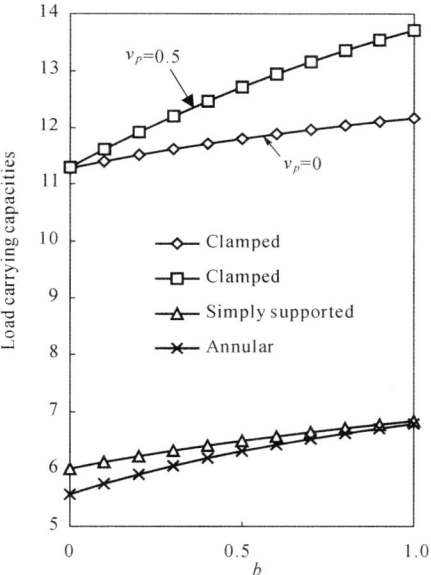

Fig. 7.6. Plastic limit loads of plates with different unified strength theory parameter b

7.4 Unified Solutions of Limit Load of Annular Plate for SD Materials

7.4.1 Unified Strength Theory

The unified strength theory (UST) has been described in Chapter 3. It has a unified model and a simple unified mathematical expression. It is applicable for various materials. In the mathematical expression of UST, α is the ratio of tensile strength σ_t to compressive strength σ_c, i.e., $\alpha = \sigma_t/\sigma_c$; b is the material strength parameter that reflects the influences of the intermediate principal stress on the yield of material ($0 \leqslant b \leqslant 1$). The UST could represent other prevailing yield or strength criteria with specific values of b. When $b = 0$, UST becomes the Mohr-Coulomb theory; when $b = 1$, it is the twin-shear strength theory (Yu et al., 1985); when $\alpha = 1$, it is the unified yield criterion (UYC); and when $b = 0$, $b = 0.5$ and $b = 1$, the Tresca yield criterion, linear approximation of the Huber-von Mises yield criterion, and the twin-shear stress yield criterion (Yu, 1961) are obtained respectively.

7.4.2 Basic Equations for Annular Plate Based on the UST

The unified strength theory (UST) can be expressed in terms of generalized stresses M_r, M_θ,

$$M_r - \frac{\alpha b}{1+b} M_\theta = \pm M_P, \quad M_r - \frac{\alpha}{1+b} M_\theta = \pm M_P,$$

$$\frac{\alpha}{1+b}(b M_r + M_\theta) = \pm M_P, \quad \frac{1}{1+b} M_r - \alpha M_\theta = \pm M_P, \tag{7.21}$$

$$\frac{\alpha}{1+b}(M_r + b M_\theta) = \pm M_p, \quad \frac{b}{1+b} M_r - \alpha M_\theta = \pm M_P.$$

It can be rewritten as

$$m_\theta = a_i m_r + b_i, \quad (i = 1, ..., 4), \tag{7.22}$$

where a_i and b_i are constants and the values are shown in Table 7.2. From Eqs.(7.2) and (7.22),

$$m_r = \frac{b_i + \frac{1}{2}p\beta^2}{1 - a_i} - \frac{p r^2}{2(3 - a_i)} + c_i r^{-1+a_i}, \quad (i = 1, ..., 4). \tag{7.23}$$

The values of a_i and b_i are listed in Table 7.1 for L_i ($i = 1, ..., 4$) (lines AB, BC, CD, DE) in Fig.7.7, respectively.

The velocity field expression based on UST is the same as that with UYC. The coefficients a_i and b_i are listed in Table 7.2.

Table 7.2. Constants a_i and b_i in the unified strength theory

	AB ($i = 1$)	BC ($i = 2$)	CD ($i = 3$)	DE ($i = 4$)
a_i	$-b$	$\alpha \frac{b}{1+b}$	$\frac{\alpha}{1+b}$	$\alpha(1 + b)$
b_i	$1 + b$	1	1	$1 + b$

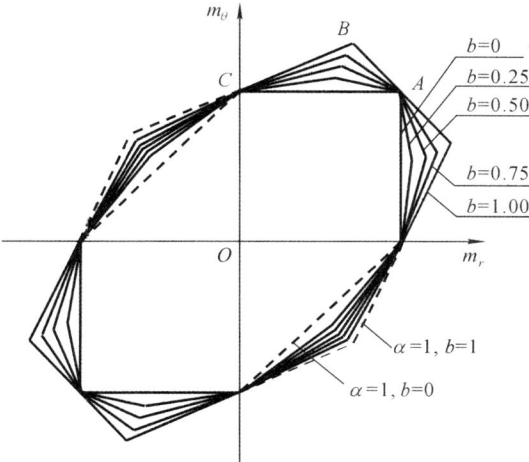

Fig. 7.7. Yield loci of UST in generalized stresses m_r-m_θ space

7.4.3 Limit Analysis

Case (1)
When $\beta \geqslant \beta_0$, m_r and m_θ correspond to $C \to B \to C$. Referring to Eq.(7.23) and the boundary conditions, the moment field can be derived as

$$m_r = \frac{1+b}{1+b-\alpha b}(1 + \frac{1}{2}p\beta^2)(1 - r^{-\frac{1+b-\alpha b}{1+b}}) - \frac{p(1+b)}{6+6b-2\alpha b}(r^2 - r^{-\frac{1+b-\alpha b}{1+b}}),$$
$$(7.24)$$

$$m_\theta = \frac{\alpha b}{1+b}m_r + 1, \tag{7.25}$$

and the limit load coefficient can be expressed as

$$p = \frac{(6+6b-2\alpha b)(\beta^{-\frac{1+b-\alpha b}{1+b}} - 1)}{2(1+b)\beta^2 + (1+b-\alpha b)\beta^{-\frac{1+b-\alpha b}{1+b}} - (3+3b-\alpha b)\beta^{\frac{1+b+\alpha b}{1+b}}}. \tag{7.26}$$

The velocity field can be derived from Eq.(7.8) with application of the boundary condition,

$$\dot{w} = \dot{w}_0 \frac{1 - r^{\frac{1+b-\alpha b}{1+b}}}{1 - \beta^{\frac{1+b-\alpha b}{1+b}}}. \tag{7.27}$$

When $r = r_0$, m_r is the maximum from Eq.(7.24). When m_r and m_θ are on the line CB, they satisfy the relation of

$$m_r(r = r_0) \leqslant \frac{1 + b}{1 + b + \alpha} \quad \text{and} \quad \left. \frac{\partial m_r}{\partial r} \right|_{r=r_0} = 0. \tag{7.28}$$

From Eqs.(7.24), (7.12) and (7.13),

$$r_0 = \left[\frac{2p(1 + b)}{(1 + \frac{1}{2}p\beta^2)(6 + 6b - 2\alpha b) - p(1 + b - \alpha b)} \right]^{-\frac{1+b}{3+3b-\alpha b}}. \tag{7.29}$$

When $\beta = \beta_0$, m_r $(r = r_0)$ lies at point B,

$$m_r(r = r_0, \beta = \beta_0) = \frac{1 + b}{1 + b + \alpha}. \tag{7.30}$$

The critical variable β_0 can be deduced from Eqs.(7.29), (7.30), (7.26) and (7.24).

Case (2)

When $\beta \leqslant \beta_0$, m_r and m_θ correspond to $C \rightarrow B \rightarrow A \rightarrow B \rightarrow C$. Assuming r_1 is the dimensionless radius when point B is reached for the first time $(C \rightarrow B)$, r_2 is the dimensionless radius when point B is reached for the second time $(C \rightarrow B \rightarrow A \rightarrow B)$. And referring to Eq.(7.23),

$$m_r = \begin{cases} \dfrac{b_2 + \frac{1}{2}p\beta^2}{1 - a_2} - \dfrac{pr^2}{2(3 - a_2)} + c_1 r^{-1+a_2}, & \beta \leqslant r \leqslant r_1, \\[2mm] \dfrac{b_1 + \frac{1}{2}p\beta^2}{1 - a_1} - \dfrac{pr^2}{2(3 - a_1)} + c_2 r^{-1+a_1}, & r_1 \leqslant r \leqslant r_2, \\[2mm] \dfrac{b_2 + \frac{1}{2}p\beta^2}{1 - a_2} - \dfrac{pr^2}{2(3 - a_2)} + c_3 r^{-1+a_2}, & r_2 \leqslant r \leqslant 1. \end{cases} \tag{7.31}$$

The integration constants c_1, c_2, c_3 and the variables p, r_1, r_2 can be derived from the following continuity and boundary conditions: (1) $m_r(r = \beta) = 0$; (2) $m_r(r = r_1, \beta \leqslant r \leqslant r_1) = \frac{1+b}{1+b+\alpha}$; (3) $m_r(r = r_1, r_1 \leqslant r \leqslant r_2) = \frac{1+b}{1+b+\alpha}$; (4) $m_r(r = r_2, r_1 \leqslant r \leqslant r_2) = \frac{1+b}{1+b+\alpha}$; (5) $m_r(r = r_2, r_2 \leqslant r \leqslant 1) = \frac{1+b}{1+b+\alpha}$; (6) $m_r(r = 1) = 0$.

From Eq.(7.8), the velocity field can be expressed as

$$\dot{w} = \dot{w}_0 \begin{cases} c_{11} r^{1-a_2} + c_{21}, & \beta \leqslant r \leqslant r_1, \\ c_{12} r^{1-a_1} + c_{22}, & r_1 \leqslant r \leqslant r_2, \\ c_{13} r^{1-a_2} + c_{23}, & r_2 \leqslant r \leqslant 1. \end{cases} \tag{7.32}$$

The continuity and boundary conditions can be written as: (1) $\dot{w}(r = 0) = \dot{w}_0$; (2) $d\dot{w}/dr$ $(r = r_i,\ i=1, 2)$ and $\dot{w}$ are continuous; (3) $\dot{w}(r = 1) = 0$. The integration constants c_{ij} $(i=1, 2,\ j=1, 2, 3)$ can be derived from these conditions.

Special Case

When the parameter b is equal to 0 in Eqs.(7.24),(7.27), the moment fields, velocity fields, and plastic limit loads of the annular plate for SD materials with $0 < \alpha \leqslant 1$ in the two different cases are the same, which is the exact solution in terms of the Mohr-Coulomb strength theory. The moments are

$$m_r = \left(1 + \frac{1}{2}p\beta^2\right)(1 - r^{-1}) - \frac{1}{6}p(r^2 - r^{-1}), \qquad (7.33)$$

$$m_\theta = 1. \qquad (7.34)$$

The corresponding plastic limit load coefficient is

$$p = \frac{6}{1 + \beta - 2\beta^2}. \qquad (7.35)$$

The velocity field is

$$\dot{w} = \frac{\dot{w}_0(1 - r)}{(1 - \beta)}. \qquad (7.36)$$

7.4.4 Results and Discussions

Figs.7.8(a) and 7.8(b) show the relations of the plastic limit load to the unified strength theory parameter b and α for Case (1) and Case (2), respectively. For a specific value of α, the plastic limit load increases with the parameter b.

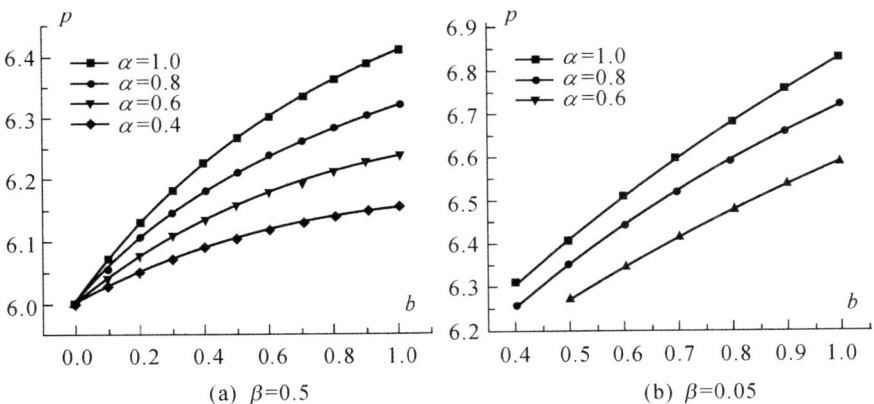

Fig. 7.8. Plastic limit loads of annular plates with different unified strength theory parameter b

Figs.7.9(a) and 7.9(b) illustrate the radial and circumferential moment fields with the radius ratio $\beta=0.5$ and 0.05, $b = 1$, which are representatives for Case (1) and Case (2), respectively. There are two peak values of m_θ (Fig.7.9(b)) for Case (2), which occur at the radii $r = r_1$ and $r = r_2$; for Case (1), there is only one peak value of m_θ (Fig.7.4(a)). The influence of different strength criteria on the moment distribution is higher for Case (2) than for Case (1). Table 7.3 shows the relationship between the plastic limit load coefficient p and the parameters α, b, and β.

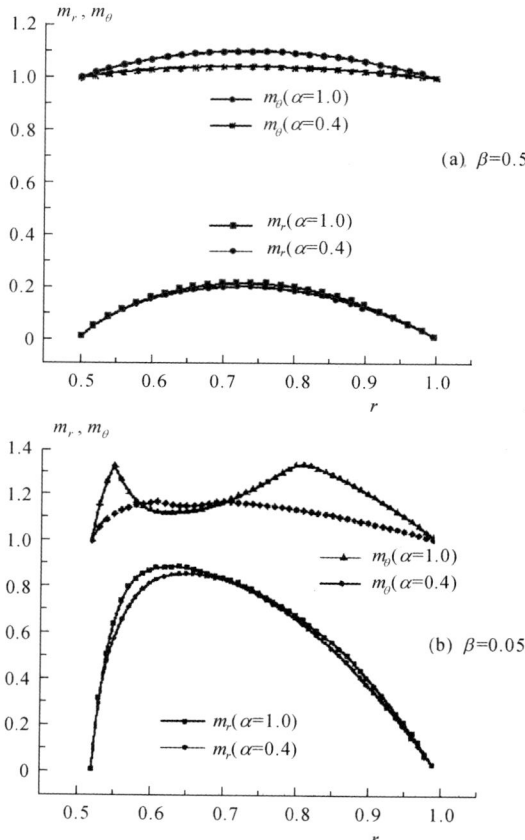

Fig. 7.9. Moment field when $b = 1$

7.5 Summary

The unified solutions of the limit-bearing capacity of an annular plate for non-SD materials (Ma et al. 1994; 1995; 1998) and SD materials (Wei and Yu, 2001; 2002) have the following characteristics:

Table 7.3. Plastic limit load having different cohesion α and b

β	α	$b = 0$	$b = 0.2$	$b = 0.4$	$b = 0.6$	$b = 0.8$	$b = 1.0$
	0.7	5.459	5.816	6.005	6.210	6.391	6.532
0.1	0.8	5.503	5.769	6.094	6.276	6.438	6.563
	1.0	5.517	5.591	6.175	6.319	6.385	6.650
	0.2	6.000	6.026	6.044	6.058	6.069	6.077
	0.4	6.000	6.051	6.088	6.117	·6.139	6.157
0.5	0.6	6.000	6.077	6.134	6.177	6.211	6.238
	0.8	6.000	6.103	6.180	6.238	6.284	6.322
	1.0	6.000	6.130	6.226	6.301	6.360	6.408
	0.2	9.600	9.616	9.628	9.636	9.643	9.648
	0.4	9.600	9.632	9.655	9.673	9.686	9.697
0.75	0.6	9.600	9.648	9.683	9.710	9.730	9.747
	0.8	9.600	9.665	9.711	9.747	9.774	9.797
	1.0	9.600	9.681	9.740	9.784	9.819	9.847
	0.2	21.429	21.441	21.450	21.457	21.463	21.467
	0.4	21.429	21.454	21.472	21.486	21.497	21.505
0.9	0.6	21.429	21.467	21.494	21.515	21.531	21.544
	0.8	21.429	21.480	21.516	21.544	21.566	21.583
	1.0	21.429	21.493	21.539	21.573	21.600	21.622

(1) It is the first time we use the unified strength theory to analyze the plastic limit load, moment field and velocity field for annular plates. A series of limit solutions from the single-shear failure criteria (Tresca criterion and Mohr-Coulomb criterion) to the twin-shear yield criterion (Yu, 1961), and the twin-shear strength theory (Yu et al., 1985) are specific cases of the unified solutions from UST with the specific values of the parameters α and b. And the upper bound and lower bound solutions can be obtained with $b = 0$ and $b = 1$, respectively. The solutions satisfy the static and dynamic conditions, namely the equilibrium equations, stress boundary conditions, yield conditions, associate flow rule, and the velocity boundary conditions. The solutions can be applied to all kinds of isotropic materials.

(2) The unified strength theory parameter b can be considered as the identifier for different strength criteria and different yield criteria. The mutual quantificational relationships are implemented in the unified strength theory.

(3) The solutions in the references for the non-SD materials are special case solutions of the results in this chapter. The tensile-compressive strength ratio has a great influence on the plastic limit load.

7.6 Problems

Problem 7.1 Determine the limit bearing capacity of annular plate by using the Tresca criterion.

Problem 7.2 Determine the limit bearing capacity of annular plate by using the unified yield criterion $(b = 0)$.

Problem 7.3 Determine the limit bearing capacity of annular plate by using the unified yield criterion $(b = 0.5)$.

Problem 7.4 Determine the limit bearing capacity of annular plate by using the unified yield criterion $(b = 0.8)$.

Problem 7.5 Determine the limit bearing capacity of annular plate by using the unified yield criterion $(b = 1.0)$.

Problem 7.6 Determine the limit bearing capacity of annular plate by using the unified strength theory $(b = 0)$.

Problem 7.7 Determine the limit bearing capacity of annular plate by using the unified strength theory $(b = 0.5)$.

Problem 7.8 Determine the limit bearing capacity of annular plate by using the unified strength theory $(b = 0.8)$.

Problem 7.9 Determine the limit bearing capacity of annular plate by using the unified strength theory $(b = 1.0)$.

References

Aryanpour M, Ghorashi M (2002) Load carrying capacity of circular and annular plates using an arbitrary yield criterion. Computer & Structures, 80:1757-1762

Casey J, Sullivan TD (1985) Pressure dependency, strength-differential effect and plastic volume expansion in metals, Int. J. Plasticity, 1:39-61

Chait R (1972) Factors influencing the strength differential of high strength steels, Metal. Trans., 3:365-371

Drucker DC (1973) Plasticity theory, strength differential (SD) phenomenon and volume expansion in metals and plastics, Metal. Trans., 4:667-673

Hodge PG Jr (1959) Plastic analysis of structures. McGraw-Hill, New York

Hodge PG Jr (1963) Limit analysis of rotationally symmetric plates and shells. Prentice-Hall, New Jersey

Hopkins HG, Prager W (1953) The load carrying capacities of circular plates. J. Mech. Phys. Solids, 2:1-13

Hopkins HG, Wang AJ (1954) Load carrying capacities of circular plates of perfectly plastic material with arbitrary yield condition. J. Mech. Phys. Solids, 3:117-129

Huang WB, Zeng GP (1989) Solving some plastic problems by twin shear stress criterion. Acta Mechanica Sinica, 21(2):249-256 (in Chinese)

Li YM (1988) Elasto-plastic limit analyses with a new yielding criterion. J. Mech. Strength, 10(3):70-74 (in Chinese)

Ma G, Hao H, Iwasaki S (1999) Unified plastic limit analysis of circular plates under arbitrary load. J. Appl. Mechanics, ASME, 66(6):568-570

Ma GW, He LN (1994) Unified solution to plastic limit of simply supported circular plate. Mechanics and Practice, 16(6):46-48 (in Chinese)

Ma GW, Iwasaki S (1998) Plastic analyses of circular plates with respect to unified criterion. Int. J. Mech. Sci., 40:963-976

Ma GW, Yu MH, Iwasaki S, et al. (1994) Plastic analysis of circular plate on the basis of the Twin shear unified yield criterion. In: Lee PKK, Tham LG, Cheung YK (eds.) Proc. Int. Conf. on Computational Methods in Structural and Geotechnical Engineering. China Translation and Printing Ltd., Hong Kong, 3:930-935

Ma GW, Yu MH, Iwasaki S, Miyamoto Y (1995) Unified elasto-plastic solution to rotating disc and cylinder. J. Struct. Eng. (Japan), 41A:79-85

Ma GW, Yu MH, Miyamoto Y, et al.(1995) Unified plastic limit solution to circular plate under portion uniform load. J. Structural Engineering (in English, JSCE), 41A:385-392

Mroz Z and Sawczuk A (1960) The load carrying capacity of annular plates. Bulletin of Science Akademic of SSSR, Moskva, 3:72-78 (in Russian)

Rauch GC, Leslie WC(1972) The extent and nature of the strength differential effect in steel. Metal. Trans, 3:365-371

Save MA, Massonnet CE (1972) Plastic analysis and design of plates, shells and disks. North-Holland, Amsterdam

Save MA, Massonnet CE, Saxce G de (1997) Plastic limit analysis of plates, shells and disks. Elsevier, Amsterdam

Wei XY, Yu MH (2001) Unified plastic limit of clamped circular plate with strength differential effect in tension and compression. Chinese Quart. Mechanics, 22(1):78-83 (in Chinese, English abstract)

Wei XY, Yu MH (2002) Unified solution for plastic limit of annular plate. J. Mech. Strength, 24(1):140-143 (in Chinese, English abstract)

Yu MH (1961) General behavior of isotropic yield function. Res. Report of Xi'an Jiaotong University, Xi'an, China (in Chinese)

Yu MH (2002) Advance in strength theory of materials under complex stress state in 20th century. Applied Mechanics Reviews, ASME, 55(3):169-218

Yu MH (2004) Unified strength theory and its applications. Springer, Berlin

Yu MH, He LN (1991) A new model and theory on yield and failure of materials under complex stress problems. In: Mechanical Behavior of Materials-6, Pergamon Press, Oxford, 3:841-846

Yu MH, He LN, Song LY (1985) Twin shear stress theory and its generalization. Scientia Sinica (Science in China), English edition, Series A, 28(12):1174-1183

Zyczkowski M (1981) Combined loadings in the theory of plasticity. Polish Scientific Publishers, PWN and Nijhoff

8

Plastic Limit Analyses of Oblique, Rhombic, and Rectangular Plates

8.1 Introduction

Plate structures are widely used in aerospace, shipping, civil, and mechanical engineering. Plastic limit analyses of flat plates with different geometries can approximately estimate the load-bearing capacities of the plates. A lot of analytical solutions for flat plates have been reported by Wood (1961), Sawczuk and Jaeger (1963), Save and Massonnet (1972), Golley (1997), Mishra et al. (1996), Moen et al. (1998). Their solutions are mainly based on the Tresca yield criterion, the Huber-von Mises yield criterion, or the Mohr-Coulomb strength criterion. The maximum principal stress criterion has also been applied for simplicity.

The Tresca-Mohr-Coulomb strength theory is a single-shear strength theory. It ignores the effect of the intermediate principal stress. The Tresca yield criterion and the Huber-von Mises yield criterion can be effectively applied for the analyses of the non-SD materials. The maximum principal stress criterion considers only one of the three principal stresses, which may be deficient in yielding valid analytical results.

The unified strength theory (UST) has attracted more and more attention in engineering applications. In this chapter the load-bearing capacity for simply supported plates of different geometries will be given. Amongst them, the unified solution to the load-bearing capacity for a simply supported oblique plate was presented by Li and Yu (2000).

In terms of the principal stresses, the mathematical expression of the UST is

$$
\begin{cases}
f = \sigma_1 - \dfrac{\alpha}{1+b}(b\sigma_2 + \sigma_3) = \sigma_t, \text{when } \sigma_2 \leqslant \dfrac{\sigma_1 + \alpha\sigma_3}{1+\alpha}, \\
f' = \dfrac{\alpha}{1+b}(\sigma_1 + b\sigma_2) - \sigma_3 = \sigma_t, \text{when } \sigma_2 \geqslant \dfrac{\sigma_1 + \alpha\sigma_3}{1+\alpha},
\end{cases} \tag{8.1}
$$

where f and f' are yield functions; σ_1, σ_2, and σ_3 are the maximum principal stress, the intermediate principal stress, and the minimum principal stress, respectively; σ_t and σ_c are the tensile and compressive strengths; α the tensile to compressive strength ratio, i.e., $\alpha = \sigma_t/\sigma_c$; b is a coefficient which reflects the relative effect of the intermediate principal stress and the intermediate principal shear stress. It is the parameter specifying the failure criterion in the unified strength theory. The unified strength theory parameter b can be obtained via the tensile strength σ_t, the compressive strength σ_c and the shear strength τ_0,

$$b = \frac{1 + \alpha - \sigma_t/\tau_0}{\sigma_t/\tau_0 - 1}.$$

The twin-shear strength theory (Yu et al., 1985) and the single-shear strength theory (Mohr-Coulomb, 1900) can be derived from Eq.(8.1) with $b = 1$ and $b = 0$, respectively. For the plane stress problem ($\sigma_2 = 0$) the UST can be simplified as

$$\begin{cases} f = \sigma_1 - \dfrac{\alpha}{1+b}\sigma_3 = \sigma_t, & \text{where} \quad 0 \leqslant \tfrac{1}{2}(\sigma_1 + \alpha\sigma_3), \\[2mm] f' = \dfrac{1}{1+b}\sigma_1 - \alpha\sigma_3 = \sigma_t, & \text{where} \quad 0 \geqslant \dfrac{1}{2}(\sigma_1 + \alpha\sigma_3). \end{cases} \tag{8.2}$$

The limit loci of the UST in the plane stress state and in the deviatoric plane are shown in Figs.8.1 and 8.2 respectively. The twelve mathematical expressions of the unified yield criterion in plane stress state are

$$\sigma_1 - \frac{\alpha b}{1+b}\sigma_2 = \sigma_t, \qquad \sigma_2 - \frac{\alpha b}{1+b}\sigma_1 = \sigma_t, \tag{8.3a}$$

$$\frac{1}{1+b}\sigma_1 + \frac{b}{1+b}\sigma_2 = \sigma_t, \qquad \frac{1}{1+b}\sigma_2 + \frac{b}{1+b}\sigma_1 = \sigma_t, \tag{8.3b}$$

$$\sigma_1 - \frac{\alpha}{1+b}\sigma_2 = -\sigma_t, \qquad \sigma_2 - \frac{\alpha}{1+b}\sigma_1 = -\sigma_t, \tag{8.3c}$$

$$\frac{1}{1+b}\sigma_1 - \alpha\sigma_2 = -\sigma_t, \qquad \frac{1}{1+b}\sigma_2 - \alpha\sigma_1 = -\sigma_t, \tag{8.3d}$$

$$-\frac{\alpha}{1+b}(b\sigma_1 + \sigma_2) = \sigma_t, \qquad -\frac{\alpha}{1+b}(b\sigma_2 + \sigma_1) = \sigma_t, \tag{8.3e}$$

$$\frac{b}{1+b}\sigma_1 - \alpha\sigma_2 = \sigma_t, \qquad \frac{b}{1+b}\sigma_2 - \alpha\sigma_1 = \sigma_t, \tag{8.3f}$$

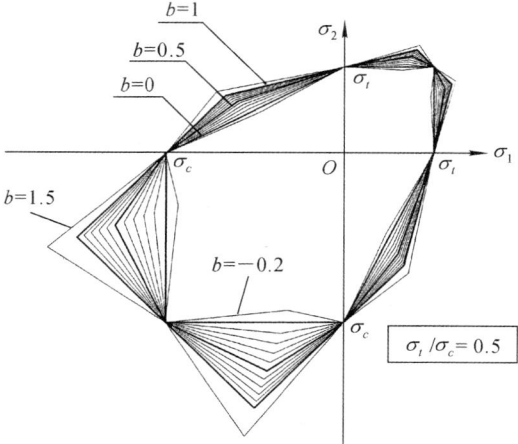

Fig. 8.1. Yield loci of the UST in the plane stress

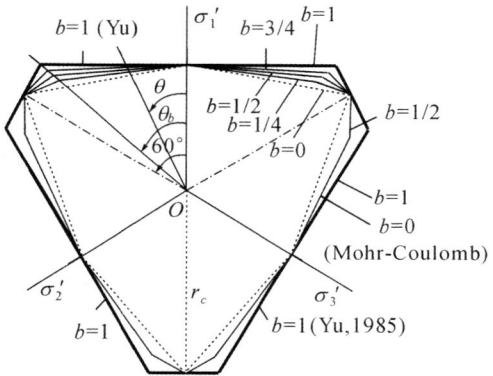

Fig. 8.2. Yield loci of the UST in the deviatoric plane

8.2 Equations for Oblique Plates

8.2.1 The Equilibrium Equation in Ordinary Coordinate System

For the oblique plate in Fig.8.3 with distribution of internal forces in Fig.8.4, u and v denote the ordinary coordinate axes; θ is the angle between the ordinary coordinate axes; $M_{n,1}$, $M_{n,2}$, and M_t are two positive bending moments and a shear moment per unit length of the oblique plate respectively. The unit of the moments is in Nm/m; $2l_1$ and $2l_2$ are respectively the total length of the two sides of the oblique plate; q is a transverse load over the plate.

The equilibrium equation of the plates in the Cartesian coordinate system is

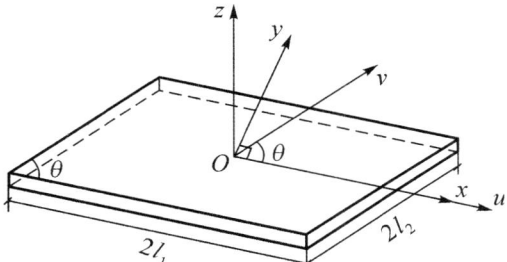

Fig. 8.3. Coordinate of the oblique plate

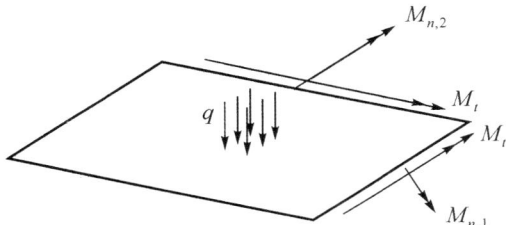

Fig. 8.4. Distribution of internal forces

$$\frac{\partial^2 M_{n,x}}{\partial x^2} + 2\frac{\partial^2 M_{n,xy}}{\partial x\partial y} + \frac{\partial^2 M_{n,y}}{\partial y^2} + q = 0, \tag{8.4}$$

where $M_{n,x}$, $M_{n,y}$, and $M_{n,xy}$ are the normal moments and shear moment per unit length in the rectangular Cartesian coordinate system.

The transformation between the rectangular Cartesian coordinate system and ordinary coordinate system can be expressed as

$$x = u + v\cos\theta, \quad y = v\sin\theta, \tag{8.5}$$

or

$$u = x - y\cot\theta, \quad v = y/\sin\theta. \tag{8.6}$$

The equilibrium equation of plates in oblique coordinate system can be derived from Eq.(8.4),

$$\frac{\partial^2 M_{n,x}}{\partial u^2} + 2\left(-\cot\theta\frac{\partial^2 M_{n,xy}}{\partial u^2} + \frac{1}{\sin\theta}\frac{\partial^2 M_{n,xy}}{\partial u\partial v}\right) + \cot^2\theta\frac{\partial^2 M_{n,y}}{\partial u^2}$$
$$- 2\frac{\cos\theta}{\sin^2\theta}\frac{\partial^2 M_{n,y}}{\partial u\partial v} + \frac{1}{\sin^2\theta}\frac{\partial^2 M_{n,y}}{\partial v^2} + q = 0. \tag{8.7}$$

8.2.2 Field of Internal Motion

Assuming that the oblique plate is simply supported around the four outer edges and subjected to a transverse load q, the functions of the internal forces for the oblique plates are

$$
\begin{cases}
M_{n,1} = c_1(l_1^2 - u^2), \\
M_{n,2} = c_2(l_2^2 - v^2), \\
M_t = c_3 uv,
\end{cases}
$$

where c_1, c_2, and c_3 are three coefficients to be determined.

According to the transformation of the internal moments between the ordinary coordinate system and the rectangular Cartesian coordinate system,

$$
\begin{cases}
M_{n,x} = -c_1\dfrac{\cos 2\theta}{\sin^2 \theta}(l_1^2 - u^2) + c_2 \cot^2 \theta(l_2^2 - v^2) + c_3\dfrac{\sin 2\theta}{\sin^2 \theta}uv, \\
M_{n,y} = c_2(l_2^2 - v^2), \\
M_{n,xy} = -c_1 \cot \theta(l_1^2 - u^2) - c_2 \cot \theta(l_2^2 - v^2) - c_3 uv.
\end{cases}
$$

The uniform transverse load can then be derived from Eqs.(8.5) and (8.7),

$$
q_l = [-2c_1(1 + 2\cos 2\theta) + 2c_2 + 2c_3 \sin \theta]/\sin^2 \theta. \tag{8.8}
$$

8.2.3 Moment Equation Based on the UST

The internal moments $M_{n,x}$, $M_{n,y}$, and $M_{n,xy}$ can be integrated form the stresses σ_x, σ_y and τ_{xy},

$$
\begin{cases}
M_{n,x} = \displaystyle\int_{-h}^{h} \sigma_x z \mathrm{d}z = \sigma_x h^2, \\
M_{n,y} = \displaystyle\int_{-h}^{h} \sigma_y z \mathrm{d}z = \sigma_y h^2, \\
M_{n,xy} = \displaystyle\int_{-h}^{h} \tau_{xy} z \mathrm{d}z = \tau_{xy} h^2,
\end{cases} \tag{8.9}
$$

where $2h$ is the thickness of the oblique plate. The UST can be rewritten for the plane stress state as

$$f = \frac{1}{1+b}\left[\frac{1+b-\alpha}{2}(\sigma_x + \sigma_y) + (1+b+\alpha)\sqrt{\left(\frac{\sigma_x - \sigma_y}{2}\right)^2 + \tau_{xy}^2}\right] = \sigma_t,$$

when $(1+\alpha)\dfrac{\sigma_x + \sigma_y}{2} + (1-\alpha)\sqrt{\left(\dfrac{\sigma_x - \sigma_y}{2}\right)^2 + \tau_{xy}^2} \geqslant 0,$

(8.10a)

$$f' = \frac{1}{1+b}\left[\frac{1-\alpha b-\alpha}{2}(\sigma_x + \sigma_y) + (1+\alpha b+\alpha)\sqrt{\left(\frac{\sigma_x - \sigma_y}{2}\right)^2 + \tau_{xy}^2}\right] = \sigma_t$$

when $(1+\alpha)\dfrac{\sigma_x + \sigma_y}{2} + (1-\alpha)\sqrt{(\dfrac{\sigma_x - \sigma_y}{2})^2 + \tau_{xy}^2} \leqslant 0.$

(8.10b)

Eqs.(8.10a) and (8.10b) can be expressed in terms of $M_{n,x}$, $M_{n,y}$, and $M_{n,xy}$ as

$$(2+b)^2(M_{n,x} - M_{n,y})^2 + 4(2+b)^2 M_{n,xy}^2 - b^2(M_{n,x} + M_{n,y})^2$$
$$= 4(1+b)M_p[(1+b)M_p - b(M_{n,x} + M_{n,y})],$$

when $(1+\alpha)\dfrac{M_{n,x} + M_{n,y}}{2} + (1-\alpha)\sqrt{\left(\dfrac{M_{n,x} + M_{n,y}}{2}\right)^2 + M_{n,xy}^2} \geqslant 0,$

(8.11a)

$$(2+b)^2(M_{n,x} - M_{n,y})^2 + 4(2+b)^2 M_{n,xy}^2 - b^2(M_{n,x} + M_{n,y})^2$$
$$= 4(1+b)M_p[(1+b)M_p + b(M_{n,x} + M_{n,y})],$$

when $(1+\alpha)\dfrac{M_{n,x} + M_{n,y}}{2} + (1-\alpha)\sqrt{\left(\dfrac{M_{n,x} + M_{n,y}}{2}\right)^2 + M_{n,xy}^2} \leqslant 0.$

(8.11b)

where M_p is the limit bending moment of the plate.

The limit loci of generalized stresses in terms of the unified strength theory for the plane plate are illustrated schematically in Fig.8.5.

8.3 Unified Solution of Limit Analysis of Simply Supported Oblique Plates

When the inequality $(1+\alpha)\dfrac{M_{n,x}+M_{n,y}}{2} + (1-\alpha)\sqrt{\left(\dfrac{M_{n,x}-M_{n,y}}{2}\right)^2 + M_{n,xy}} \geqslant 0$ is satisfied, the unified yield function for u and v can be derived by substituting Eq.(8.7) into Eq.(8.11a). With calculus regarding u and v, the limit

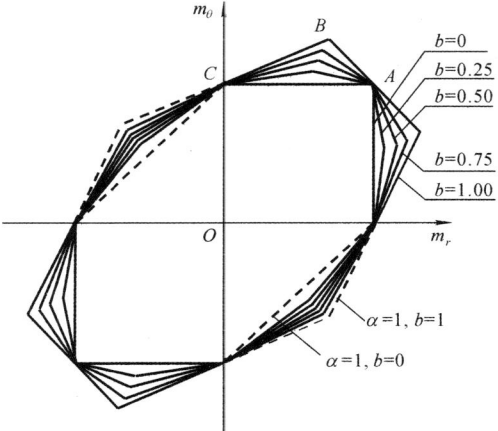

Fig. 8.5. Generalized unified yield criterion in plane stress state

loading at the points $(0,0)$, $(\pm l_1, 0)$, $(0, \pm l_2)$, and $(\pm l_1, \pm l_2)$ of the plate can then be derived. Defining

$$G = 4(1+b)^2 M_p^2, \tag{8.12a}$$

$$E_1 = \left[(1+b+\alpha)^2 \frac{\cos^2 2\theta}{\sin^4 \theta} + 4(1+b+\alpha)^2 \cot^2 \theta - (1+b-\alpha)^2 \frac{\cos^2 2\theta}{\sin^4 \theta}\right] l_1^4, \tag{8.12b}$$

$$E_2 = \left[(1+b+\alpha)^2 \frac{\cos^2 2\theta}{\sin^4 \theta} + 4(1+b+\alpha)^2 \cot^2 \theta - (1+b-\alpha)^2 \csc^4 \theta\right] l_2^4, \tag{8.12c}$$

$$E_3 = \left[(1+b+\alpha)^2 \frac{\sin^2 2\theta}{\sin^4 \theta} + 4(1+b+\alpha)^2 - (1+b-\alpha)^2 \frac{\sin^2 2\theta}{\sin^4 \theta}\right] l_1^2 l_2^2, \tag{8.12d}$$

$$F_1 = 4(1+b)(1+b-\alpha) M_p \frac{\cos 2\theta}{\sin^2 \theta} l_1^2, \tag{8.12e}$$

$$F_2 = 4(1+b)(1+b-\alpha) M_p \csc^2 \theta l_2^2, \tag{8.12f}$$

$$F_3 = 4(1+b)(1+b-\alpha) M_p \frac{\sin 2\theta}{\sin^2 \theta} l_1 l_2. \tag{8.12g}$$

The limit load is derived from Eq.(8.8),

$$q_l = [-2c_1(1 + 2\cos 2\theta) + 2c_2 + 2c_3 \sin \theta]/\sin^2 \theta, \qquad (8.13)$$

where c_1, c_2, and c_3 are

$$\begin{cases} c_1 = \left(F_1 + \sqrt{F_1^2 + 4E_1G} \right)/(2E_1), \\ c_2 = \left(-F_2 + \sqrt{F_2^2 + 4E_2G} \right)/(2E_2), \\ c_3 = \left(-F_3 + \sqrt{F_3^2 + 4E_3G} \right)/(2E_3). \end{cases} \qquad (8.14)$$

When $(1+\alpha)\frac{M_{n,x}+M_{n,y}}{2} + (1-\alpha)\sqrt{(\frac{M_{n,x}-M_{n,y}}{2})^2 + M_{n,xy}} \leqslant 0$, the limit load q_l can be derived from Eq.(8.11b) with the same form of Eq.(8.13), while the coefficients c_1, c_2, and c_3 are given as

$$c_1 = \left(-F_1 + \sqrt{F_1^2 + 4E_1G} \right)/(2E_1), \qquad (8.15a)$$

$$c_2 = \left(F_2 + \sqrt{F_2^2 + 4E_2G} \right)/(2E_2), \qquad (8.15b)$$

$$c_3 = \left(F_3 + \sqrt{F_3^2 + 4E_3G} \right)/(2E_3). \qquad (8.15c)$$

The limit load q_l for a parallelogram plate with $\theta = \pi/3$ can be derived from Eq.(8.13),

$$q_l = \left[\frac{2}{l_2^2} + \frac{4(1+b)}{9(1+b)+3\alpha} \cdot \frac{1}{l_1 l_2} \right] M_p. \qquad (8.16)$$

When $\theta = \pi/4$, the limit load q_l becomes

$$q_l = \left[\frac{-4(1+b)}{(1+b+\alpha)} \frac{1}{l_1^2} + \frac{2(1+b)}{\alpha} \frac{1}{l_2^2} \right. \\ \left. + \frac{(4+2\sqrt{2})(1+b)^2 + (4-2\sqrt{2})\alpha(1+b)}{(1+b)^2 + \alpha^2 + 6\alpha(1+b)} \frac{1}{l_1 l_2} \right] M_p. \qquad (8.17)$$

The relations between the limit load q_l and the unified strength theory parameter b of the UST for various oblique plates ($\theta = \pi/3$, $l_1 = 1.5l_2$; $\theta = \pi/3$, $l_1 = 2l_2$; $\theta = \pi/4$, $l_1 = 1.5l_2$; $\theta = \pi/4$, $l_1 = 2l_2$) are given in Fig.8.6 to Fig.8.9.

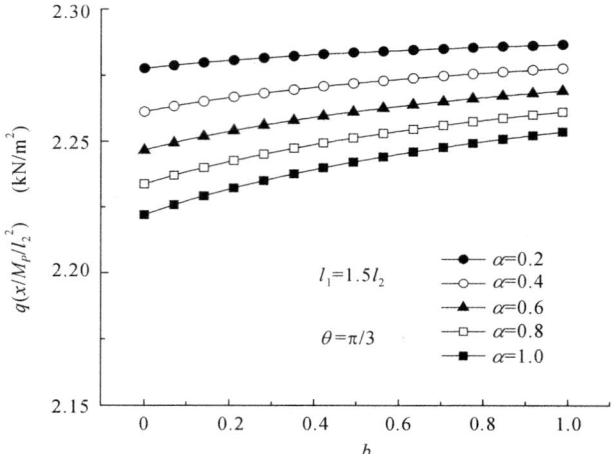

Fig. 8.6. Relations between q_l and unified strength theory parameter b ($\theta = \pi/3$, $l_1 = 1.5l_2$)

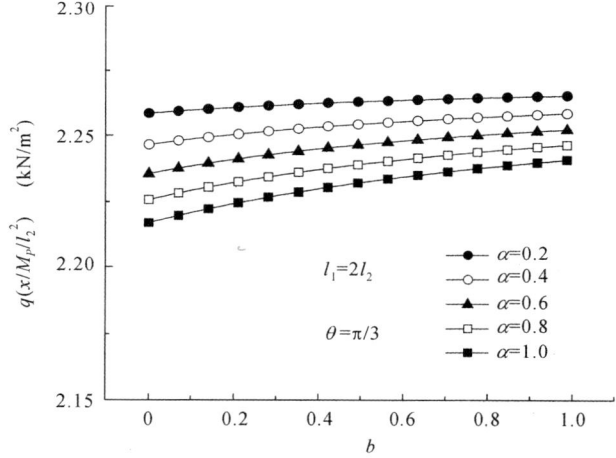

Fig. 8.7. Relations between q_l and unified strength theory parameter b ($\theta = \pi/3$, $l_1 = 2l_2$)

8.4 Limit Load of Rhombic Plates

For $l_1 = l_2$, the limit load can be obtained from Eqs.(8.16) and (8.17) for $\theta = \pi/3$,

$$q_l = \left[2 + \frac{4(1+b)}{9(1+b) + 3\alpha}\right] \frac{M_p}{l_2^2} \quad \text{when} \quad \theta = \frac{\pi}{3}. \tag{8.18}$$

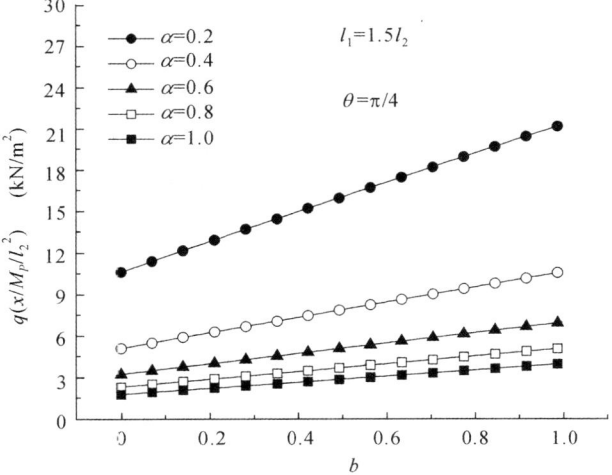

Fig. 8.8. Relations between q_l and unified strength theory b ($\theta = \pi/4$, $l_1 = 1.5l_2$)

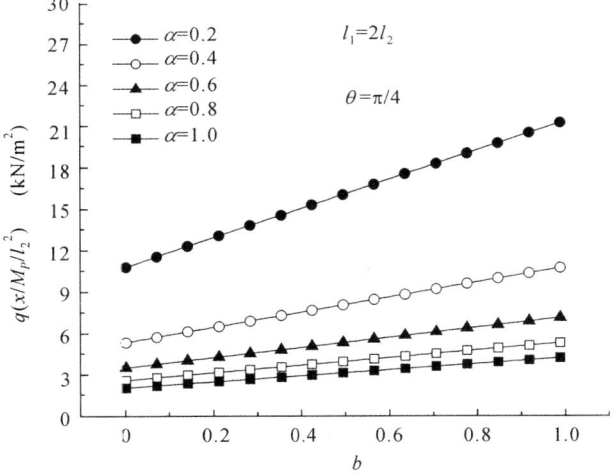

Fig. 8.9. Relations between q_l and unified strength theory b ($\theta = \pi/4$, $l_1 = 2l_2$)

The relations between limit load q_l and the unified strength theory parameter b are shown in Fig.8.10 and Fig.8.11 for $\theta = \pi/3$ and $\theta = \pi/4$ respectively.

8.5 Limit Load of Rectangular Plates

When $\theta = \pi/2$, the limit load for rectangular plates can be derived from Eq.(8.13),

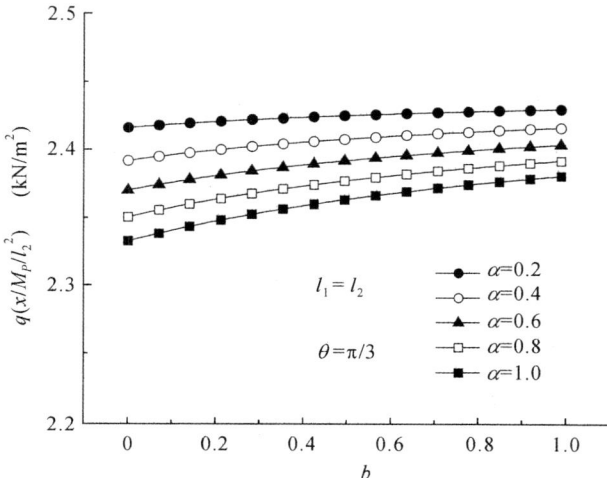

Fig. 8.10. Relations between limit load q_l and unified strength theory parameter b ($\theta = \pi/3$, $l_1 = l_2$)

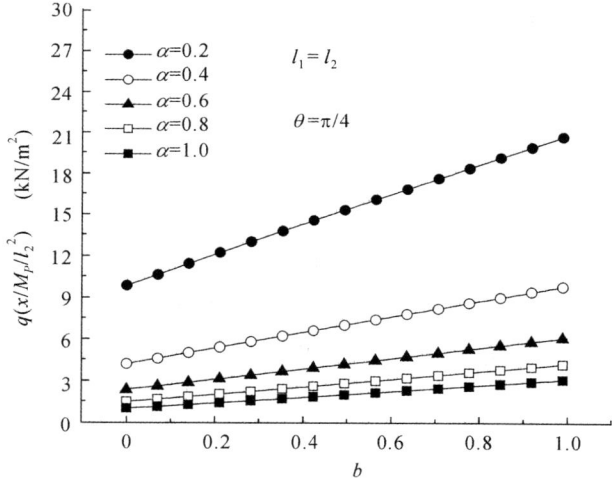

Fig. 8.11. Relations between limit load q_l and unified strength theory parameter b ($\theta = \pi/4$, $l_1 = l_2$)

$$q_l = \left[2 \left(\frac{1}{l_1^2} + \frac{1}{l_2^2} \right) + \frac{2(1+b)}{(1+b+\alpha)l_1 l_2} \right] M_p. \qquad (8.19)$$

If the plate consists of non-SD material ($\alpha = 1$), the limit load in Eq.(8.19) can be simplified as

$$q_l = \left[2 \left(\frac{1}{l_1^2} + \frac{1}{l_2^2} \right) + \frac{2(1+b)}{(2+b)l_1 l_2} \right] M_p. \tag{8.20}$$

Figs.8.12 to 8.14 show the limit load q_l versus the unified strength theory parameter b for different rectangular plates.

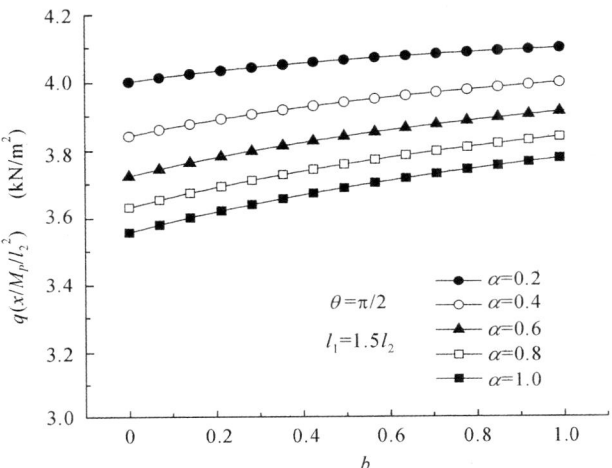

Fig. 8.12. Relations between q_l and unified strength theory parameter b ($\theta = \pi/2$, $l_1 = 1.5l_2$)

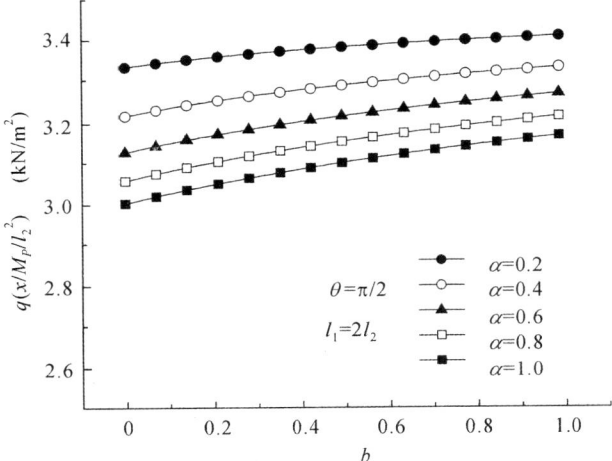

Fig. 8.13. Relations between and unified strength theory parameter b ($\theta = \pi/2$, $l_1 = 2l_2$)

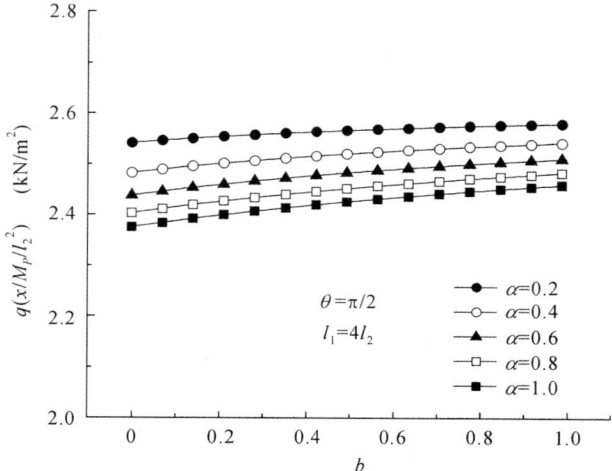

Fig. 8.14. Relations between q_l and unified strength theory parameter b ($\theta = \pi/2$, $l_1 = 4l_2$))

It is seen that both α and b have significant influences on the limit load. For a given value of α, the limit load increases with increasing parameter b. On the other hand, for a given value of b, the limit load decreases with the increase of α.

The limit load q_l versus the ratio is shown in Figs.8.15 to 8.17 for different parameter b.

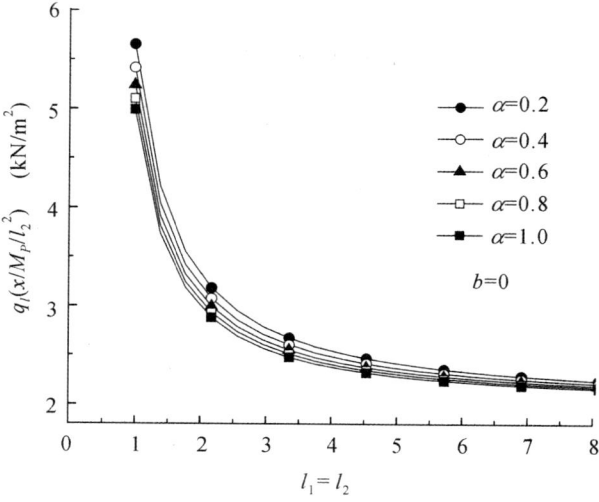

Fig. 8.15. Variation of q_l for different rectangular plates ($b = 0$)

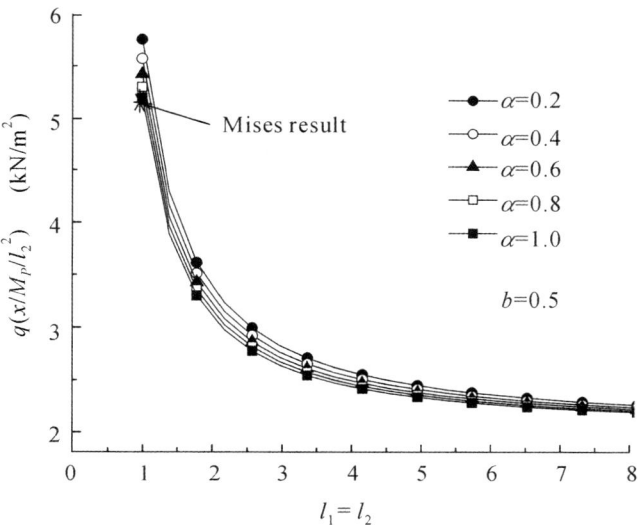

Fig. 8.16. Variation of q_l for different rectangular plates ($b = 0.5$)

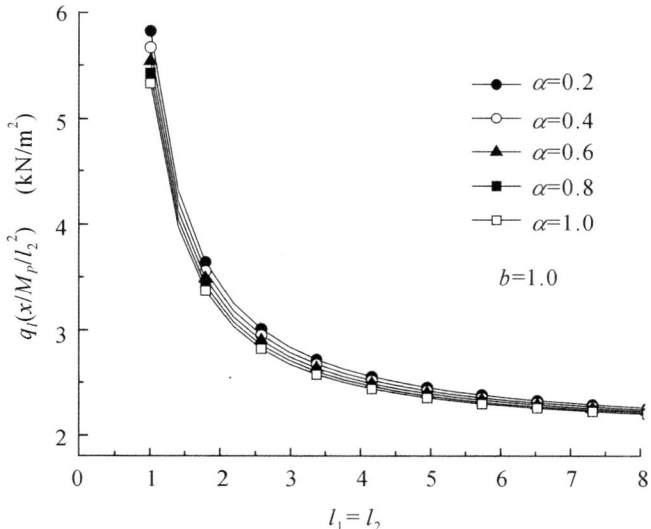

Fig. 8.17. Variation of q_l for different rectangular plates ($b = 1.0$)

It is seen that the limit load q_l decreases with the increase of the ratio l_1/l_2 and α for a given value of b. On the other hand, $q_l(M_p/l_2^2)$ approaches a constant of 2 (kN/m^2), and is independent of b and α.

8.6 Unified Limit Load of Square Plates

The limit load of square plates can be obtained by further simplifying the limit load solution in Eq.(8.10) with $l_1 = l_2$

$$q_l = 2 \left[2 + \frac{(1+b)}{1+b+\alpha} \right] \frac{M_p}{l_2^2}. \tag{8.21}$$

When $\alpha = 1$, the limit load based on the twin shear yield criterion can be derived as

$$q_l = 2 \left[2 + \frac{(1+b)}{2+b} \right] \frac{M_p}{l_2^2}, \tag{8.22}$$

and

$$q_l = 5.155 \frac{M_p}{l_2^2}, \quad \text{when} \quad b = \frac{1}{1+\sqrt{3}}, \tag{8.23}$$

which is identical to the following solution based on the von Mises criterion (Wang, 1998)

$$q_l = 5.2 \frac{M_p}{l_2^2}, \quad \text{when} \quad b = \frac{1}{2}. \tag{8.24}$$

Fig. 8.18 shows the limit load q_l with respect to the parameter b for square plates.

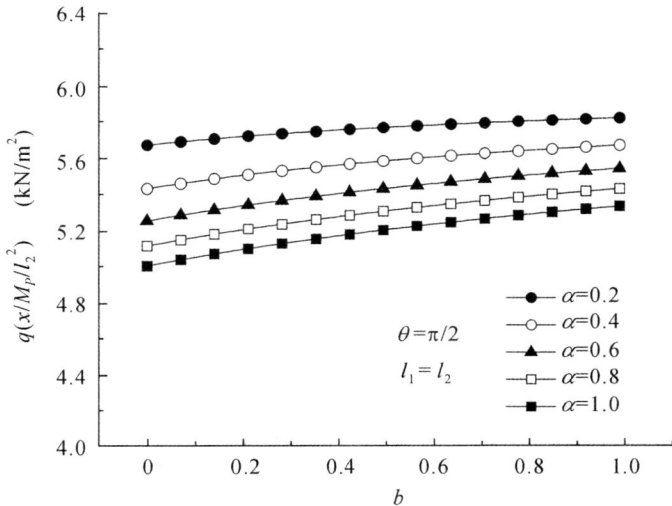

Fig. 8.18. Relations of q_l to unified strength theory parameter b ($\theta = \pi/2$, $l_1 = l_2$)

8.7 Tabulation of the Limit Load for Oblique, Rhombic and Square Plates

For convenient comparison and easier reference, the relations of the limit load $q_l(M_p/l_2^2)$ to the unified strength theory parameter b, the SD ratio α, the angle θ for the oblique plate, and the length ratio of l_1/l_2 are tabulated in Table 8.1 to Table 8.4.

Table 8.1. Relation of $q_l(M_p/l_2^2)$ to b and α with $\theta = 45°$

	α	$b{=}0$	$b{=}0.25$	$b{=}0.5$	$b{=}0.75$	$b{=}1.00$
	0.2	8.820	12.585	15.313	18.009	20.678
	0.4	4.193	5.603	7.015	8.422	8.820
$l_1/l_2 = 1.0$	0.6	2.352	3.262	4.193	5.132	6.074
	0.8	1.484	2.130	2.803	3.493	4.193
	1.0	1.000	1.484	1.998	2.531	3.078
	0.2	10.621	13.323	15.993	18.637	21.260
	0.4	5.097	6.492	7.879	8.255	10.621
$l_1/l_2 = 1.5$	0.6	3.235	4.164	5.097	6.028	6.955
	0.8	2.316	3.003	3.699	4.397	5.097
	1.0	1.778	2.316	2.865	3.420	3.978
	0.2	10.743	13.405	16.039	18.652	21.248
	0.4	5.311	6.684	8.047	8.400	10.743
$l_1/l_2 = 2.0$	0.6	3.468	4.390	5.311	6.227	7.140
	0.8	2.547	3.237	3.929	4.620	5.311
	1.0	2.000	2.547	3.099	3.652	4.206

The limit load q_l versus the ratio for different lengths of rectangular plates and the unified strength theory parameter b are listed in Table 8.4. The limit load for square plate and rectangular plates with $l_1/l_2 = 2$, $l_1/l_2 = 4$, $l_1/l_2 = 7$, $l_1/l_2 = 10$, and $l_1/l_2 = \infty$ are given. It is seen that the limit load q_l decreases with the increase of the ratio l_1/l_2 and α for a given value of b, and the limit load q_l increases with the increase in the unified strength theory parameter b in any case.

In Table 8.4, the result of $\alpha = 1$ and $b = 0$ is the same as the result for Tresca material; the result for $\alpha = b = 1$ is the same as the result for twin-shear yield criterion material; the result of $\alpha = 1$ and $b = 1/2$ is the

Table 8.2. Relation of $q_l(M_p/l_2^2)$ to b and α with $\theta = 60°$

	α	$b=0$	$b=0.25$	$b=0.5$	$b=0.75$	$b=1.00$
	0.2	2.417	2.422	2.426	2.428	2.430
	0.4	2.392	2.402	2.408	2.413	2.417
$l_1/l_2 = 1.0$	0.6	2.370	2.383	2.392	2.399	2.404
	0.8	2.351	2.366	2.377	2.386	2.392
	1.0	2.333	2.351	2.364	2.373	2.381
	0.2	2.278	2.281	2.284	2.285	2.287
	0.4	2.261	2.268	2.272	2.275	2.278
$l_1/l_2 = 1.5$	0.6	2.247	2.255	2.261	2.266	2.269
	0.8	2.234	2.244	2.252	2.257	2.261
	1.0	2.222	2.234	2.242	2.249	2.254
	0.2	2.208	2.211	2.213	2.214	2.215
	0.4	2.196	2.201	2.204	2.206	2.208
$l_1/l_2 = 2.0$	0.6	2.185	2.192	2.196	2.199	2.202
	0.8	2.175	2.183	2.189	2.193	2.196
	1.0	2.167	2.175	2.182	2.187	2.190

linear approximation of the result of the Huber-von Mises criterion. Because the Huber-von Mises criterion has a nonlinear mathematical expression, it is relatively complicated to derive analytical solutions in structural plasticity. For practical application, approximating the numerical solutions based on the Huber-von Mises criterion is adopted. UST with $\alpha = 1$ and $b = 1/2$ can be considered as a linear approximation of the Huber-von Mises yield criterion which is more suitable for the derivation of analytic solutions. The results of the Huber-von Mises criterion and the unified strength theory with $\alpha = 1$ and $1/(1+\sqrt{3})$ are very close with a percentage difference less that 3%, which is even as low as 0.87% for the plastic limit load of the square plate.

For a given value of α, the limit load q_l can be derived from Eq.(8.13) and Eq.(8.2) for different materials. As a result the material properties of the plates can be taken into account more appropriately if the UST is applied.

8.8 Summary

Based on the unified strength theory, the plastic limit analyses for oblique plates are carried out and the unified limit load is derived. It gives a se-

Table 8.3. Relation of $q_l(M_p/l_2^2)$ to b and α with $\theta = 90°$

	α	$b=0$	$b=0.25$	$b=0.5$	$b=0.75$	$b=1.00$
	0.2	5.667	5.724	5.765	5.795	5.818
	0.4	5.429	5.515	5.579	5.628	5.667
$l_1/l_2 = 1.0$	0.6	5.250	5.351	5.429	5.489	5.538
	0.8	5.111	5.220	5.304	5.373	5.429
	1.0	5.000	5.111	5.200	5.273	5.333
	0.2	4.556	4.718	4.845	4.948	5.032
	0.4	4.317	4.485	4.620	4.731	4.824
$l_1/l_2 = 1.5$	0.6	4.139	4.304	4.441	4.556	4.654
	0.8	4.000	4.160	4.295	4.411	4.511
	1.0	3.889	4.043	4.175	4.289	4.389
	0.2	4.167	4.387	4.569	4.722	4.853
	0.4	3.929	4.139	4.318	4.472	4.605
$l_1/l_2 = 2.0$	0.6	3.750	3.949	4.122	4.272	4.405
	0.8	3.611	3.799	3.963	4.109	4.239
	1.0	3.500	3.676	3.833	3.974	4.100

ries of solutions covering those from the single-shear theory (Mohr-Coulomb strength theory) to the twin-shear strength theory (Yu, 1985). The unified solution of the limit load for the oblique, rhombic, rectangular, and square plates encompasses the solutions as special cases as reported by other researchers as well as a series of new solutions.

The parameter b has a significant influence on the load-bearing capacities of oblique plates and the influences vary for different conditions. The influences vary with the state of stress. When the intermediate principal stress σ_2 is close to the minimum principal stress σ_3 the difference in the limit load based on different strength criteria is minimal. However, when the intermediate principal stress σ_2 is close to $\sigma_2 = (\sigma_1 + \sigma_3)/2$, the influence on the limit load is significant.

The limit load q_l for different materials and structures can be obtained when α and b vary and when the different l_1/l_2 and θ are adopted.

Table 8.4. Relations of limit load $q_l(M_p/l_2^2)$ for different rectangular plates

b	α	$l_1/l_2 = 1.0$ (square)	$l_1/l_2 = 2$	$l_1/l_2 = 4$	$l_1/l_2 = 7$	$l_1/l_2 = 10$	$l_1/l_2 = \infty$
	0.2	5.667	3.333	2.542	2.279	2.187	2.0
	0.4	5.429	3.214	2.482	2.245	2.163	2.0
0	0.6	5.250	3.125	2.438	2.219	2.145	2.0
	0.8	5.111	3.056	2.403	2.200	2.131	2.0
	1.0	5.000	3.000	2.375	2.184	2.120	2.0
	0.2	5.765	3.382	2.566	2.293	2.196	2.0
	0.4	5.579	3.289	2.520	2.266	2.178	2.0
0.5	0.6	5.429	3.214	2.482	2.245	2.163	2.0
	0.8	5.304	3.152	2.451	2.227	2.150	2.0
	1.0	5.200	3.100	2.425	2.212	2.140	2.0
$\frac{1}{1+\sqrt{3}}$	1.0	5.155(Mises)	3.077	2.414	2.206	2.135	2.0
1.0	0.2	5.818	3.409	2.580	2.301	2.202	2.0
(twin-	0.4	5.667	3.333	2.542	2.279	2.187	2.0
shear)	0.6	5.538	3.269	2.510	2.261	2.174	2.0
	0.8	5.429	3.214	2.482	2.245	2.163	2.0
	1.0	5.333	3.167	2.458	2.231	2.153	2.0

8.9 Problems

Problem 8.1 Try your hand at an application of the unified yield criterion for limit analysis of a square plate.

Problem 8.2 Try your hand at an application of the unified yield criterion for limit analysis of a rectangular plate with different l_1, l_2 and θ.

Problem 8.3 Try your hand at an application of the unified yield criterion for limit analysis of a rhombic plate with different θ.

Problem 8.4 Try your hand at an application of the unified yield criterion for limit analysis of an oblique plate with different l_1, l_2 and θ.

Problem 8.5 Why does the solution obtained by using the unified yield criterion contain all the solutions of the Tresca yield criterion, the von Mises yield criterion, the twin-shear yield criterion and other possible yield criteria adopted for those materials with the same yield stress in tension and in compression?

Problem 8.6 Write a paper regarding the plastic analysis of an oblique plate with different l_1, l_2 and θ using the unified yield criterion.

Problem 8.7 Can you introduce a unified plastic solution for an oblique plate with different l_1, l_2 and θ using the unified strength theory? The ratio of tensile strength σ_t to compressive strength σ_c is $\alpha = \sigma_t/\sigma_c = 0.8$.

Problem 8.8 A high-strength alloy has the strength ratio in tension and compression $\alpha = 0.9$. Find the unified solution for a square plate made of this alloy.

Problem 8.9 A high-strength alloy has the strength ratio in tension and compression $\alpha = 0.9$. Find the unified solution for a rectangular plate made of this alloy.

Problem 8.10 A high-strength alloy has the strength ratio in compression and tension $\alpha = 0.9$. Find the unified solution for a rhombic plate made of this alloy.

Problem 8.11 A high-strength alloy has the strength ratio in compression and tension $\alpha = 0.9$. Find the unified solution for an oblique plate made of this alloy.

Problem 8.12 Compare the plastic solutions of a square plate using the unified yield criterion and the unified strength theory with $\alpha = 0.8$.

Problem 8.13 Compare the plastic solutions of a rectangular plate using the unified yield criterion and the unified strength theory with $\alpha = 0.8$.

Problem 8.14 Compare the plastic solutions of a rhombic plate using the unified yield criterion and the unified strength theory with $\alpha = 0.8$.

Problem 8.15 Compare the plastic solutions of an oblique plate using the unified yield criterion and the unified strength theory with $\alpha = 0.8$.

References

Golley BW (1997) Semi-analytical solution of rectangular plates. J. Eng. Mech., ASCE, 123(7):669-677

Grigorendo YM, Kryukov NN (1995) Solution of problems of the theory of plates and shells with spline functions (Survey). Int. Appl. Mech., 31(6):413-434

Hodge PG (1963) Limit analysis of rotationally symmetric plates and shells. Prentice-Hall, Englewood Cliffs, NJ

Krys'ko VA, Komarov SA, Egurnov NV (1996) Buckling of flexible plates under the influence of longitudinal and transverse loads. Int. Appl. Mech., 32(9):727-733

Li JC, Yu MH, Fan SC (2000a) A unified solution for limit load of simply supported oblique plates, rhombus plates, rectangular plates and square plates. China Civil Engineering Journal, 33(6):76-89 (in Chinese, English abstract)

Li JC, Yu MH, Xiao Y (2000b) Unified limit solution for metal oblique plates. Chinese J. Mechanical Engineering, 36(8):25-28 (in Chinese, English abstract)

Mishra RC and Chakrabarti SK (1996) Rectangular plates tensionless elastic foundation: Some new results. J. Eng. Mech., ASCE, 122(4):385-387

Moen LA, Langseth N, Hopperstad OS (1998). Elasto-plastic buckling of anisotropic aluminum plate elements. J. Struct. Eng., ASCE, 124(6):712-719

Save MA, Massonnet CE (1972) Plastic analysis and design of plates, shells and disks. North-Holland, Co., Amsterdam

Sawczuk A, Jaeger T (1963) Limit design theory of plates. Springer, Berlin

Wood RH (1961) Plastic and elastic design of slabs and plates. London, Thames and Hudson

Yu MH (1992) A new system of strength theory. Xi'an Jiaotong University Press, Xi'an (in Chinese)

Yu MH (1998) Twin shear theory and its application. Science Press, Beijing (in Chinese)

Yu MH (2004) Unified strength theory and its application. Springer, Berlin

Yu MH, He LN (1991) A new model and theory on yield and failure of materials under the complex stress state. In: Mechanical Behavior of Materials-6, Pergamon Press, Oxford, 3:841-846

Yu MH, He LN, Song LY (1985) Twin shear stress theory and its generalization, Scientia Sinica (Science in China), English ed. Series A28(11):1174-1183; Chinese ed., Series A, 28(12):1113-1120

9

Plastic Limit Analysis of Pressure Vessels

9.1 Introduction

Thin-walled vessels and thick-walled cylinders are applied widely in industry as pressure vessels, pipes, gun barrels, cylinders of rockets, etc. The limit analyses of thick-walled hollow spheres and cylinders under internal pressure were discussed in detail by Hill (1950), and Johnson and Mellor (1962). Further studies on this subject were reported by Derrington and Johnson (1958), Johnson and Mellor (1962), Tuba (1965), and Zyczkowski (1981). The Tresca yield criterion or the Huber-von Mises yield criterion is usually applied for the design of thin-walled pressure vessels. The result using the Huber-von Mises yield criterion for a spherical vessel is similar to that using the Tresca yield criterion. These solutions are applicable only for non-SD materials. It can be seen in the textbook of plasticity.

For SD materials two-parameter failure criteria have to be used (Drucker, 1973; Richmond et al. 1980). The limit pressure of a thick-walled hollow cylinder with material, following the Mohr-Coulomb strength theory, was discussed by Xu and Liu (1995). The limit pressures of the thick-walled hollow sphere and cylinder with material following the twin-shear strength theory were reported by Ni et al. (1998) and Zhuang (1998). Application of the twin-shear strength theory in the strength-calculation of gun barrels was given by Liu et al. (1998) and Li et al. (2007).

The elastic limit and plastic limit of the thin-walled vessel and thick-walled cylinder were studied with respect to the unified strength theory by Wang and Fan (1998), a series of unified solutions of limit loads for pressure vessels were given (Wang and Fan, 1998). Zhao et al. (1999), Feng et al. (2004a; 2004b) and recently Li et al. (2007) also give some results of unified solutions for pressure vessels. The unified strength theory is also applied to the unified limit load solution for fiber-reinforced concrete cylinder taking into consideration the strain softening of material by Chen et al. (2006). The effects of failure criterion on the elastic limit and plastic limit loads of the thin-walled pressure vessel and thick-walled pressure vessel were summarized by Yu (2004).

In most of the applications the thickness of the cylinder is constant and the cylinder is subjected to a uniform internal pressure p. The deformations of the cylinder are symmetrical with respect to the symmetric axis of the cylinder. The deformations at a cross section sufficiently far from the junction of the cylinder and its end caps are independent of the axial coordinate z. In particular, if the cylinder is open-ended (no end caps) and unconstrained, it undergoes axisymmetric deformations due to pressure p which is independent of z. If the deformation of cylinder is constrained by end caps, the displacements and stresses at cylinder cross sections near the end cap junctions differ from those at sections far away from the end cap junctions, if axially symmetrical loads and constraints are considered. Thus the solution is axisymmetrical; the solutions are functions of the radial coordinate r only. In the case of a thin-walled cylinder, the difference in stresses at the inner wall and outer wall is small if the thickness t is much less than the vessel diameter. The internal stresses can then be assumed to be independent of the radial coordinate r. Relationships between the internal pressure p, the dimensions of the thin-walled vessel, circumferential and axial stresses in a pressure vessel, can be found in textbooks of *Mechanics of Materials* or *Strength of Materials*.

The systematic results of elastic and plastic limit loads for thin-walled and thick-walled pressure vessels will be described in this chapter.

9.2 Unified Solution of Limit Pressure of Thin-walled Pressure Vessel

Considering the stresses in a thin-walled pressure vessel subjected to an internal pressure as shown in Fig.9.1, the pressure incurs a circumferential stress (or hoop stress) σ_1 and a longitudinal stress σ_m or σ_2 that can be expressed as

$$\sigma_1 = \frac{pD}{2t}, \; \sigma_2 = \sigma_m = \frac{pD}{4t}, \; \sigma_3 = 0. \tag{9.1}$$

Based on the unified strength theory,

$$F = \sigma_1 - \frac{\alpha}{1+b}(b\sigma_2 + \sigma_3) = \sigma_t, \quad \text{when} \quad \sigma_2 \leqslant \frac{\sigma_1 + \alpha\sigma_3}{1+\alpha}, \tag{9.2a}$$

$$F' = \frac{1}{1+b}(\sigma_1 + b\sigma_2) - \alpha\sigma_3 = \sigma_t, \quad \text{when} \quad \sigma_2 \leqslant \frac{\sigma_1 + \alpha\sigma_3}{1+\alpha}. \tag{9.2b}$$

The stresses of a thin-walled vessel satisfy the inequity condition $\sigma_2 = \frac{1}{2}(\sigma_1 + \sigma_3) \leqslant \frac{\sigma_1 + \alpha\sigma_3}{1+\alpha}$, therefore the first formula of the unified strength theory

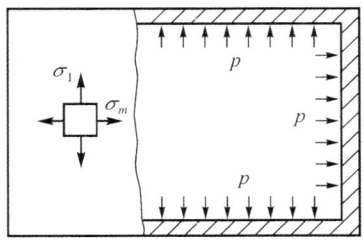 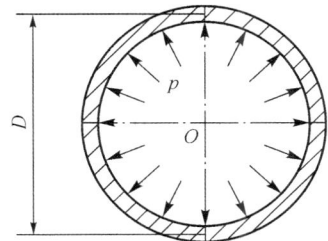

Fig. 9.1. Stresses in thin-walled pressure vessel

Eq.(9.2a) is valid for the yield condition of the vessel. Substituting Eq.(9.1) into Eq.(9.2a), the yield condition for a thin walled cylinder obeying the unified strength theory is obtained

$$F = \sigma_1 - \frac{\alpha}{1+b}(b\sigma_2 + \sigma_3) = \frac{pD}{2t} - \frac{\alpha b}{1+b}\frac{pD}{4t} = \sigma_t. \tag{9.3}$$

The elastic limit pressure can be derived as

$$p_e = \frac{1+b}{2+2b-\alpha b}\frac{4t}{D}\sigma_t. \tag{9.4}$$

If the material has an allowable tensile stress of $[\sigma] = \sigma_t/n$, where n is the factor of safety, the allowable limit pressure is

$$[p] = \frac{1+b}{2+2b-\alpha b}\frac{4t}{D}[\sigma]. \tag{9.5}$$

If the internal pressure p and allowable stress $[\sigma]$ are given, the wall thickness should satisfy

$$t \geqslant \frac{2+2b+\alpha b}{1+b}\frac{pD}{4[\sigma]}. \tag{9.6}$$

The relationship between the limit pressure and wall thickness and the unified strength theory parameter b are illustrated in Fig.9.2 and Fig.9.3, respectively.

The unified solution with $b = 0$ and $\alpha \neq 1$ is with respect to the Mohr-Coulomb material and the solution for the Tresca material is a special case of the unified solution with $b = 0$ and $\alpha = 1$. The unified solution with $b = 1$ and $\alpha \neq 1$ corresponds to the generalized twin-shear criterion (Yu, 1983), and the unified solution with $b = \alpha = 1$ is the same as the solution of the twin-shear stress criterion (Yu, 1961) or the maximum deviatoric stress criterion (Haythorthwaite, 1960). Other solutions are new which can be applied to different materials. Therefore the unified solution can be adopted for analysis of structures made of various materials.

It is worth noting that:

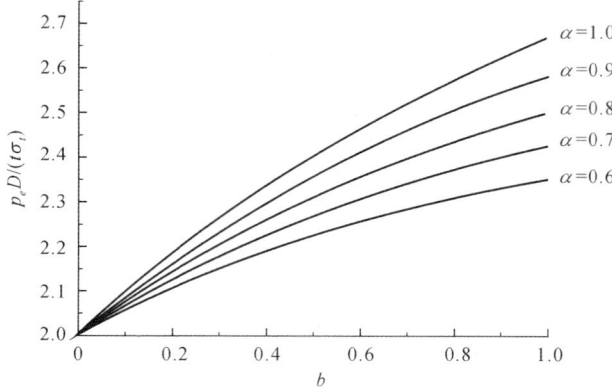

Fig. 9.2. Elastic limit pressure versus the unified strength theory parameter b

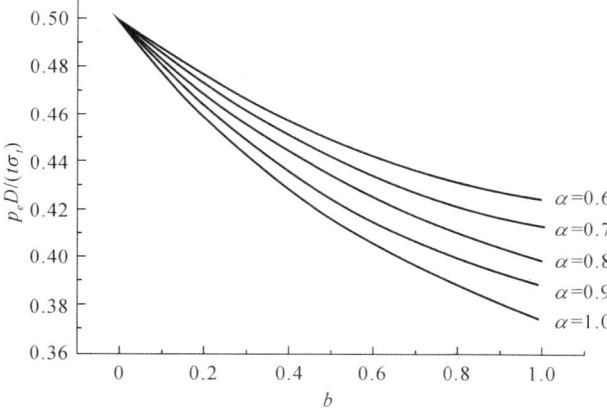

Fig. 9.3. Minimum wall thickness versus the unified strength theory parameter b

(1) The traditional solution is a single solution (with respect to $b = 0$ in the unified strength theory), which can be adopted only for one kind of material. The unified solution, however, gives a serial solution, which can be adopted for various materials and structures.

(2) The solution for Tresca material ($b = 0$ and $\alpha = 1$) is identical to the solution for the Mohr-Coulomb material ($b = 0$ and $\alpha \neq 1$). It means that the SD effect of materials ($\alpha \neq 1$) cannot be considered by the Mohr-Coulomb strength theory in this case.

(3) All the solutions of the bearing capacity of structures with $b > 0$ are higher than the solution of the traditional Tresca or Mohr-Coulomb criterion. All the solutions of the required wall thickness of pressure vessels with $b > 0$ are lower than the solution of the traditional Tresca or Mohr-Coulomb criterion.

(4) The application of the unified strength theory and the unified solutions are more economical in terms of the effective use of materials and energy.

Example 9.1 Design of Space Shuttle

The satellite carrier rockets launched by the United States, Europe and China have very large diameters, for example the diameter of a satellite carrier rocket is 3.5 m with a pressurized body length of 15 m. Under an internal pressure p, the stresses of the rocket wall are

$$\sigma_1 = \frac{pD}{2t}, \ \sigma_2 = \frac{pD}{4t}, \ \sigma_3 = 0.$$

The stresses of a thin-walled vessel satisfy the inequity condition,

$$\sigma_2 = \frac{1}{2}(\sigma_1 + \sigma_3) \leqslant \frac{\sigma_1 + \sigma_3}{2}.$$

Therefore any one of the two expressions of the unified yield criterion Eq.(9.2a) or Eq.(9.2b) is valid for the yield condition of the vessel. The calculated formulas of thickness of a missile body under inner pressure is

$$t \geqslant \frac{2 + 2b + \alpha b}{1 + b} \frac{pD}{4[\sigma]} \quad \text{(for SD materials),}$$

$$t \geqslant \frac{2 + 3b}{1 + b} \frac{pD}{4[\sigma]} \quad \text{(for non-SD materials).}$$

From the above results, for a specific allowable stress $[\sigma]$, factor of safety n, internal pressure p and diameter D, the required wall thickness of the pressure vessels depends on the parameter b in the unified yield criterion. When $b = 0$ which corresponds to the single shear criterion, the required thickness is the largest; when $b = 1$ with respect to the twin shear stress criterion, the thickness is the smallest. The difference between the two required thicknesses is 33.3%.

A carrier rocket is a tool to launch a satellite and itself is not the target to be launched and positioned in space. Given the certain capacity of the launching system, a reduction in the rocket's selfweight is beneficial for increasing the satellite weight. Based on Zhang (1998) and Xia (1999), the cost of launching a satellite per ton mass is as high as tens of thousands of US dollars. The transportation cost per unit effective mass is about US$22,000 even if the satellite is positioned on a lower track (Xia, 1999; Zhang, 1998).

The application of new materials is an effective way of reducing the self-weight of the rocket. A new ultra-high-strength Al-Cu-Li 2195 alloy has been successfully used to fabricate large-scale tanks for the space shuttle, and the weight of the tanks decreases by 3405 kg.

On the other hand, using a more accurate analysis method may be an alternative way of making a cost-effective design of the carrier rocket. The current design of the rocket strength is based on the Tresca criterion which is a single shear criterion. If the unified yield criterion for $b = 1$ or $b = 1/2$

is adopted, the required wall thickness of the rocket can be reduced by more than 16.5%. A rough estimation shows that the body weight of the rocket may reduce by 440 kg if the wall thickness is 1 mm thinner. Thus the economic benefit is as high as US$9.2 million if the unified yield criterion with $b = 1$ is used.

9.3 Limit Pressure of Thick-walled Hollow Sphere

If a thick-walled sphere with inner radius r_a and outer radius r_b is subjected to an internal pressure p, as shown in Fig.9.4. The sphere will deform symmetrically about the center; the radial and any two orthogonal tangential directions will be the principal directions.

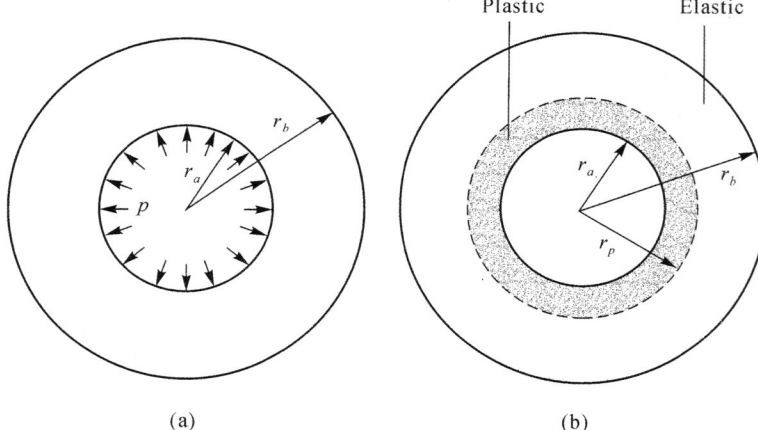

Fig. 9.4. Thick-walled sphere shell

The three corresponding principal strains are ε_r, ε_θ, ε_φ and $\varepsilon_\theta = \varepsilon_\varphi$. The equilibrium equation is

$$\frac{\mathrm{d}\sigma_r}{\mathrm{d}r} = 2\frac{\sigma_\theta - \sigma_r}{r}. \tag{9.7}$$

The elastic stress-strain relations are

$$\begin{cases} \varepsilon_r = \dfrac{1}{E}(\sigma_r - 2\nu\sigma_\theta), \\[2mm] \varepsilon_\theta = \dfrac{1}{E}[(1 - \nu)\sigma_\theta - \nu\sigma_r)]. \end{cases} \tag{9.8}$$

or

$$\begin{cases} \sigma_\theta = \sigma_\varphi = \dfrac{E}{(1+\nu)(1-2\nu)}(\varepsilon_\theta + \nu\varepsilon_r), \\[4mm] \sigma_r = \dfrac{E}{(1+\nu)(1-2\nu)}[(1-\nu)\varepsilon_r + 2\nu\varepsilon_\theta. \end{cases} \tag{9.9}$$

The compatibility equation has the form of

$$\frac{\mathrm{d}\varepsilon_\theta}{\mathrm{d}r} + \frac{\varepsilon_\theta - \varepsilon_r}{r} = 0. \tag{9.10}$$

9.3.1 Elastic Limit Pressure of Thick-walled Sphere Shell

The Lame solutions of the elastic stress distribution had been given (Johnson and Mellor, 1962) as follows:

$$\sigma_r = \frac{pr_a^3}{r^3}\frac{(r_b^3 - r^3)}{(r_a^3 - r_b^3)}, \tag{9.11}$$

$$\sigma_\theta = \sigma_\varphi = \frac{pr_a^3}{2r^3}\frac{(2r^3 + r_b^3)}{(r_b^3 - r_a^3)}. \tag{9.12}$$

For convenience of formulation, the following dimensionless quantities are introduced:

$$K = \frac{r_b}{r_a}, \quad \rho = \frac{r}{r_a}, \quad \alpha = \frac{\sigma_t}{\sigma_c}, \tag{9.13}$$

where σ_t is the yield strength in uniaxial tension.

The stress expressions can then be written as

$$\sigma_r = \frac{\rho^3 - K^3}{\rho^3(K^3 - 1)}p, \quad \sigma_\theta = \sigma_\varphi = \frac{2\rho^3 + K^3}{2\rho^3(K^3 - 1)}p. \tag{9.14}$$

The magnitude of these stresses in elastic range is limited by the yield criterion. When the unified strength theory Eq.(9.2a) and Eq.(9.2b) are used, because $\sigma_\theta = \sigma_\varphi > \sigma_r$, i.e., $\sigma_1 = \sigma_\theta$ (or σ_φ), $\sigma_2 = \sigma_\varphi$ (or σ_θ), $\sigma_3 = \sigma_r$, $\tau_{12} = 0$, $\tau_{13} = \tau_{23}$, there is

$$\sigma_2 \geqslant \frac{\sigma_1 + \alpha\sigma_3}{1 + \alpha}.$$

Thus Eq.(9.2b) should be used as the yield condition when the unified strength theory is applied. The inner surface of the spherical shell yields first at the elastic limit pressure. Substituting the stress components at the inner surface into Eq.(9.2b), the elastic limit pressure p_e in terms of the unified strength theory is derived as

$$p_e = \frac{K^3 - 1}{(1 - \alpha) + K^3(\frac{1}{2} + \alpha)} \sigma_t. \tag{9.15}$$

The relation of elastic limit pressure with the ratios of the outer radius to the inner radius $K = r_b/r_a$ is shown in Fig. 9.5. It is worth noting that, as K approaches infinity, the elastic limit pressure approaches a specific value. If $\alpha = 1$, this pressure is equal to 2/3 of the yield stress σ_y. When the ratio K is larger than 2, the increment of the limit pressure is the minimum in spite of different α.

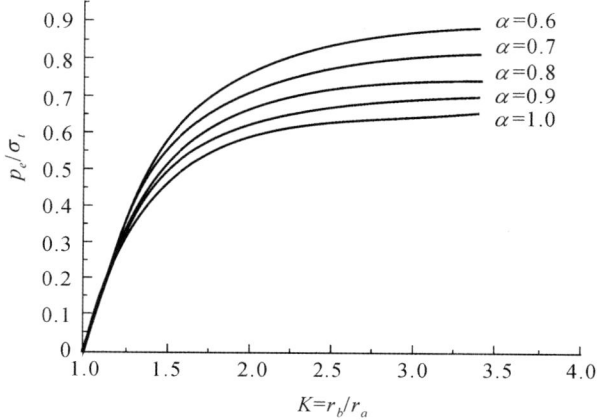

Fig. 9.5. Relation of elastic limit pressure of sphere shell to $K = r_b/r_a$

The elastic limit pressure for non-SD materials are a special case in Eq. (9.15) with $\alpha = 1$,

$$p_e = \frac{2}{3} \frac{K^3 - 1}{K^3} \sigma_y. \tag{9.16}$$

This result is identical with the previous result (Johnson and Mellor, 1962), which is a special case for results obtained based on the unified yield criterion.

The limit pressure of a thick-walled hollow sphere is independent of the strength parameter b. The reason is that the stress state of a spherical shell is spherically symmetrical about the center, and the three principal stresses satisfy $\sigma_1 = \sigma_2 > \sigma_3$. All the limit loci of the unified strength theory with different parameter b intersect each other for this stress state.

9.3.2 Plastic Limit Pressure of Thick-walled Sphere Shell

When the internal pressure p reaches the elastic limit pressure p_e, the inner surface of the hollow sphere shell yields. As the internal pressure increases, the plastic zone spreads towards the outer surface. Denoting the outer radius of the plastic zone as r_p, and assuming the material is perfectly plastic, the failure condition of the unified strength theory for the spherical shell can be simplified as

$$\sigma_\theta - \alpha\sigma_r = \sigma_t. \tag{9.17}$$

The stresses in the plastic region $(r_a \leqslant r \leqslant r_p)$ can then be derived from the equilibrium equation (Eq.(9.9)) with application of the boundary condition of $r = r_a$, $\sigma_r = -p$,

$$\begin{cases} \sigma_r^p = \dfrac{\sigma_t}{1-\alpha}\left[1-\left(\dfrac{r_a}{r}\right)^{2(1-\alpha)}\right] - p\left(\dfrac{r_a}{r}\right)^{2(1-\alpha)}, \\[12pt] \sigma_\theta^p = \sigma_\varphi^p = \dfrac{\sigma_t}{1-\alpha}\left[1-\alpha\left(\dfrac{r_a}{r}\right)^{2(1-\alpha)}\right] - \alpha p\left(\dfrac{r_a}{r}\right)^{2(1-\alpha)}. \end{cases} \tag{9.18}$$

Since no stress-strain relation is required to derive the stresses, it is a statically determinate problem.

At the plastic zone boundary of $r = r_p$, the radial stress σ_r^p can be calculated by substituting the boundary condition into Eq.(9.18). The elastic part of the sphere is then considered as a new sphere with an inner radius of r_p and an outer radius of r_b with an internal pressure of σ_r^p at $r = r_p$. The stresses in the elastic region $(r_p \leqslant r \leqslant r_b)$ can be written as

$$\begin{cases} \sigma_r^e = \dfrac{r_p^3(1-\frac{r_b^3}{r^3})}{r_p^3(1-\alpha)+r_b^3(\frac{1}{2}+\alpha)}\sigma_t, \\[14pt] \sigma_\theta^e = \sigma_\varphi^e = \dfrac{r_p^3(1+\frac{r_b^3}{2r^3})}{r_p^3(1-\alpha)+r_b^3(\frac{1}{2}+\alpha)}\sigma_t. \end{cases} \tag{9.19}$$

The pressure at the elastic-plastic boundary and the radius of the plastic zone can be derived from the stress continuous condition at the elasto-plastic boundary,

$$p_{ep} = \left\{ -\frac{1}{1-\alpha}\left[\left(\frac{r_p}{r_a}\right)^{2(1-\alpha)} - 1\right] + \frac{(r_b^3 - r_p^3)\left(\frac{r_p}{r_a}\right)^{2(1-\alpha)}}{r_p^3(1-\alpha)+r_b^3(\frac{1}{2}+\alpha)} \right\} \sigma_t. \tag{9.20}$$

When r_p is equal to the outer radius of the sphere r_b, the sphere shell is completely plastic. The plastic limit of the internal pressure p_p can be derived as

$$p_p = \frac{\sigma_t}{1-\alpha}(K^{2(1-\alpha)} - 1). \tag{9.21}$$

Eq.(9.21) is the same as the result based on the twin-shear strength theory (Zhuang, 1998) and the result obtained by using the Mohr-Coulomb strength theory. The relationship of the plastic limit pressure with different ratios of the outer radius to the inner radius $K = r_b/r_a$ is shown in Fig.9.6. The plastic limit pressure increases with the increase of the ratio.

The plastic limit pressure of a thick-walled hollow sphere shell of non-SD materials can be calculated from Eq.(9.21) with $\alpha = 1$,

$$p_p = 2 \ln K. \tag{9.22}$$

Eq.(9.22) is the same as the result based on the Tresca criterion (Johnson and Mellor, 1962).

The stresses in the plastic region $(r_p \leqslant r \leqslant r_b)$ can be expressed as

$$\sigma_r^p = \frac{\sigma_t}{1-\alpha} \left[1 - \left(\frac{r_b}{r} \right)^{2(1-\alpha)} \right], \tag{9.23}$$

$$\sigma_\theta^p = \sigma_\varphi^p = \frac{\sigma_t}{1-\alpha} \left[1 - \alpha \left(\frac{r_b}{r} \right)^{2(1-\alpha)} \right]. \tag{9.24}$$

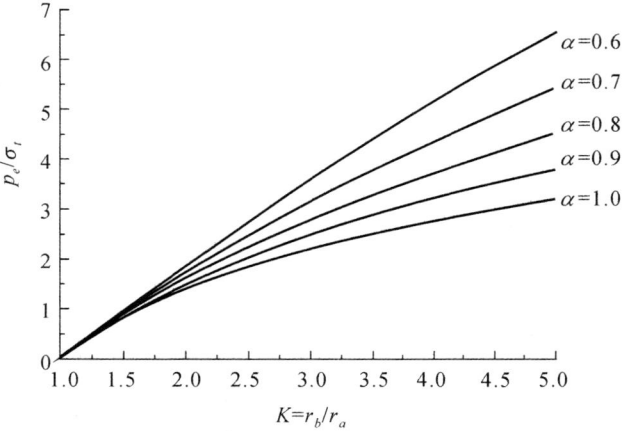

Fig. 9.6. Relation of plastic limit pressure of sphere shell to the ratio of $K = r_b/r_a$

9.4 Unified Solution of Elastic Limit Pressure of Thick-walled Cylinder

Considerable works on elasto-plastic analysis for a thick-walled cylinder under internal pressure have been reported by Turner (1909), Nadai (1931), Manning (1945), Allen and Sopwith (1951), and Crossland and Bones (1958). Discussions in depth on this subject can also be found in the books by Hill (1950), Johnson and Mellor (1962), Mendelson (1968), and Chakrabarty (1987).

The twin-shear yield criterion proposed by Yu (1961; 1983) was applied to study the limit pressure of a thick-walled cylinder by Li (1998), Huang and Zeng (1989). The generalized twin-shear strength theory (Yu et al., 1985) was also used to derive the limit pressure of a thick-walled cylinder and hollow sphere shell by Zhuang (1998), Zhao et al. (1999), and Ni et al. (1998). It was also applied to gun barrels by Liu et al. (1998). The elastic limit pressure, plastic limit pressure, and autofrettage pressure in an autofretted gun barrel were reported by Liu et al. (1998).

Nowadays, gun barrels are made of high-strength steel having different strength in tension and compression. Therefore, the results with respect to the generalized twin-shear strength theory, which takes into account the SD effect of materials, should be more appropriate. The unified yield criterion (Yu and He, 1991) was used to drive the limit pressure for thick-walled tubes with different end conditions, e.g., the open-end condition, the closed-end condition, and the plane strain condition (Wang and Fan, 1998). For pressure-sensitive materials, a failure criterion should take into account the SD effect of material. The unified strength theory takes all the stress components into account and is suitable for both non-SD and SD materials. In this chapter the effects of yield criteria on elastic and plastic limit pressures for thick-walled tubes using the unified strength theory are summarized and discussed.

Considering a thick-walled cylinder with the inner and outer radii of the cylinder r_a and r_b under an internal pressure p and a longitudinal force P, the radius of the cylinder is assumed to be so large that the plane transverse sections remain on the plane during the expansion. It implies that the longitudinal strain ε_z is independent of the radius. Since the stresses and strains in a cross-section that is sufficiently far away from the ends do not vary along the length of the cylinder, the equation of equilibrium can be written as

$$\frac{\mathrm{d}\sigma_r}{\mathrm{d}r} = \frac{\sigma_\theta - \sigma_r}{r}. \tag{9.25}$$

It should be mentioned that the z axis of the cylindrical coordinates (r, θ, z) is the longitudinal axis of the tube. Based on the generalized Hooke's law, the longitudinal stress in the elastic state can be written as

$$\sigma_z = E\varepsilon_z + \nu(\sigma_r + \sigma_\theta), \tag{9.26}$$

where E is Young's modulus and ν the Poisson's ratio. The radial strain ε_r and the circumferential strain ε_θ are

$$\varepsilon_r = -\nu\varepsilon_z + \frac{1+\nu}{E}\left[(1-\nu)\sigma_r - \nu\sigma_\theta\right], \tag{9.27a}$$

$$\varepsilon_\theta = -\nu\varepsilon_z + \frac{1+\nu}{E}\left[(1-\nu)\sigma_\theta - \nu\sigma_r\right]. \tag{9.27b}$$

The compatibility equation is $\frac{d}{dr}(\sigma_r + \sigma_\theta) = 0$, which indicates that $\sigma_r + \sigma_\theta$ have constant values at each stage of the elastic expansion. Integrating Eq. (9.25) and applying the boundary conditions of $\sigma_r = 0$ at $r = r_b$, and $\sigma_r = -p$ at $r = r_a$, the stresses are given as

$$\sigma_r = -p\left(\frac{\frac{r_b^2}{r^2} - 1}{\frac{r_b^2}{r_a^2} - 1}\right), \quad \sigma_\theta = p\left(\frac{\frac{r_b^2}{r^2} + 1}{\frac{r_b^2}{r_a^2} - 1}\right). \tag{9.28}$$

This is Lame's solution given by Lame (1852). If the resultant longitudinal load is denoted by P, the axial stress σ_z is $P/\left[\pi(r_b^2 - r_a^2)\right]$ since this stress is constant over the cross section. In particular, $p = 0$ represents an open-end condition and $P = \pi r_a^2 p$ represents a closed-end condition. For a plane strain condition ($\varepsilon_z = 0$), σ_z can be directly derived from Eqs.(9.26) and (9.28),

$$\sigma_z = \frac{p}{K^2 - 1}, \quad \text{closed end}, \tag{9.29a}$$

$$\sigma_z = 0, \quad \text{open end}, \tag{9.29b}$$

$$\sigma_z = \frac{2\nu p}{K^2 - 1}. \quad \text{plane strain.} \tag{9.29c}$$

The corresponding axial strains are

$$\varepsilon_z = \frac{(1-2\nu)p}{(K^2-1)E}, \quad \text{closed end}, \tag{9.30a}$$

$$\varepsilon_z = 0, \quad \text{open end}, \tag{9.30b}$$

$$\varepsilon_z = \frac{-2\nu p}{(K^2-1)E}, \quad \text{plane strain.} \tag{9.30c}$$

In all the three cases, σ_z is the intermediate principal stress. For the closed-end condition, σ_z is the average or the mean value of the other two principal stresses. If the material is assumed to be incompressible in both the elastic and plastic ranges, σ_z of the plane strain condition is identical to that of the closed-end condition. There are $\sigma_1 = \sigma_\theta$, $\sigma_2 = \sigma_z$, $\sigma_3 = \sigma_r$, and

$$\sigma_2 = \frac{1}{2}(\sigma_1 + \sigma_3) \leqslant \frac{\sigma_1 + \alpha\sigma_3}{1+\alpha}. \tag{9.31}$$

Thus the first equation of the unified strength theory should be applied as the yield conditions,

$$\sigma_1 - \frac{\alpha}{1+b}(b\sigma_2 + \sigma_3) = \sigma_t. \tag{9.32}$$

Substituting Eq.(9.31) into Eq.(9.32), the yield condition in the case of thick-walled cylinder with closed-end and plane strain condition can be expressed as

$$\frac{2 + (2 - \alpha)\alpha}{2(1 + b)}\sigma_\theta - \frac{\alpha(2 + b)}{2(1 + b)}\sigma_r = \sigma_t. \tag{9.33}$$

For on open-end cylinder it is

$$\sigma_\theta - \frac{\alpha}{1 + b}\sigma_r = \sigma_t. \quad \text{(open end)} \tag{9.34}$$

Substituting Eq.(9.28) into Eqs.(9.33) and (9.34), we get

$$[2 + (2 - \alpha)b]\frac{p}{K^2 - 1}\left(\frac{r_b^2}{r^2} + 1\right) + \alpha(2 + b)\frac{p}{K^2 - 1}\left(\frac{r_b^2}{r^2} - 1\right) = 2(1 + b)\sigma_t. \tag{9.35}$$

The elastic limit pressure in terms of the unified strength theory can be derived as

$$p_e = \frac{(1 + b)(K^2 - 1)\sigma_t}{K^2(1 + b + \alpha) + (1 + b)(1 - \alpha)}, \quad \text{closed end} \tag{9.36}$$

$$p_e = \frac{(1 + b)(K^2 - 1)\sigma_t}{(1 + b)(K^2 + 1) + \alpha(K^2 - 1)}, \quad \text{open end} \tag{9.37}$$

$$p_e = \frac{(1 + b)(K^2 - 1)\sigma_t}{K^2(1 + b + \alpha) + (1 + b)(1 - \alpha)}. \quad \text{plane strain} \tag{9.38}$$

The limit pressure for the closed-end cylinder based on the Mohr-Coulomb strength theory (single-shear theory) is

$$p_e = \frac{K^2 - 1}{(1 + \alpha)K^2 + (1 - \alpha)}\sigma_t. \quad \text{(Mohr-Coulomb strength theory)} \tag{9.39}$$

The limit pressure of a thick-walled cylinder in terms of twin-shear strength theory was reported by Zhuang (1998) and Ni et al. (1998) as

$$p_e = \frac{2(K^2 - 1)}{(2 + \alpha)K^2 + 2(1 - \alpha)}\sigma_t. \quad \text{(twin shear strength theory)} \tag{9.40}$$

These limit pressures are specific cases of the solutions in terms of the unified solution with $b = 0$ and $b = 1$ respectively.

For non-SD materials, i.e., $\alpha = 1$ or $\sigma_t = \sigma_c = \sigma_y$, Eqs.(9.36)~(9.38) are simplified as

$$p_e = \frac{(1+b)(K^2-1)}{K^2(2+b)}\sigma_y, \quad \text{closed end} \tag{9.41}$$

$$p_e = \frac{(1+b)(K^2-1)}{K^2(2+b)+b}\sigma_y, \quad \text{open end} \tag{9.42}$$

$$p_e = \frac{(1+b)(K^2-1)}{K^2(2+b)+b(1-2\nu)}\sigma_y. \quad \text{plane strain} \tag{9.43}$$

These results are identical to the solutions with Yu unified yield criterion (Wang and Fan, 1998).

The elastic limit pressure for the Tresca material at closed end, open end, and plane strain conditions can be obtained from Eqs.(9.41)∼(9.43) with $\alpha = 1$, $b = 0$. The solutions for different conditions are identical,

$$p_e = \frac{K^2-1}{2K^2}\sigma_y. \tag{9.44}$$

The elastic limit pressure for the Huber-von Mises material can be approximately derived from the unified solution with $\alpha = 1$, $b=1/2$,

$$p_e = \frac{3(K^2-1)}{5K^2}\sigma_y, \quad \text{closed end} \tag{9.45}$$

$$p_e = \frac{3(K^2-1)}{5K^2+(1-2\nu)}\sigma_y. \quad \text{plane strain} \tag{9.46}$$

The classical solutions for Huber-von Mises material are

$$p_e = \frac{K^2-1}{\sqrt{3K^2}}\sigma_y, \quad \text{closed end} \tag{9.47}$$

$$p_e = \frac{K^2-1}{\sqrt{3K^4+1}}\sigma_y, \quad \text{open end} \tag{9.48}$$

$$p_e = \frac{K^2-1}{\sqrt{3K^4+(1-2\nu)^2}}\sigma_y. \quad \text{plane strain} \tag{9.49}$$

The percentage difference between the approximated elastic limit pressure with regard to the unified solution with $\alpha = 1$, $b = 1/2$, and the exact solution based on the Huber-von Mises criterion is as low as 0.38%.

The elastic limit pressure in terms of the twin-shear yield criterion can be derived from the unified solution with $\alpha = 1$, $b = 1$,

$$p_e = \frac{2(K^2-1)}{3K^2}\sigma_y, \quad \text{closed end} \tag{9.50}$$

$$p_e = \frac{2(K^2-1)}{3K^2+1}\sigma_y, \quad \text{open end} \tag{9.51}$$

$$p_e = \frac{2(K^2 - 1)}{3K^2 + (1 - 2\nu)}\sigma_y. \quad \text{plane strain} \tag{9.52}$$

The percentage difference of the solutions between the Tresca material and the twin-shear material is as high as 33.4%.

It can be noted from the above derivation that all the solutions with regard to the prevailing yield criteria can be approximated or deduced from the unified solution in view of the unified strength theory with specific values of α and b. The variations of the unified solution regarding different values of α and b are illustrated in Figs.9.7 and 9.8.

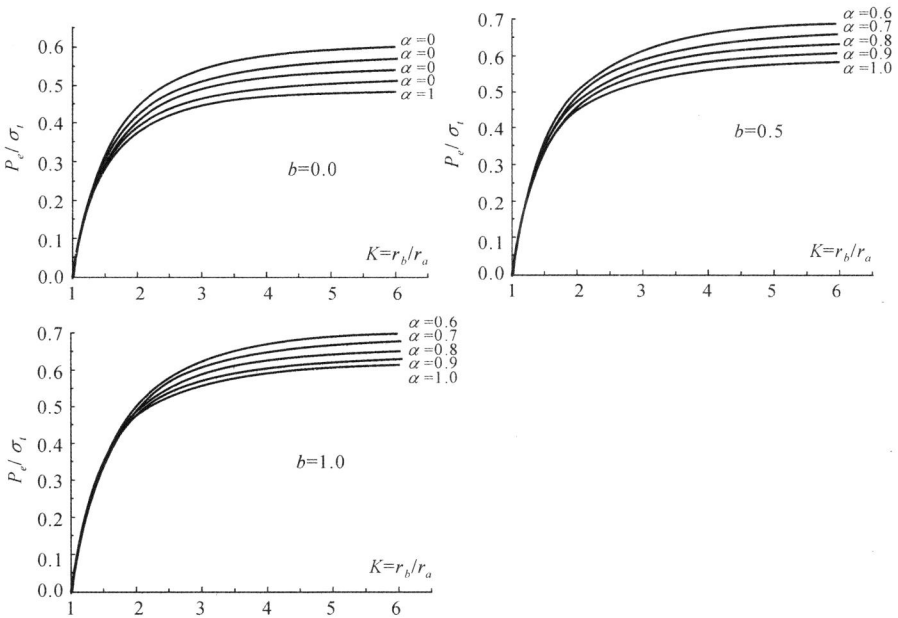

Fig. 9.7. Relation of elastic pressure with $K = r_b/r_a$

The results of the elastic limit pressures of a thick-walled cylinder for closed end and open end in view of different yield criteria are summarized in Table 9.1 and Table 9.2.

When a uniform pressure p is applied externally to a thick-walled cylinder with wall ratio r_b/r_a, the elastic stress distribution of σ_r and σ_θ can be derived from Eq.(9.28) by exchanging the positions of r_a and r_b in the formulation. In this case both the stresses are compressive, and the magnitude of σ_θ is higher than that of σ_r.

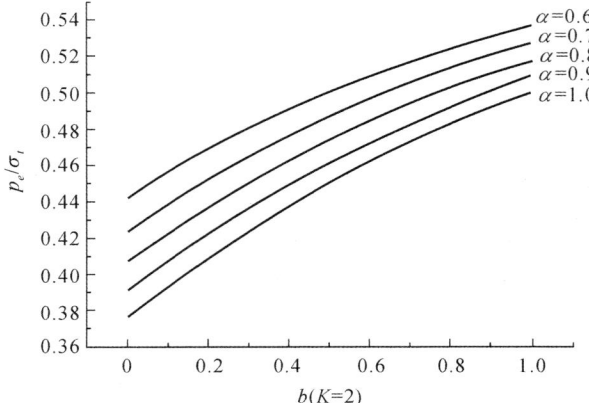

Fig. 9.8. Relation of elastic pressure with the unified strength theory parameter b

Table 9.1. Summary of elastic limit pressures for closed end

	Materials	Elastic limit pressures	Failure criterion used
1	SD material $\alpha \neq 1$	$p_e = \dfrac{(1+b)(K^2-1)\sigma_t}{K^2(1+b+\alpha)+(1+b)(1-\alpha)}$	Unified strength theory
2	SD material $\alpha \neq 1$	$p_e = \dfrac{K^2-1}{(1+\alpha)K^2+(1-\alpha)}\sigma_t$	Unified strength theory $b=0$, Mohr-Coulomb
3	SD material $\alpha \neq 1$	$p_e = \dfrac{2(K^2-1)}{K^2(2+\alpha)+2(1-\alpha)}\sigma_t$	Unified strength theory $b=1$, twin-shear theory
4	$\alpha = 1$ materials	$p_e = \dfrac{(1+b)(K^2-1)}{K^2(2+b)}\sigma_y$	Unified yield criterion
5	$\alpha = 1$ materials	$p_e = \dfrac{K^2-1}{2K^2}\sigma_y$	Unified yield criterion $b=0$, Tresca criterion
6	$\alpha = 1$ materials	$p_e = \dfrac{K^2-1}{\sqrt{3K^2}}\sigma_y$	von Mises yield criterion
7	$\alpha = 1$ materials	$p_e = \dfrac{3(K^2-1)}{5K^2}\sigma_y$	Unified yield criterion $b=1/2$
8	$\alpha = 1$ materials	$p_e = \dfrac{2(K^2-1)}{3K^2}\sigma_y$	Unified yield criterion $b=1$, twin-shear criterion

9.5 Unified Solution of Plastic Limit Pressure of Thick-walled Cylinder

9.5.1 Stress Distribution

When the internal pressure exceeds p_e, a plastic zone starts at the inner surface and spreads towards the outer surface. If the outer radius of the

Table 9.2. Summary of elastic limit pressures for open end

	Materials	Elastic limit pressures	Failure criterion used
1	SD material $\alpha \neq 1$	$p_e = \frac{(1+b)(K^2-1)\sigma_t}{(1+b)(K^2+1)+\alpha(K^2-1)}$	Unified strength theory
2	SD material $\alpha \neq 1$	$p_e = \frac{K^2-1}{(1+\alpha)K^2+(1-\alpha)}\sigma_t$	Unified strength theory $b = 0$, Mohr-Coulomb
3	SD material $\alpha \neq 1$	$p_e = \frac{2(K^2-1)}{K^2(2+\alpha)+2(1-\alpha)}\sigma_t$	Unified strength theory $b = 1$, twin-shear theory
4	$\alpha = 1$ materials	$p_e = \frac{(1+b)(K^2-1)}{K^2(2+b)+b}\sigma_y$	Unified yield criterion
5	$\alpha = 1$ materials	$p_e = \frac{K^2-1}{2K^2}\sigma_y$	Unified yield criterion $b = 0$, Tresca criterion
6	$\alpha = 1$ materials	$p_e = \frac{K^2-1}{\sqrt{3K^4+1}}\sigma_y$	von Mises yield criterion
7	$\alpha = 1$ materials	$p_e = \frac{3(K^2-1)}{5K^2+1}\sigma_y$	Unified yield criterion $b = 1/2$
8	$\alpha = 1$ materials	$p_e = \frac{2(K^2-1)}{3K^2+1}\sigma_y$	Unified yield criterion $b = 1$, twin-shear criterion

elastic-plastic boundary is denoted as r_c, in the elastic region ($r_c \leqslant r \leqslant r_b$), the radial and circumferential stresses can be derived from Lame's equations with application of the boundary conditions of $\sigma_r = 0$ at $r = r_b$, and the stresses at $r = r_c$ satisfying the yield conditions. The pressure reaches its maximum value when the plastic zone reaches the outer surface of the thick-walled tube.

The elastic part of the elastic-plastic thick-walled tube can be considered as a new tube with the inner radius r_c, outer radius r_b and an internal pressure p_e. The stress distribution in the elastic region for incompressible material can be written as

$$\sigma_\theta = \frac{p_e r_c^2}{r_b^2 - r_c^2}\left(1 + \frac{r_b^2}{r^2}\right), \tag{9.53}$$

$$\sigma_r = \frac{p_e r_c^2}{r_b^2 - r_c^2}\left(1 - \frac{r_b^2}{r^2}\right), \tag{9.54}$$

$$\sigma_z = \frac{\nu}{2}(\sigma_\theta + \sigma_r), \tag{9.55}$$

where

$$p_e = \frac{2(1 + b)(r_b^2 - r_c^2)}{(2 + 2b - \alpha b)(r_b^2 + r_c^2) + \alpha(2 + b)(r_b^2 - r_c^2)} \sigma_t. \tag{9.56}$$

9.5.2 Plastic Zone in the Elasto-plastic Range

In the plastic zone, for elastic-perfectly-plastic material, the stress state satisfies Eq.(9.2a) or Eq.(9.2b) when the unified strength theory is adopted. According to the stress state condition of Eq. (9.3a), the first equation of the unified strength theory, i.e., Eq. (9.2a), should be applied as the yield condition,

$$\frac{2 + (2 - \alpha)b}{2(1 + b)} \sigma_\theta - \frac{\alpha(2 + b)}{2(1 + b)} \sigma_r = \sigma_t. \tag{9.57}$$

Substituting Eq.(9.57) into the equilibrium equation in Eq.(9.25), we get

$$\frac{d\sigma_r}{dr} + \frac{2(1 + b)(1 - \alpha)}{2 + (2 - \alpha)b} \frac{\sigma_r}{r} - \frac{2(1 + b)}{2 + (2 - \alpha)b} \frac{\sigma_t}{r} = 0. \tag{9.58}$$

The general solution to this differential equation is

$$\sigma_r = \frac{c}{r^{\frac{2(1+b)(1-\alpha)}{2+(2-\alpha)b}}} + \frac{\sigma_t}{1 - \alpha}. \tag{9.59}$$

The integration constant can be determined by the boundary condition of $r = r_a$, $\sigma_r = -p$ as $-p = \frac{c}{r_a^{\frac{2(1+b)(1-\alpha)}{2+(2-\alpha)b}}} + \frac{\sigma_t}{1-\alpha}$, which gives

$$c = (-p - \frac{\sigma_t}{1 - \alpha}) A^{\frac{2(1+b)(1-\alpha)}{2+(2-\alpha)b}}. \tag{9.60}$$

Therefore, the stress distribution in the plastic region ($r_a \leqslant r \leqslant r_c$) is

$$\sigma_r = - \left(p + \frac{\sigma_t}{1 - \alpha} \right) \left(\frac{r_a}{r} \right)^{\frac{2(1+b)(1-\alpha)}{2+(2-\alpha)b}} + \frac{\sigma_t}{1 - \alpha}, \tag{9.61}$$

$$\sigma_\theta = \frac{2(1 + b)\sigma_t}{2 + (2 - \alpha)b} - \frac{\alpha(2 + b)}{2 + (2 - \alpha)b} \left[\left(p + \frac{\sigma_t}{1 - \alpha} \right) \left(\frac{r_a}{r} \right)^{\frac{2(1+b)(1-\alpha)}{2+(2-\alpha)b}} + \frac{\sigma_t}{1 - \alpha} \right], \tag{9.62}$$

$$\sigma_z = \frac{1}{2}(\sigma_r + \sigma_\theta). \tag{9.63}$$

Eqs.(9.61)~(9.63) give the stresses of a thick-walled cylinder at the plastic region. Since no stress-strain relation is required to derive the stresses, the problem is considered statically determinate.

9.5.3 Plastic Zone Radius in the Elasto-plastic Range

The pressure on the elastic and plastic zone boundary satisfies Eq.(9.57) of the elastic zone solution. Assuming that the radius of the plastic zone is r_c, for a given internal pressure p, the plastic zone radius r_c can be determined from Eq.(9.57). When pressure increases, the plastic zone radius r_c increases gradually from r_a to r_b.

The stress continuity of radial stress σ_r across $r = r_c$ gives

$$\sigma_{r=r_c}(\text{elastic zone}) = \sigma_{r=r_c}(\text{plastic zone}).$$

Substituting the radial stress in Eq.(9.55) and the radial stress in Eq.(9.60) into the stress continuity condition, the relation of pressure p to plastic zone radius is derived,

$$
\begin{aligned}
p = & \left(\frac{r_c}{r_a}\right)^{\frac{2(1+b)(1-\alpha)}{2+(2-\alpha)b}} \left[\frac{2(1+b)(r_b^2 - r_c^2)}{(2+2b-\alpha b)(r_b^2 + r_c^2) + \alpha(2+b)(r_b^2 - r_c^2)}\right. \\
& \left. + \frac{1}{1-\alpha}\right]\sigma_t - \frac{\sigma_t}{1-\alpha}.
\end{aligned}
\tag{9.64}
$$

As an example, the relation of pressure versus the plastic zone radius is illustrated schematically in Fig.9.9 for the ratio of the external radius r_b to the internal radius r_a, $K = r_b/r_a = 2$.

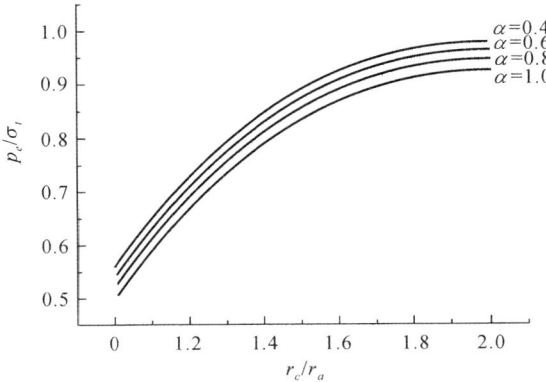

Fig. 9.9. Plastic zone radius versus internal pressure for different α ($K = 2, b = 1.0$)

9.5.4 Plastic Limit Pressure

9.5.4.1 Plastic Limit Pressure for SD Materials

When r_c is equal to r_b, the thick-walled tube is completely plastic. The plastic limit pressure for the thick-walled cylinder is

$$p_p = \frac{\sigma_t}{1 - \alpha}\left(K^{\frac{2(1+b)(1-\alpha)}{2+2b-\alpha b}} - 1\right). \tag{9.65}$$

The solution in Eq.(9.65) is for a thick-walled cylinder with closed end or plane strain conditions. It can be referred to as the unified solution of plastic limit pressure for thick-walled cylinder.

When $b = 0$, the plastic limit pressure in terms of the Mohr-Coulomb theory is deduced from the unified solution,

$$p_p = \frac{\sigma_t}{1 - \alpha}(K^{(1-\alpha)} - 1). \tag{9.66}$$

When $b = 1$, the unified solution becomes the plastic limit pressure in terms of the twin-shear strength theory,

$$p_p = \frac{\sigma_t}{1 - \alpha}(K^{\frac{4(1-\alpha)}{4-\alpha}} - 1). \tag{9.67}$$

9.5.4.2 Plastic Limit Pressure for Non-SD Materials

The unified solution for non-SD materials can be derived from the unified solution with $\alpha = 1$. The plastic limit pressure of a thick-walled cylinder based on the unified yield criterion can be expressed as

$$p_p = \frac{2(1 + b)\sigma_t}{2 + b} \ln K. \tag{9.68}$$

The limit pressure in terms of the Tresca yield criterion can be derived from the unified solution with $b = 0$,

$$p_p = \sigma_t \ln K, \tag{9.69}$$

which is identical to the classical solution based on the Tresca yield criterion.

The plastic limit pressure in terms of the linear Huber-von Mises yield criterion can be approximately obtained with the unified solution with $b = 1/2$,

$$p_p = \frac{6}{5}\sigma_t \ln K. \tag{9.70}$$

The plastic limit pressure in terms of the twin-shear yield criterion can be obtained from the unified solution with $b = 1$,

$$p_p = \frac{4}{3}\sigma_t \ln K. \tag{9.71}$$

Eqs.(9.70) and (9.71) are identical to the plastic limit pressure based on the twin-shear strength theory (Zhuang, 1998).

The relation of the plastic limit pressure with respect to the different parameter b and different thickness of cylinder ($K = r_b/r_a = 1.8$, $K = 2.0$, $K = 2.5$, $K = 3.0$) are shown in Fig.9.10 to Fig.9.13. From these figures the effect of failure criteria is prominent.

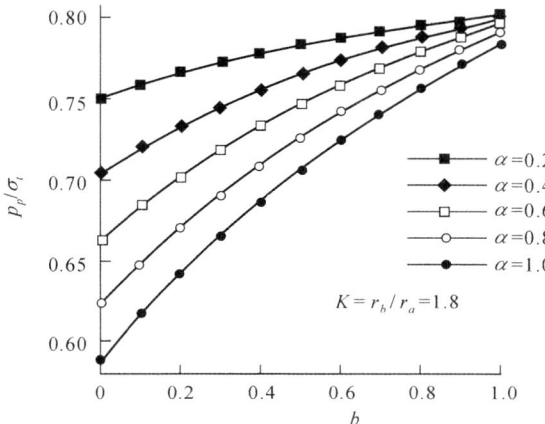

Fig. 9.10. Relation of plastic limit pressure to the unified strength theory parameter b when $K = 1.8$

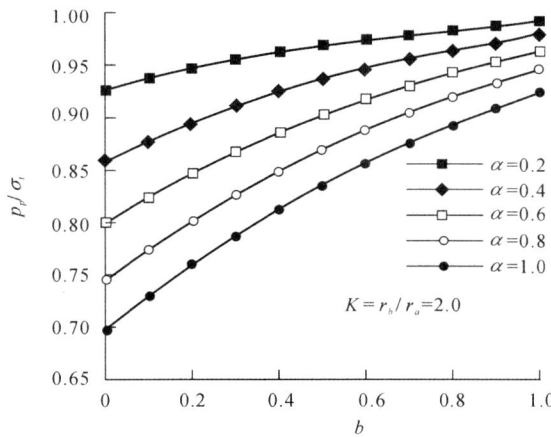

Fig. 9.11. Relation of plastic limit pressure to the unified strength theory parameter when $K = 2.0$

From Figs.9.13, 9.14 and Table 9.3, the elastic limit pressure in terms of the unified strength theory increases monotonically with increasing b for all the three end conditions.

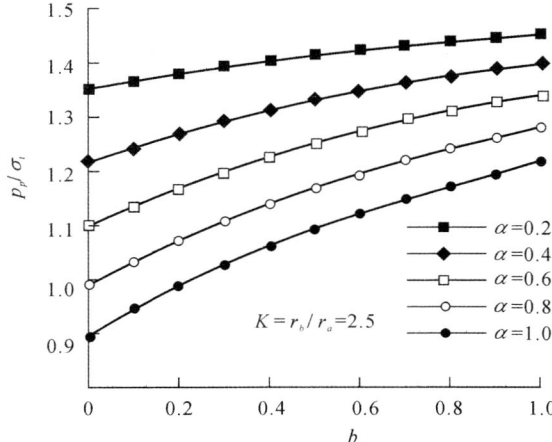

Fig. 9.12. Relation of plastic limit pressure to the unified strength theory parameter when $K = 2.5$

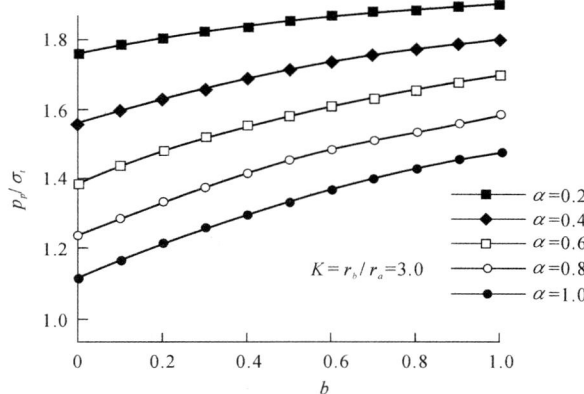

Fig. 9.13. Relation of plastic limit pressure to the unified strength theory parameter when $K = 3.0$

The elastic limit pressure in terms of the Tresca criterion is equal to that of the unified strength theory with $b = 0$ and $\alpha = 1$. The elastic limit pressure in terms of the von Mises criterion is equal to that of the unified strength theory with $b \simeq 4$. Therefore it can be concluded that the Huber-von Mises and the Tresca criteria are encompassed in the unified strength theory with regard to the elastic limit pressure. The maximum elastic limit pressure in terms of the twin-shear yield criterion is obtained with $b = 1$. It is 33.4% and 15.5% higher than those obtained from the Tresca criterion and the Huber-von Mises criterion respectively. It was also found that the higher values obtained from the unified strength theory were insensitive to the variations

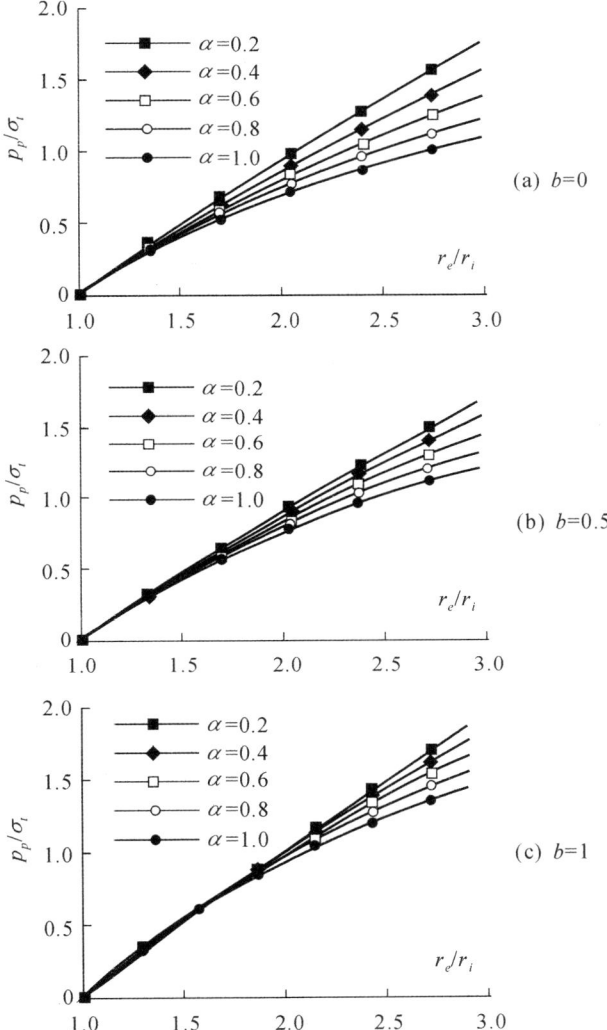

Fig. 9.14. Relation of p_p/σ_t and r_e/r_i with different b and α

of the inner-to-outer-radius ratio regardless of the different end conditions. For the plastic limit pressure, similar statements can be made.

The results of plastic limit pressures of a thick-walled cylinder under the closed end condition in terms of different yield criteria are summarized in Table 9.3.

The elastic limit pressure and plastic limit pressure, two important parameters in the design of a cylinder, have been derived using the unified strength theory. It was found that the percentage difference of the elastic-plastic limit

Table 9.3. Summary of elastic limit pressures for open end

	Materials	Elastic limit pressures	Failure criterion used
1	SD material $\alpha \neq 1$	$p_p = \frac{\sigma_t}{1-\alpha}(K^{\frac{2(1+b)(1-\alpha)}{2+2b-\alpha b}} - 1)$	Unified strength theory
2	SD material $\alpha \neq 1$	$p_p = \frac{\sigma_t}{1-\alpha}(K^{(1-\alpha)} - 1)$	Unified strength theory $b = 0$, Mohr-Coulomb
3	SD material $\alpha \neq 1$	$p_p = \frac{\sigma_t}{1-\alpha}(K^{\frac{4(1-\alpha)}{4-\alpha}} - 1)$	Unified strength theory $b = 1$, twin-shear theory
4	$\alpha = 1$ materials	$p_p = \frac{2(1+b)\sigma_t}{2+b} \ln K$	Unified yield criterion $\alpha = 1$
5	$\alpha = 1$ materials	$p_p = \sigma_t \ln K$	Tresca yield criterion $\alpha = 1$, $b = 0$
6	$\alpha = 1$ materials	$p_p = \frac{6}{5}\sigma_t \ln K$	Unified yield criterion $b = 1/2$
7	$\alpha = 1$ materials	$p_p = \frac{4}{3}\sigma_t \ln K$	Twin-shear yield criterion $\alpha = 1$, $b = 1$

pressures derived from different criteria could differ from one from another as much as 33.4%. If the unified strength criterion is used in the design instead of the Tresca or the Huber-von Mises criterion, it could lead to a substantial saving of material.

9.6 Summary

The limit analysis of the thick-walled hollow sphere and cylinder under pressure was discussed in detail in literature. The Tresca yield criterion or the Huber-von Mises yield criterion has been applied for analysis and design purposes. The solution is adopted only for one kind of material.

In the last decade the elastic limit and plastic limit of thin-walled vessels and thick-walled cylinders were studied by researchers with respect to the unified strength theory. Unified limit solutions for a thick-wall cylinder subject to external and internal pressure are given. The unified solution is a series of results which can be adopted for more materials. They are described in this chapter.

The application of the unified strength theory is also extended from ideal elasto-plastic materials to hardening material. The unified limit analysis of a

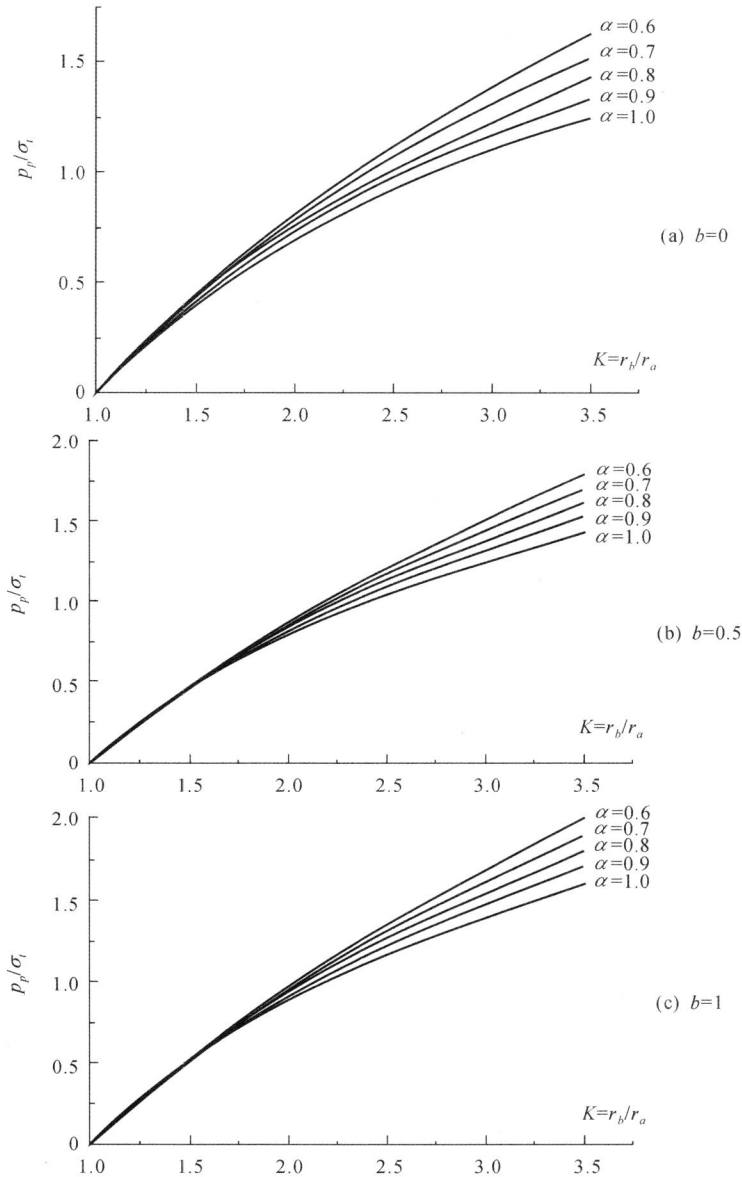

Fig. 9.15. Relation of plastic limit pressure with the thickness of cylinder

thick-wall cylinder of linearly strengthened material is derived by Ma (2004). It was also extended recently to brittle materials such as concrete or rock. Based on Yu unified strength theory with a material's strain softening property considered, a unified strength criterion for strain softening materials has also been proposed to find out the bearing capacity of a thick-walled cylinder subject to external pressure.

A unified solution for a cylinder is generalized to take into account the strain softening material, elasto-brittle-plastic materials and fiber-reinforced concrete. The bearing capacity analysis for a thick walled cylinder, take into account elasto-brittle-plastic and strain softening, is presented by Xu and Yu (2004), Chen et al. (2006a; 2006b).

9.7 Problems

Problem 9.1 Derive the elastic limit pressure equation for a spherical shell under internal pressure in terms of the Mohr-Coulomb theory.

Problem 9.2 Derive the elastic limit pressure equation for a spherical shell under internal pressure in terms of the twin-shear strength theory.

Problem 9.3 Explain why we would expect the Mohr-Coulomb strength theory, the twin-shear strength theory and the unified strength theory to coincide in the case of a spherical shell with spherical symmetry.

Problem 9.4 Derive the elastic limit pressure equation for a thick-walled cylinder under internal pressure using the Mohr-Coulomb theory

$$p_e = \frac{K^2 - 1}{K^2(1 + \alpha)(1 - \alpha)} \sigma_t, \ K = r_b/r_a.$$

Problem 9.5 Derive the elastic limit pressure equation for a thick-walled cylinder under internal pressure by using the twin-shear strength theory

$$p_e = \frac{K^2 - 1}{K^2(1 + \alpha/2)(1 - \alpha)} \sigma_t, \ K = \frac{r_b}{r_a}.$$

Problem 9.6 Compare the results of Problem 9.4 with those of Problem 9.5.

Problem 9.7 A uniform pressure p is applied externally to a thick-walled cylinder of wall ratio r_b/r_a. In this case both the stresses are negative, where σ_θ is more compressive than σ_r. Introduce the elastic limit external pressure equation for a thick-walled cylinder under external pressure using the Mohr-Coulomb strength theory.

Problem 9.8 A uniform pressure p is applied externally to a thick-walled cylinder of wall ratio r_b/r_a. In this case both the stresses are negative, where σ_θ is more compressive than σ_r. Introduce the elastic limit external pressure equation for a thick-walled cylinder under external pressure using the twin-shear strength theory.

Problem 9.9 A uniform pressure p is applied externally to a thick-walled cylinder of wall ratio r_b/r_a. In this case both the stresses are negative, where σ_θ is more compressive than σ_r. Introduce the plastic limit external pressure equation for a thick-walled cylinder under external pressure using the unified strength theory.

Problem 9.10 Compare the results obtained for Problems 9.7, 9.8 and 9.9.

Problem 9.11 A uniform pressure p is applied externally to a thick-walled cylinder of wall ratio r_b/r_a. In this case both the stresses are negative, where σ_θ is more compressive than σ_r. Introduce the elastic limit external pressure equation for a thick-walled cylinder under external pressure using the Mohr-Coulomb strength theory.

Problem 9.12 A uniform pressure p is applied externally to a thick-walled cylinder of wall ratio r_b/r_a. In this case both the stresses are negative, where σ_θ is more compressive than σ_r. Introduce the plastic limit external pressure equation for a thick-walled cylinder under external pressure using the twin-shear strength theory.

Problem 9.13 A uniform pressure p is applied externally to a thick-walled cylinder of wall ratio r_b/r_a. In this case both the stresses are negative, where σ_θ is more compressive than σ_r. Introduce the plastic limit external pressure equation for a thick-walled cylinder under external pressure using the unified strength theory.

Problem 9.14 Compare the results obtained in Problems 9.11, 9.12 and 9.13.

Problem 9.15 Explain why we have to determine the stress state condition $\sigma_2 \leqslant \frac{\sigma_1 + \alpha\sigma_3}{1+\alpha}$ or $\sigma_2 \geqslant \frac{\sigma_1 + \alpha\sigma_3}{1+\alpha}$ using the unified strength theory.

Problem 9.16 How do you choose between the two equations in the unified strength theory?

Problem 9.17 What is the result if you use the second equation of the unified strength theory for the stress state of $\sigma_2 \leqslant \frac{\sigma_1 + \alpha\sigma_3}{1+\alpha}$?

Problem 9.18 What is the result if you use the first equation of the unified strength theory for the stress state of $\sigma_2 \geqslant \frac{\sigma_1 + \alpha\sigma_3}{1+\alpha}$?

Problem 9.19 Complete discussions of the effects of pressure and temperature on yielding of thick-walled spherical shells given by Johnson and Mellor (1962), Mendelson (1968), and Chakrabarty (1987). The Tresca yield criterion was used in these studies. Can you obtain a more complete study on this subject using the unified yield criterion ($\alpha = 1$)?

Problem 9.20 A complete discussion of the effects of pressure and temperature on yielding of thick-walled spherical shells was given by Johnson and Mellor (1962), Mendelson (1968), and Chakrabarty (1987). The Tresca yield criterion was used in these studies. Can you obtain a more complete study on this subject using the unified strength theory ($\alpha \neq 1$)?

Problem 9.21 Complete discussions of the effects of pressure and temperature on yielding of thick-walled cylinder given by Johnson and Mellor (1962), Mendelson (1968), and Chakrabarty (1987). The Tresca yield criterion was used in these studies. Can you obtain a more complete study on this subject using the unified yield criterion ($\alpha = 1$)?

Problem 9.22 Complete discussions of the effects of pressure and temperature on yielding of thick-walled cylinder given by Johnson and Mellor (1962), Mendelson (1968), and Chakrabarty (1987). The Tresca yield criterion was used in these studies. Can you obtain a more complete study on this subject using the unified strength theory ($\alpha \neq 1$)?

Problem 9.23 The unified yield criterion can be used in many fields. Write an article regarding the application of the unified yield criterion.

Problem 9.24 The unified strength theory can be used in many fields. Write an article regarding the application of the unified strength theory.

References

Allen DN de, Sopwith DG (1951) The stresses and strains in a partially plastic thick tube under internal pressure and end-load. Proceedings of Royal Society, London, A205:69-83

American Society of Mechanical Engineers (1995) Cases of ASME Boiler and Pressure Vessel Code, Case N-47. ASME, New York

Chakrbarty J (1987) Theory of Plasticity, McGraw-Hill. New York

Chen CF, Xiao SJ, Yang Y (2006a) Unified limit solution of thick wall cylinder subject to external pressure considering strain softening. Journal of Hunan University (Natural Sciences), 33 (2):1-5 (in Chinese, English abstract)

Chen CF, Yang Y, Gong XN (2006b) The unified limit load solution of fiber-reinforced concrete cylinders considering the strain softening of material. Journal of Basic Science and Engineering, 14(4):496-505 (in Chinese, English abstract)

Crossland B, Bones JA (1958) Behavior of thick-walled steel cylinders subjected to internal pressure. Proc. Inst. Mech. Eng., 172:777-785

Derrington MG, Johnson W (1958) The onset of yield in a shell subject to internal pressure and a uniform heat flow. Appl. Sci. Res., Series A, 7:408-415

Drucker DC (1973) Plasticity theory, strength differential (SD) phenomenon and volume expansion in metals and plastics. Metal. Trans., 4:667-673

Feng JJ, Zhang JY, Zhang P, Han LF (2004a) Plastic limit load analysis of thick-walled tube based on twin-shear unified strength theory. Acta Mechanica Solida Sinica, 25(2):208-212 (in Chinese, English abstract)

Feng JJ, Zhang JY, Zhang P, Tan YQ, Han LF (2004b) Limit analysis of thick-walled tubes in complex stress state. Engineering Mechanics, 21(5): 188-192 (in Chinese, English abstract)

Heyman J (1971) Plastic design of frames: applications. Cambridge University Press, London

Hill R (1950) The mathematical theory of plasticity, Oxford University Press, London

Hodge PG Jr (1959) Plastic analysis of structures. McGraw-Hill, New York

Horne MR (1979) Plastic theory of structures. Pergamon Press, Oxford

Huang WB, Zeng GP (1989) Solving some plastic problems by using the Twin shear stress criterion. Acta Mechanica Sinica, 21(2):249-256 (in Chinese, English abstract)

Johnson W, Mellor B (1962) Engineering plasticity. Van Nostrand Reinhold, London

Kameniarzh JA (1996) Limit analysis of solids and structures. CRC Press, Boca Raton, FL

Koiter WT (1960) General theorems for elastic plastic solids. In: Sneddon JN, Hill R (eds.) Progress in Solid Mechanics. North-Holland Publ. Co., Amsterdam, 1:165-221

Li BF, Liu XQ, Ni XH (2007) Application of the twin shear strength theory in strength — calculation of gun barrels. Science Technology and Engineering, 7(13):3235-3237 (in Chinese)

Li YM (1998) Elasto-plastic limit analysis with a new yield criterion (twin-shear yield criterion). J. Mech. Strength, 10(3):70-74 (in Chinese)

Liu XQ, Ni XH, Yan S, et al. (1998) Application of the twin shear strength theory in strength-calculation of gun barrels. In: Yu MH, Fan SC (eds.) Strength Theory: Applications, Developments and Prospects for the 21st Century. Science Press, New York, Beijing, 1039-1042

Ma JH (2004) Unified limit analysis of thickwall cylinder of linear strengthened material. Mechanics and Practice, 26(1):49-51 (in Chinese)

Mendelson A (1968) Plasticity: theory and application. Macmillan, New York

Neal BG (1956) The plastic methods of structural analysis. Wiley, New York (2nd ed. 1963)

Ni XH, Liu XQ, Liu YT, Wang XS (1998) Calculation of stable loads of strength-differential thick cylinders and spheres by the twin shear strength theory. In: Yu MH, Fan SC (eds.) Strength Theory: Applications, Developments and Prospects for the 21st Century. Science Press, New York, Beijing, 1043-1046

Richmond O, Spitzig WA(1980) Pressure dependence and dilatancy of plastic flow. Theoretical and Applied Mechanics, 15th, ICTAM, 377-386

Save MA, Massonnet CE (1972) Plastic analysis and design of plates, shells and disks. North-Holland Publ. Co., Amsterdam

Tuba IS(1965) Elastic-plastic analysis for hollow spherical media under uniform radial loading, J. Franklin Inst., 280:343-355

Wang F (1998) Non-linear analysis of RC plate and shell using unified strength theory. Ph.D. Thesis, Nanyang Technological University, Singapore

Wang F, Fan SC (1998) Limit pressures of thick-walled tubes using different yield criteria. In: Yu MH, Fan SC (eds.) Strength Theory: Applications, Developments and Prospects for the 21st Century. Science Press, New York, Beijing, 1047-1052

Xia DS (1999) Technological research of tank structural material for launch vehicle. Missiles and Space Vehicles, (3):32-38

Xu BY, Liu XS (1995) Applied elasto-plasticity. Tsinghua University Press, Beijing (in Chinese)

Xu SQ, Yu MH (2004) Elasto brittle-plastic carrying capacity analysis for a thick walled cylinder under unified theory criterion. Chinese Quart. Mechanics, 25(4):490-495

Yu MH (1961a) General behavior of isotropic yield function. Res. Report of Xi'an Jiaotong University, Xi'an (in Chinese)

Yu MH (1961b) Plastic potential and flow rules associated singular yield criterion. Res. Report of Xi'an Jiaotong University, Xi'an (in Chinese)

Yu MH (1983) Twin shear stress yield criterion. Int. J. Mech. Sci., 25(1):71-74

Yu MH (1998) Twin shear theory and its application. Science Press, Beijing (in Chinese)

Yu MH (2002a) Advances in strength theories for materials under complex stress state in the 20th century. Applied Mechanics Reviews, ASME, 55(3):169-218

Yu MH (2002b) New system of strength theory . Xi'an Jiaotong University Press, Xi'an (in Chinese)

Yu MH (2004) The effects of failure criterion on structural analysis. In: Unified Strength Theory and Its Applications, Springer, Berlin, 241-264

Yu MH, He LN (1991) A new model and theory on yield and failure of materials under complex stress state. Proceedings, Mechanical Behavior of Materials-6, Pergamon Press, Oxford, 3:851-856

Yu MH, He LN (1983) Non-Schmid effect and twin shear stress criterion of plastic deformation in crystals and polycrystalline metals. Acta Metallurgica Sinica, 19(5):190-196 (in Chinese, English abstract)

Yu MH, He LN, Song LY (1985) Twin shear stress theory and its generalization. Scientia Sinica, English ed. Series A, 28(11):1174-1183

Zhang QW (1998) Launch vehicle status and development. Science & Technology Reviews, 1998(2):3-6, 16 (in Chinese)

Zhao JH, Zhang YQ, Li JC (1999) The solutions of some plastic problems in plane strain by using the unified strength theory and the unified slip field theory. Chinese J. Mechanical Engineering 35(6):61-66 (in Chinese)

Zhao JH, Zhang YQ, Liao HJ (2000) Unified limit solutions of thick wall cylinder and thick wall spherical shell with unified strength theory. Chinese J. Applied Mechanics, 17(1):157-161 (in Chinese)

Zhuang JH (1998) The analysis of limit inner pressure of thick-walled tube and sphere shell with different strength in tension and compression by theory of generalized twin-shear strength theory. Mechanics and Practices, 29(4):39-41 (in Chinese)

Zyczkowski M (1981) Combined loadings in the theory of plasticity. Polish Scientific Publishers, PWN and Nijhoff

10

Dynamic Plastic Response of Circular Plate

10.1 Introduction

The dynamic elastic response of plates and beams has been studied thoroughly. On the other hand, it is complicated to derive a dynamic plastic response for structures because of the plastic constitutive model. The first solution to the dynamic response of a rigid and simply supported circular plate was derived by Hopkins and Prager in 1954. Over the past fifty years very many research efforts have been concentrated on this subject by considering various boundary, loading conditions, and plastic flow assumptions (Florence, 1977; Jones and Oliveira, 1980; Jones, 1968; 1971; Stronge and Yu, 1993). Membrane mode solutions for impulsively loaded circular plates were derived by Symonds and Wierzbicki (1979). The dynamic response and failure of fully clamped circular plates under impulsive loading was studied by Shen and Jones (1993).

So far most of the studies are based on the Tresca yield criterion or the maximum stress criterion. Little attention has been paid to the influence of different yield criteria on the dynamic plastic behavior of structures. In fact strength behavior varies from material to material. A yield criterion with a single material parameter is not effective for materials with SD effect, such as stainless and refractory steels of high strength, high strength aluminum alloys, rocks and concrete materials (Casey and Sulivan, 1985; Chait, 1972; Drucker, 1973).

The unified yield criterion (UYC), a special case of the unified strength theory (UST) was adopted by Ma et al. (1998) to analyze the dynamic plastic responses of simply supported circular plates under moderate partial uniformly distributed impulsive loading. The UYC unifies the yield criteria with a single material parameter that is suitable for the non-SD materials.

Unified solutions of dynamic plastic response for the clamped circular plate and simply supported circular plate under moderate partial uniformly distributed impulsive loading applying the unified strength theory (UST) were reported by Wei et al. (2001) and Wang et al. (2005). The results based on UYC, the static and dynamic results from the Tresca, the Huber-von Mises yield criteria and the twin-shear yield criterion are special cases of the unified solutions. A series of new solutions are also given.

Analyses of the dynamic plastic response of structures are useful for the design of vehicles, engines and various structures sustaining an impact load. The unified solutions to dynamic plastic load-bearing capacities, moment fields and velocity fields of a simply supported circular plate are introduced in this chapter. The strength difference in tension and compression and the effect of different yield criteria are taken into account by application of the unified strength theory. Upper bound and lower bound plastic responses of the plate under moderate partial-uniformly distributed impulsive loading are derived. Static and kinetic admissibility of the dynamic plastic solutions are discussed. The unified solutions of static plastic load-bearing capacities, moment fields and velocity fields of a simply supported circular plate can also been derived from the dynamic solutions using the unified strength theory. The solutions are suitable for many materials, with or without strength difference in tension and compression, and they are able to reflect the effect of the intermediate principal stress.

The solutions based on the Tresca, the Huber-von Mises, the Mohr-Coulomb theory, the twin-shear strength theory, and the unified yield criterion are special cases of the present unified solutions. The influence of the unified strength theory parameter b and the tension-to-compression strength ratio α on the dynamic and static solutions are discussed. It is shown that the effects of unified strength theory parameter b and the strength ratio α on the load-bearing capacity of the plate in the dynamic plastic limit state are more significant than in the static plastic limit state.

10.2 Dynamic Equations and Boundary Conditions of Circular Plate

For a simply supported circular plate, as shown in Fig.10.1, with radius of a, the thickness of h and the mass per unit area of $\tilde{\mu}$, we denote R the radial distance from the plate center.

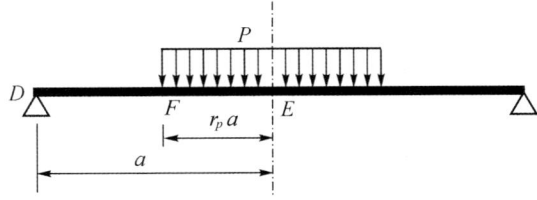

Fig. 10.1. Simply supported circular plate under rectangular impulsive loading

The generalized stresses are M_r, M_θ, Q_{rz}; M_0 is the ultimate bending moment; W is the bending deflection response. M_r and M_θ are the radial and circumferential bending moments respectively; Q_{rz} is the transverse shear force. The impulsive rectangular load is assumed to have a peak value of P with a duration of τ.

If the circular plate is made of an elastic-perfectly-plastic material, the ultimate bending moment M_0 can be derived as

$$M_0 = \frac{\sigma_t h^2}{2(1+\alpha)}. \tag{10.1}$$

For convenience of derivation, the following dimensionless variables are defined:

$$r = R/a, \ m_\theta = M_\theta/M_0, \ m_r = M_r/M_0, \ p = Pa^2/M_0,$$
$$q = Q_{rz}a/M_0, \ \mu = \tilde{\mu}a^3/M_0, \ w = W/a.$$

Assume that the plate is subjected to a rectangular impulsive loading $P(t) = P$ ($P_s \leqslant P \leqslant 2P_s$, where P_s is the static plastic limit load) with the duration of τ. The analysis can be divided into two phases, i.e., $0 \leqslant t \leqslant \tau$ and $\tau \leqslant t \leqslant T$, where T is the duration of the plate response. The motion equations of a circular plate using the dimensionless variables can be written as

$$\partial(rm_r)/\partial r - m_\theta - rq = 0, \tag{10.2}$$

$$\partial(rq)/\partial r + rp - \mu r\ddot{w} = 0, \tag{10.3}$$

where p is a dimensionless impulsive loading. Equations of geometry are

$$\dot{k}_r = -\partial^2 \dot{w}/\partial r^2, \tag{10.4a}$$

$$\dot{k}_\theta = -(\partial \dot{w}/\partial r)/r. \tag{10.4b}$$

In the first phase it satisfies

$$p = \begin{cases} p_0, & 0 \leqslant r \leqslant r_p, \\ 0, & r_p \leqslant r \leqslant 1, \end{cases} \tag{10.5}$$

where r_p is the dimensionless loading radius and $r_p = R_p/a$.

In the second phase the impact load vanishes,

$$p = 0. \tag{10.6}$$

Fig.10.2 shows the yield loci of UST expressed with the generalized stresses m_r and m_θ.

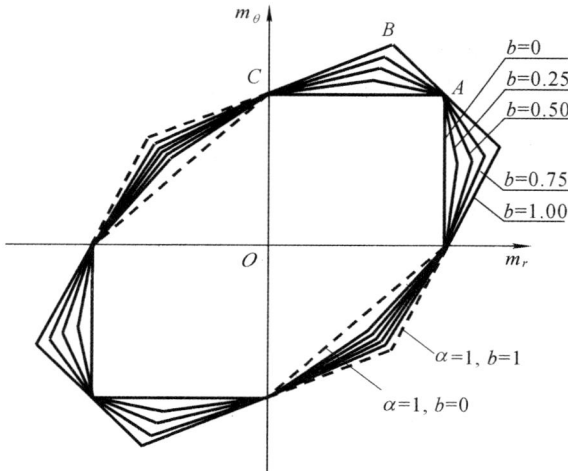

Fig. 10.2. Yield loci of UST in generalized stresses m_r-m_θ space (α=1)

In the plastic limit state the moments of the plate center ($r = 0$) satisfy $m_r = m_\theta = 1$ (point A in Fig.10.2), the simply supported edge ($r = 1$) satisfies $m_r = 0$ (point C in Fig. 10.2). Bending moments of all the points in the plate are located on segments AB and BC. The mathematical expression of the yield conditions with respect to AB and BC are

$$m_\theta = a_i m_r + b_i \ (i = 1, 2), \tag{10.7}$$

where $i = 1$ for AB, and $i = 2$ for BC. a_i and b_i are constants and there are $a_1 = -b$, $b_1 = 1 + b$, $a_2 = \alpha b/(1 + b)$, $b_2 = 1$.

According to the associated flow rule,

$$\dot{k}_r = \dot{\lambda}\partial F/\partial m_r, \quad \dot{k}_\theta = \dot{\lambda}\partial F/\partial m_\theta, \tag{10.8}$$

where F is the plastic potential function, being the same as the yield function, thus

$$\dot{k}_r = -a_i \dot{k}_\theta. \tag{10.9}$$

Eqs.(10.2), (10.3) and (10.7) give the moment governing equation,

$$\partial^2 (rm_r)/\partial r^2 - a_i \partial m_r/\partial r = -rp + \mu r\ddot{w}. \tag{10.10}$$

Governing equation of the velocity is derived by substituting Eqs.(10.4a) and (10.14b) into Eq.(10.9),

$$\partial^2 \dot{w}/\partial r^2 + a_i \partial \dot{w}/(r\partial r) = 0. \tag{10.11}$$

10.2.1 First Phase of Motion ($0 \leqslant T \leqslant \tau$)

In the first phase of motion the plate is subjected to a constant loading p_0 in the inner region $0 \leqslant r \leqslant t_p$. Integrating Eq.(10.11) twice with respect to r gives the velocity equations,

$$\dot{w} = \dot{w}_1 \begin{cases} c_{11}r^{1-a_1} + c_{21} & 0 \leqslant r \leqslant r_1 \\ c_{12}r^{1-a_2} + c_{22} & r_1 \leqslant r \leqslant 1, \end{cases} \tag{10.12}$$

where c_{1i} and c_{2i} ($i=1, 2$) are integration constants, r_1 is the dividing radius where the internal moments m_r and m_θ correspond to point B in Fig. 10.2, $\dot{w}_1$ is the velocity of the plate center which is a function of time t.

The continuity and boundary conditions of the velocity variable can be expressed as: (1) $\dot{w}(r = 0) = \dot{w}_1$; (2) $\mathrm{d}\dot{w}/\mathrm{d}r(r = r_1)$ and $\mathrm{d}\dot{w}$ $(r = r_1)$ are continuous; (3) $\dot{w}(r = 1) = 0$. With reference to these conditions, the integration constants can be derived,

$$c_{11} = -\frac{(1+b-\alpha b)r_1^{-\frac{b^2+b+\alpha b}{1+b}}}{(1+b)^2 - b(1+b+\alpha)r_1^{\frac{1+b-\alpha b}{1+b}}}, \quad c_{21} = 1,$$

$$c_{12} = -c_{22} = -\frac{(1+b)^2}{(1+b)^2 - b(1+b+\alpha)r_1^{\frac{1+b-\alpha b}{1+b}}}.$$

There are two regions of the moment response ($r_p \leqslant r_1$ and $r_p \geqslant r_1$) during the first phase of motion as shown in Fig.10.3.

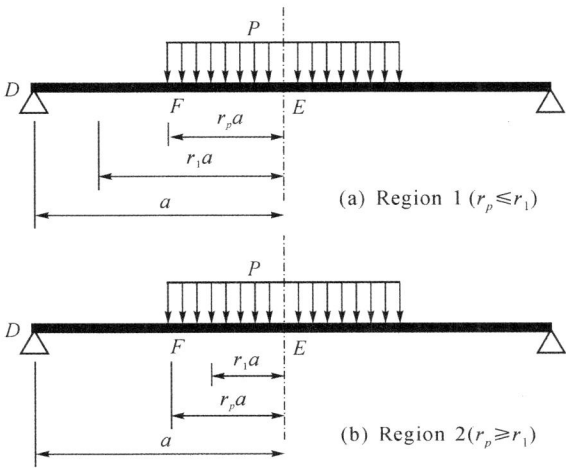

Fig. 10.3. Two regions of the moment response

The continuity and boundary conditions for the moment responses are: (4) $m_r(r=0)=1$; (5) $m_r(r=r_1)$ is continuous and equal to $(1+b)(1+b+\alpha)$; (6) $\partial m_r/\partial r(r=r_1)$ is continuous; (7) m_r $(r=r_p)$ is continuous; (8) $\partial m_r/\partial r(r=r_p)$ is continuous; (9) m_r $(r=1)=0$.

Region 1 $(r_p \leqslant r_1)$

As shown in Fig.10.3(a), the plate is subjected to a rectangular impulsive loading p_0 within the range of r_p. The moment response field can be derived as

$$
m_{r1} = \frac{-p_0 + \mu\ddot{w}_1 c_{21}}{2(3-a_1)} r^2 + \frac{\mu\ddot{w}_1 c_{11}}{(3-a_1)(4-2a_1)} r^{3-a_1}
$$
$$
+ c_{31} r^{-1+a_1} + c_{41} \quad (0 \leqslant r \leqslant r_p),
$$
(10.13a)

$$
m_{r2} = \frac{\mu\ddot{w}_1 c_{21}}{2(3-a_1)} r^2 + \frac{\mu\ddot{w}_1 c_{11}}{(3-a_1)(4-2a_1)} r^{3-a_1}
$$
$$
+ c_{32} r^{-1+a_1} + c_{42} \quad (r_p \leqslant r \leqslant r_1),
$$
(10.13b)

$$
m_{r3} = \frac{\mu\ddot{w}_1 c_{22}}{2(3-a_2)} r^2 + \frac{\mu\ddot{w}_1 c_{12}}{(3-a_2)(4-2a_2)} r^{3-a_2}
$$
$$
+ c_{33} r^{-1+a_2} + c_{43} \quad (r_1 \leqslant r \leqslant 1),
$$
(10.13c)

where c_{3i} and c_{4i} $(i=1,2,3)$ are integration constants and they are derived from the continuity and boundary conditions,

$$
c_{31} = 0, \; c_{41} = 1, \; c_{32} = \frac{p_0 r_p^{3-a_1}}{(1-a_1)(3-a_1)}, \; c_{42} = 1 - \frac{p_0 r_p^2}{2(1-a_1)},
$$

$$
c_{33} = \left(\frac{\mu\ddot{w}_1 c_{21}}{(3-a_1)(a_2-1)} - \frac{\mu\ddot{w}_1 c_{22}}{(3-a_2)(a_2-1)} \right) r_1^{3-a_2} + \frac{\mu\ddot{w}_1 c_{11} r_1^{4-a_1-a_2}}{(4-2a_1)(a_2-1)}
$$
$$
- \frac{\mu\ddot{w}_1 c_{12} r_1^{4-2a_2}}{(4-2a_2)(a_2-1)} + \frac{a_1-1}{a_2-1} c_{32} r_1^{a_1-a_2},
$$

$$
\mu\ddot{w}_1 = \frac{(3-a_1)(4-2a_1)}{(2-a_1)r_1^2 + c_{11}r_1^{3-a_1}} \left[-\frac{\alpha}{1+b+\alpha} + \frac{p_0 r_p^2}{2(1-a_1)} - \frac{p_0 r_p^{3-a_1} r_1^{-1+a_1}}{(1-a_1)(3-a_1)} \right],
$$
(10.14)

where r_1 satisfies

$$\frac{\mu\ddot{w}_1 c_{22}}{2(3-a_2)} + \frac{\mu\ddot{w}_1 c_{12}}{(3-a_2)(4-2a_2)} + c_{33} + c_{43} = 0. \tag{10.15}$$

For a given rectangular impulsive loading p_0 and loading radius r_p, r_1 can be calculated from Eqs.(10.15) and (10.14) in the range of $(r_p, 1)$.

Region 2$(r_p \geqslant r_1)$

With reference to Fig.10.3(b), the moment response field is expressed as

$$m_{r1} = \frac{-p_0 + \mu\ddot{w}_1 c_{21}}{2(3-a_1)}r^2 + \frac{\mu\ddot{w}_1 c_{11}}{(3-a_1)(4-2a_1)}r^{3-a_1}$$
$$+ c_{31}r^{-1+a_1} + c_{41} \quad (0 \leqslant r \leqslant r_1), \tag{10.16a}$$

$$m_{r2} = \frac{-p_0 + \mu\ddot{w}_1 c_{22}}{2(3-a_2)}r^2 + \frac{\mu\ddot{w}_1 c_{12}}{(3-a_2)(4-2a_2)}r^{3-a_2}$$
$$+ c_{32}r^{-1+a_2} + c_{42} \quad (r_1 \leqslant r \leqslant r_p), \tag{10.16b}$$

$$m_{r3} = \frac{\mu\ddot{w}_1 c_{22}}{2(3-a_2)}r^2 + \frac{\mu\ddot{w}_1 c_{12}}{(3-a_2)(4-2a_2)}r^{3-a_2}$$
$$+ c_{33}r^{-1+a_2} + c_{43} \quad (r_p \leqslant r \leqslant 1), \tag{10.16c}$$

where c_{3i} and c_{4i} $(i = 1, 2, 3)$ are integration constants. According to the continuity and boundary conditions, the integration constants are derived as

$$c_{31} = 0, \; c_{41} = 1,$$

$$c_{32} = \left(\frac{-p_0 + \mu\ddot{w}_1 c_{21}}{(3-a_1)(a_2-1)} - \frac{-p_0 + \mu\ddot{w}_1 c_{22}}{(3-a_2)(a_2-1)}\right)r_1^{3-a_2}$$
$$+ \frac{\mu\ddot{w}_1 c_{11}r_1^{4-a_1-a_2}}{(4-2a_1)(a_2-1)} - \frac{\mu\ddot{w}_1 c_{12}r_1^{4-2a_2}}{(4-2a_2)(a_2-1)},$$

$$c_{42} = \frac{1+b}{1+b+\alpha} - c_{32}r_1^{-1+a_2} - \frac{-p_0 + \mu\ddot{w}_1 c_{22}}{2(3-a_2)}r_1^2 - \frac{\mu\ddot{w}_1 c_{12}}{(3-a_2)(4-2a_2)}r_1^{3-a_2},$$

$$c_{33} = \frac{p_0 r_p^{3-a_2}}{(1-a_2)(3-a_2)} + c_{32},$$

$$c_{43} = c_{42} + c_{32}r_p^{-1+a_2} - \frac{p_0 r_p^2}{2(3-a_2)} - c_{33}r_p^{-1-a_2},$$

$$\mu\ddot{w}_1 = \frac{(2+b)(1+b+\alpha)p_0 r_1^2 - 2\alpha(3+b)(2+b)}{(2+b)(1+b+\alpha)r_1^2 + (1+b+\alpha)c_{11}r_1^{3+b}}, \tag{10.17}$$

where r_1 satisfies

$$\frac{\mu \ddot{w}_1 c_{22}}{2(3 - a_2)} + \frac{\mu \ddot{w}_1 c_{12}}{(3 - a_2)(4 - 2a_2)} + c_{33} + c_{43} = 0. \qquad (10.18)$$

For a given rectangular impulsive loading p_0 and loading radius r_p, r_1 can be calculated from Eqs.(10.17) and (10.18) in the range of $(0, r_p)$. The moment fields are then determined.

For a impulsive loading p_0 satisfying the static and kinetic admissibility, there is a critical state between the two cases. The corresponding radius is the critical dividing radius r_{1p}, which satisfies $r_{1p} = r_p = r_1$. Compared with the actual loading radius r_p, if $r_p \leqslant r_{1p}$, the moment response fields are calculated using the equation for Region 1, and if not, then using those for Region 2.

Eqs.(10.15) and (10.17) show that $\mu \ddot{w}_1$ is a constant during the first phase of motion. The displacement and velocity at the end of the phase for both regions are

$$w = \ddot{w}_1 \tau^2 (c_{1i} r^{1-a_i} + c_{2i})/2 \ (i = 1, 2), \qquad (10.19)$$

$$\dot{w} = \ddot{w}_1 \tau (c_{1i} r^{1-a_i} + c_{2i}) \ (i = 1, 2). \qquad (10.20)$$

Eq.(10.20) gives kinetic energy absorbed during the first phase,

$$k_e = \int_0^{2\pi} \int_0^1 \frac{1}{2} \mu M_0 a \dot{w}^2 r d\theta dr = \mu M_0 a \ddot{w}_1^2 \tau^2 \left(\int_0^{r_1} (c_{11} r^{1-a_1} + c_{21})^2 r dr \right.$$

$$\left. + \int_{r_1}^1 (c_{12} r^{1-a_2} + c_{22})^2 r dr \right) = \mu M_0 a \ddot{w}_1^2 \tau^2 K, \qquad (10.21)$$

where

$$K = \frac{c_{11}^2}{4 - 2a_1} r_1^{4-2a_1} + \frac{2c_{11}c_{21}}{3 - a_1} r_1^{3-a_1} + \frac{c_{21}^2}{2} r_1^2 + \frac{c_{12}^2}{4 - 2a_2} (1 - r_1^{4-2a_2})$$

$$+ \frac{2c_{12}c_{22}}{3 - a_2} (1 - r_1^{3-a_2}) + \frac{c_{22}^2}{2} (1 - r_1^2),$$

and k_e is the kinetic energy consumed during the second phase.

10.2.2 Second Phase of Motion $(\tau \leqslant t \leqslant T)$

The circular plate is unloaded during this phase of motion and therefore $p = 0$ on the whole plate. The plastic deformation continues to spread on the plate. The velocity profile during this phase has the same form as that of the first phase. The velocity response can be obtained by replacing r_1 in Eq.(10.12) with r_2. Similar to the first phase of motion, the moment response field can be derived as

$$m_{r1} = \frac{\mu \ddot{w}_2 c_{21}}{2(3-a_1)} r^2 + \frac{\mu \ddot{w}_2 c_{11}}{(3-a_1)(4-2a_1)} r^{3-a_1}$$
$$+ c_{31} r^{-1+a_1} + c_{41} \quad (0 \leqslant r \leqslant r_2),$$
(10.22a)

$$m_{r2} = \frac{\mu \ddot{w}_2 c_{22}}{2(3-a_2)} r^2 + \frac{\mu \ddot{w}_2 c_{12}}{(3-a_2)(4-2a_2)} r^{3-a_2}$$
$$+ c_{32} r^{-1+a_2} + c_{42} \quad (r_2 \leqslant r \leqslant 1).$$
(10.22b)

The boundary and continuity conditions can be put as $m_r(r=0) = 1$; $m_r \ (r=r_2)$ is continuous and equals $(1+b)(1+b+\alpha)$; $m_r \ (r=1) = 0$. With reference to these conditions, the integration constants are calculated as

$$c_{31} = 0, \ c_{41} = 1,$$

$$c_{32} = \left(\frac{\mu \ddot{w}_2 c_{21}}{(3-a_1)(a_2-1)} - \frac{\mu \ddot{w}_2 c_{22}}{(3-a_2)(a_2-1)} \right) r_2^{3-a_2}$$
$$+ \frac{\mu \ddot{w}_2 c_{11} r_2^{4-a_1-a_2}}{(4-2a_1)(a_2-1)} - \frac{\mu \ddot{w}_2 c_{12} r_2^{4-2a_2}}{(4-2a_2)(a_2-1)},$$

$$c_{42} = \frac{1+b}{1+b+\alpha} - c_{32} r_2^{-1+a_2} - \frac{\mu \ddot{w}_2 c_{22}}{2(3-a_2)} r_2^2 - \frac{\mu \ddot{w}_2 c_{12}}{(3-a_2)(4-2a_2)} r_2^{3-a_2}$$

$$\mu \ddot{w}_2 = \frac{-2\alpha(3+b)(2+b)}{(2+b)(1+b+\alpha)r_2^2 + (1+b+\alpha)c_{11} r_2^{3+b}},$$
(10.23)

where r_2 satisfies

$$\frac{\mu \ddot{w}_2 c_{22}}{2(3-a_2)} + \frac{\mu \ddot{w}_2 c_{12}}{(3-a_2)(4-2a_2)} + c_{32} + c_{42} = 0.$$

Eq.(10.23) implies that $\mu \ddot{w}_2$ remains constant during the second phase of motion.

Since μ and $\ddot{w}_2$ are constants, the velocity and displacement responses at the plate center with the assumption of $\ddot{w}_2 = c_1$ can be written as

$$\dot{w}_2 = \int_\tau^t c_1 dt = c_1(t-\tau) + c_2; \ w_2 = \int_\tau^t \dot{w}_2 dt = \frac{1}{2} c_1(t-\tau)^2 + c_2(t-\tau) + c_3.$$
(10.24)

According to the velocity and displacement conditions at $t = \tau$, the integration constants are obtained as $c_2 = \dot{w}_1$, $c_3 = w_1$. Therefore,

$$w_2 = \frac{1}{2}\ddot{w}_2(t - \tau)^2 + \dot{w}_1(t - \tau) + w_1. \tag{10.25}$$

Since $\ddot{w}_1$ is also a constant with reference to Eq.(10.17), the velocity and displacement responses at the plate center during the first phase with the assumption of $\ddot{w}_1 = c_4$ are $\dot{w}_1 = \int_0^t c_4 dt = c_4 t + c_5$, $w_1 = \frac{1}{2}c_4 t^2 + c_5 t + c_6$, respectively.

According to the initial conditions, i.e., $\dot{w}_1 = 0$ and $w_1 = 0$ when $t = 0$, the integration constants are obtained as $c_5 = c_6 = 0$. There are $\dot{w}_1 = \ddot{w}_1 t$, $w_1 = \frac{1}{2}\ddot{w}_1 t^2$. The displacement response at the plate center is derived by substituting $\dot{w}_1 = \ddot{w}_1 \tau$, $w_1 = \frac{1}{2}\ddot{w}_1 \tau^2$ into Eq.(10.25) at $t = \tau$,

$$w_2 = \frac{1}{2}\ddot{w}_2(t - \tau)^2 + \frac{1}{2}\ddot{w}_1\tau(2t - \tau), \tag{10.26}$$

and the velocity response is

$$\dot{w}_2 = \ddot{w}_2(t - \tau) + \ddot{w}_1\tau, \tag{10.27}$$

where $\ddot{w}_1$ and $\ddot{w}_2$ are constants determined by Eqs.(10.17) and (10.23) respectively.

Denoting the duration of response T, and $T = \eta\tau$, there is $\dot{w}_2 = 0$ at $t = T$. Eq.(10.27) gives

$$\eta = 1 - \frac{\ddot{w}_1}{\ddot{w}_2}, \tag{10.28}$$

where η is the response time factor. The transverse displacement is obtained from Eqs.(10.25) and (10.12) by replacing r_1 with r_2,

$$w_f = -\frac{1}{2}\ddot{w}_2\tau^2\eta(\eta - 1)^2(c_{1i}r^{1-a_i} + c_{2i}) \quad (i = 1, 2). \tag{10.29}$$

10.3 Static and Kinetic Admissibility

If the impulsive time is sufficiently long, and the acceleration $\ddot{w}_1$ at the plate center during the first phase is equal to 0, the dynamic solutions degenerate to static plastic limit solutions. The static limit loadings are derived from Eqs.(10.14) and (10.17) with reference to $\ddot{w}_1 = 0$,

$$p_s = \frac{2\alpha(1+b)(3+b)}{(1+b+\alpha)\left[(3+b) - 2\left(\frac{r_p}{r_s}\right)^{1+b}\right]r_p^2} \tag{10.30}$$

and

$$p_s = \frac{2\alpha(3+b)}{(1+b+\alpha)\,r_s^2}, \tag{10.31}$$

where p_s is the dimensionless static limit load, r_s is the dividing radius at which the moments m_r and m_θ correspond to point B in Fig.10.2. r_s is calculated from Eq.(10.15) or (10.18) with reference to $\ddot{w}_1 = 0$ for a static case. There is a critical state between the two cases with the dividing radius r_{sp} satisfying $r_{sp} = r_p = r_s$, if $r_p \leqslant r_{sp}$, r_s and p_s are calculated from Eqs.(10.15) and (10.30); otherwise, from Eqs.(10.18) and (10.31). The critical radius can be derived from Eq.(10.30) or (10.31) by substituting $\ddot{w}_1 = 0$ and $r_{sp} = r_p = r_s$ into Eq.(10.15) or (10.18),

$$r_{sp} = \left(\frac{2\alpha}{(2+b)\alpha - (1+b)}\right)^{\frac{1+b}{\alpha b - 1 - b}}. \tag{10.32}$$

A circular plate can sustain a short duration pressure that well exceeds the static plastic limit loading, but it is necessary to demonstrate that the foregoing theoretical solutions do not violate the yield condition and are therefore statically admissible. According to the yield conditions shown in Fig.10.2, the radial moment field is a decreasing function from point A to point B, and then to C. In order to avoid a yield violation at $r = 0$ and $r = 1$ in the bending moment distribution, at $r = 0$ there should be,

$$\partial(m_r)/\partial r \leqslant 0, \ \partial^2(m_r)/\partial r^2 \leqslant 0, \tag{10.33}$$

and at $r = 1$,

$$\partial(m_r)/\partial r \leqslant 0, \partial^2(m_r)/\partial r^2 \geqslant 0. \tag{10.34}$$

For the first phase of motion, Eq.(10.33) is satisfied automatically at the plate center. Thus we need to check only if $\partial^2 m_r/\partial r^2 \leqslant 0$ at $r = 0$. Differentiating Eq.(10.13a) or Eq.(10.16a) twice with respect to r, the static admissible dynamic loading at $r = 0$ should satisfy

$$-p_0 + \mu\ddot{w}_1 \leqslant 0, \tag{10.35}$$

which gives the maximum dynamic impulsive loading p_{d1} with reference to Eqs.(10.14) and (10.15) or Eqs.(10.17) and (10.18).

Since $\partial^2 m_r/\partial r^2 \geqslant 0$ at $r = 1$ is satisfied automatically, the following equation is derived according to Eq.(10.13c) or Eq.(10.16c) with reference to $\partial m_r/\partial r$ at $r = 1$

$$\frac{\mu \ddot{w}_1 c_{22}}{(3 - a_2)} + \frac{\mu \ddot{w}_1 c_{12}}{(4 - 2a_2)} + (-1 + a_2)c_{33} = 0. \tag{10.36}$$

The maximum dynamic impulsive loading p_{d2} can be calculated from Eqs.(10.14) and (10.15) or Eqs.(10.17) and (10.18) with reference to Eq.(10.36).

During the second phase of motion m_r is independent of p_0. Thus the static admissible loading should satisfy the inequality of

$$p_s \leqslant p_0 \leqslant p_d = \min(p_{d1}, p_{d2}), \tag{10.37}$$

where p_d is the maximum statically admissible impulsive loading.

Given $\beta = p_0/p_s$, $\beta_d = p_d/p_s$, Eq.(10.37) can be rewritten as

$$1 \leqslant \beta \leqslant \beta_d, \tag{10.38}$$

where β is defined as the statically admissible loading factor and β_d is the maximum statically admissible loading factor.

The response acceleration during both the first phase of motion and the second one is independent of time. The displacement, velocity and acceleration are continuous during the entire response time. Thus the dynamic solutions are kinetically admissible. The aforementioned theoretical analysis is statically admissible under the condition of $1 \leqslant \beta \leqslant \beta_d$. Therefore the above solutions are the exact solutions throughout the entire response for a rigid-perfectly-plastic circular plate.

10.4 Unified Solution of Dynamic Plastic Response of Circular Plate

Considering the special case of uniformly distributed impulsive loading for the plate, i.e., $r_p = 1$, the static solutions can be derived from the dynamic solutions. The relationship of the static plastic limit loading p_s to the unified strength theory parameters b and α is shown in Fig.10.4.

For a given value of α, the limit loading reaches the minimum value when $b = 0$ and maximum value when $b = 1$. The relationship of static plastic limit moment field to parameters α and b is shown in Fig.10.5.

The relationship of dynamic plastic limit loading to parameters α and b is shown in Fig.10.6.

Similarly, the dynamic plastic limit loading reaches the minimum value when $b = 0$ and maximum value when $b = 1$ for a given value of α. For a

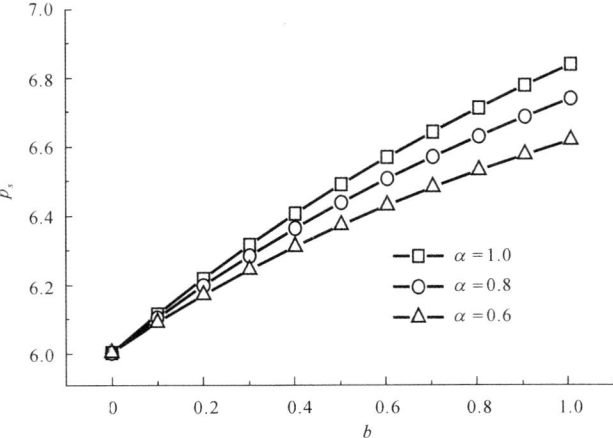

Fig. 10.4. Relations of static plastic limit loading p_s to unified strength theory parameters b and α

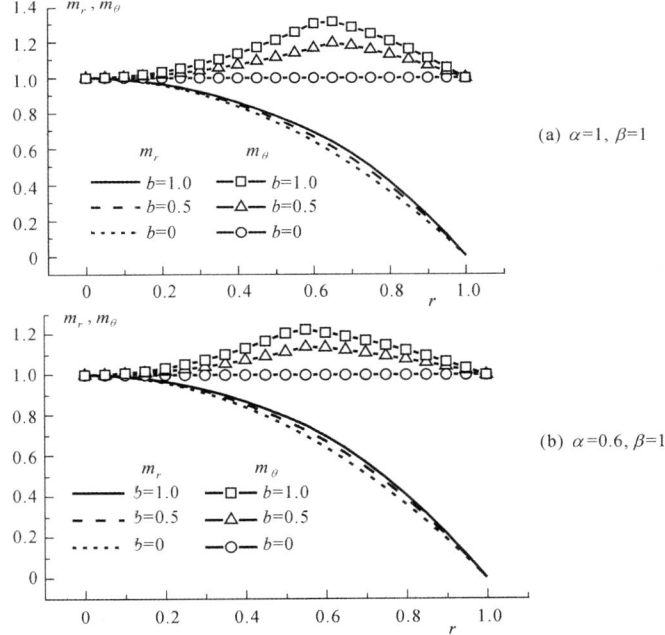

Fig. 10.5. Moment fields of static plastic limit

given value of b the limit loading increases with the increase of α (except for $b = 0$). Figs.10.4 and 10.6 demonstrate that the effect of the unified strength theory parameter b on the dynamic limit loading (70% increase from $b = 0$

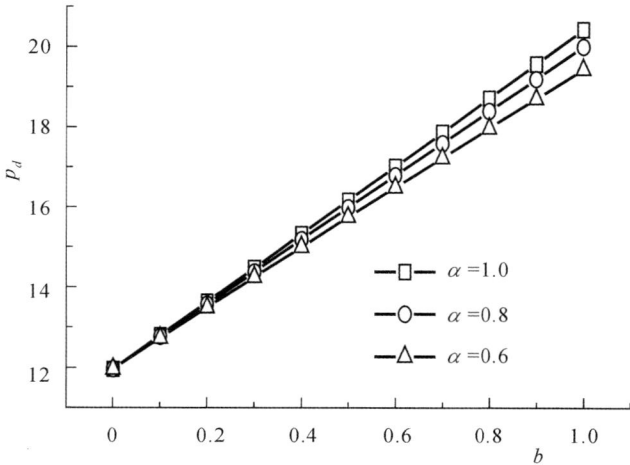

Fig. 10.6. Relationship of dynamic plastic limit loading p_d to unified strength theory parameters b and α

to $b = 1$ when $\alpha=1$) is more significant than that on the static limit loading (35% increase from $b = 0$ to $b = 1$ when $\alpha=1$).

Figs.10.7 and 10.8 show the variation of dynamic plastic limit moment fields versus parameters α and b during the first and second phases of motion respectively.

The plastic moment field during the first phase of motion is a function of the impulsive loading. On the other hand, it is independent of the impulsive loading during the second phase of motion. It is shown that the effect of the variation of unified strength theory parameter b on the circumferential bending moment is higher than that on the radial bending moment. The parameter α has a significant influence on both circumferential and radial bending moments.

Figs.10.9 and 10.10 show the displacement and velocity responses at the plate center when subjected to dynamic limit loading with varying α and b respectively. It is seen that α affects both displacement and velocity responses. For a given value of α, the displacement and velocity responses at the plate center increase with increasing unified strength theory parameter b.

Fig.10.11 shows the permanently deformed displacements of the circular plate subjected to dynamic limit loading. From Fig.10.11, both α and b have apparent influences on the plastic deformed displacement. For a given value of α, the plastic displacement increases with increasing unified strength theory parameter b and for a given value of b, the permanent displacement decreases with decreasing α.

Fig.10.12 shows the permanent displacements versus the parameter b for different α with the dynamic load $p_0 = 12$. It is seen that the variation of

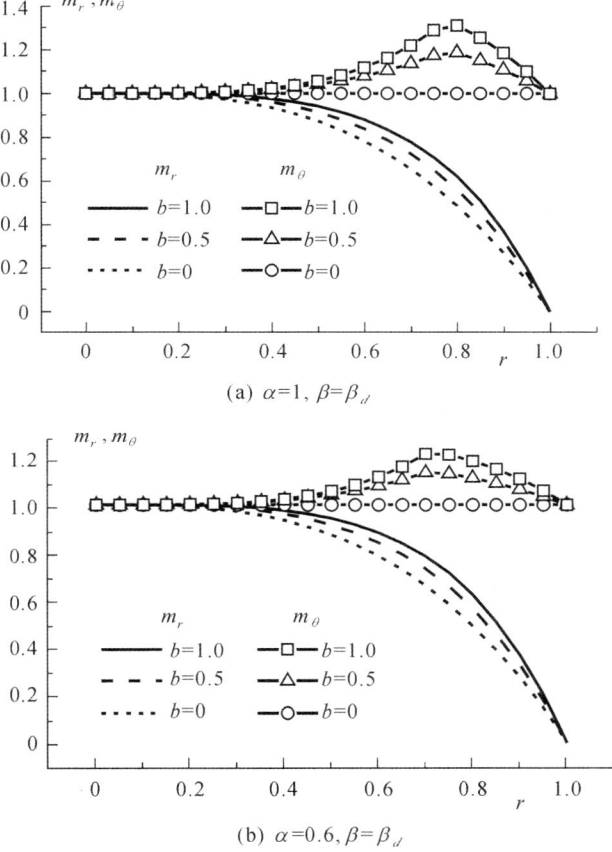

Fig. 10.7. Dynamic plastic limit moment fields during the first phase of motion

plastic deformed displacement is independent of α. The maximum permanent displacement at the plate center is obtained with $b = 0$ and the minimum one with $b = 1$.

The results of the dynamic limit loading, velocity, displacement time histories and moment responses with $\alpha=1$ are exactly the same as those derived by Ma et al. (1998) for non-SD materials.

When $\alpha=1$ the UST degenerates to UYC. The UYC becomes the Tresca yield criterion when $b = 0$ and the twin-shear yield criterion (Yu, 1961; 1983) when $b = 1$. The Huber-von Mises yield criterion can be approximated with $b = 0.5$. As seen from Figs.10.4, 10.6 and 10.12, the maximum displacement response and minimum static and dynamic plastic limit loadings with respect to the Tresca yield criterion are the lower bound responses. The minimum displacement response and the maximum static and dynamic plastic limit

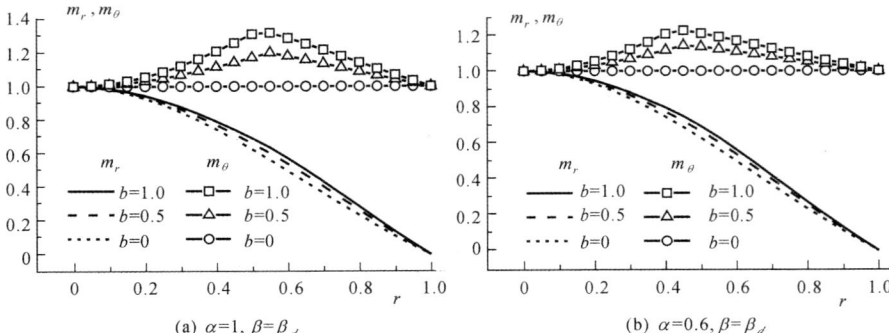

Fig. 10.8. Dynamic plastic limit moment fields during the second phase of motion

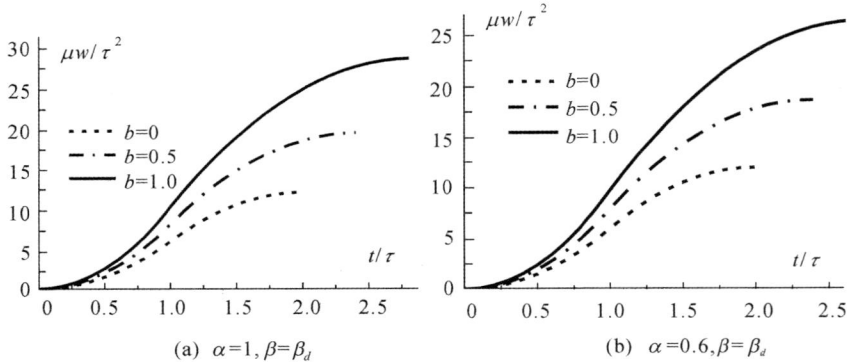

Fig. 10.9. Displacement responses at the center of plate

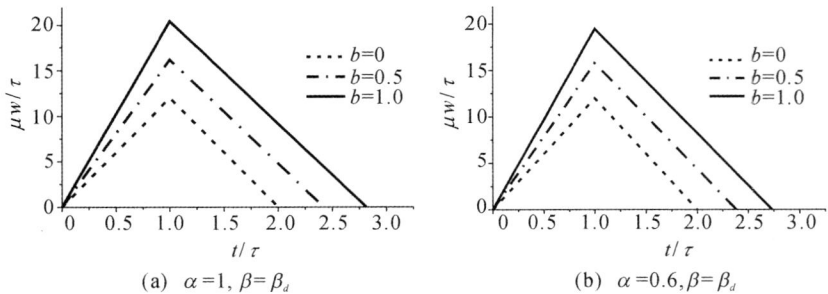

Fig. 10.10. Velocity responses at the center of plate

loadings with respect to the twin shear stress criterion are the upper bound responses.

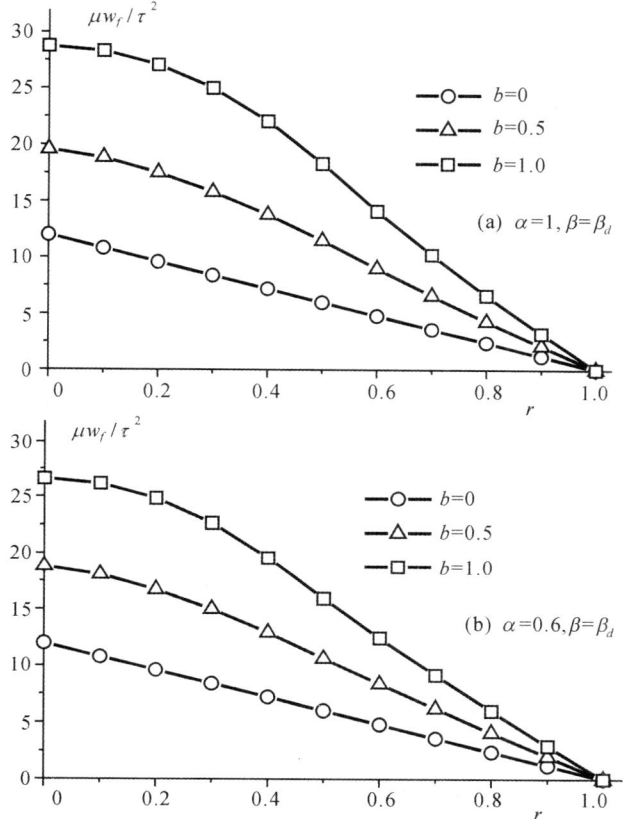

Fig. 10.11. Permanently deformed displacements of plate

10.5 Special Cases of the Unified Solutions

The solutions with respect to the Tresca criterion and the twin shear stress criterion (Yu, 1961; 1983) can be obtained from the current unified solution with a specific value of unified strength theory parameter b. Solutions with the Huber-von Mises criterion fall between those from the Tresca criterion and the twin shear stress criterion and can be approximated by the solution by using UYC with $b = 0.5$. For a given impulsive loading radius r_p and the pulse force p_0 satisfying static admissibility, r_1 and r_2 can be calculated from Eq.(10.15) or Eq.(10.18) for the first phase of motion, and from Eq.(10.24) for the second phase of motion. Substituting r_1 and r_2 into the equations for the integration constants and response quantities, moment, velocity, then the displacement response fields of the plate are derived for the two motion phases.

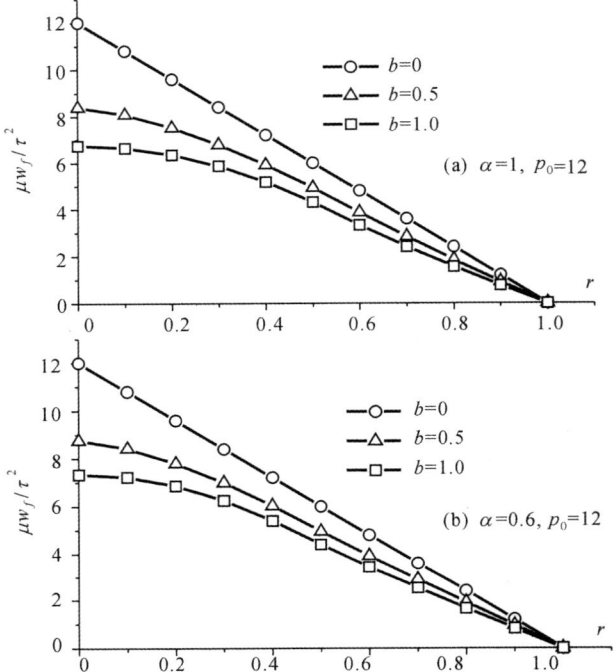

Fig. 10.12. Permanently deformed displacements of plate

Figs.10.13 and 10.14 show the moment fields for the two phases of motion for a plate subjected to a uniformly impulsive loading $(r_p = 1)$ with respective load factors $\alpha = 1$ and $\alpha = \alpha_d$ regarding three different yield criteria.

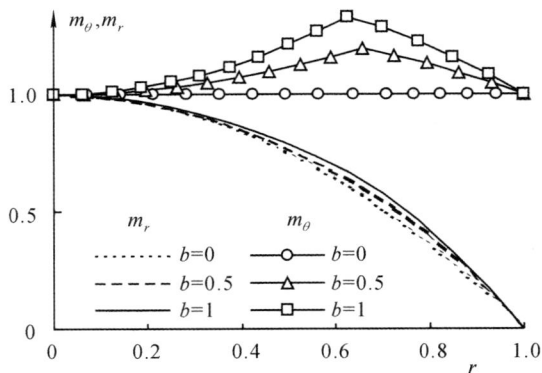

Fig. 10.13. Moment fields during the first phase of motion $(\alpha = 1)$

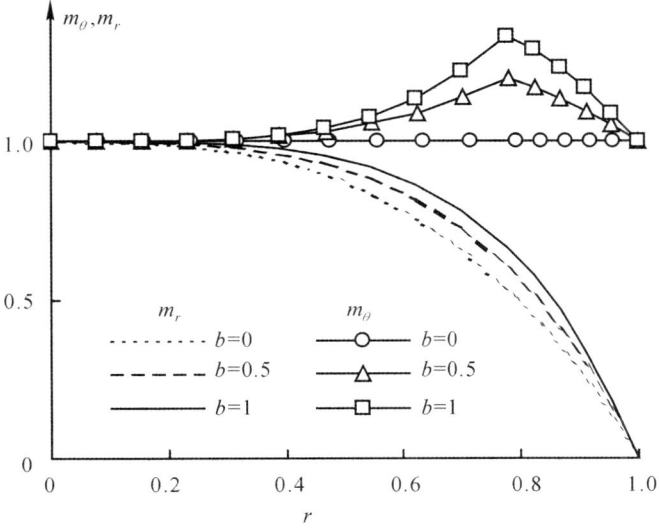

Fig. 10.14. Moment fields during the first phase of motion ($\alpha = \alpha_d$)

Fig.10.15 shows the moment fields during the second phase of motion which is independent of the loading radius r_p and loading factor α.

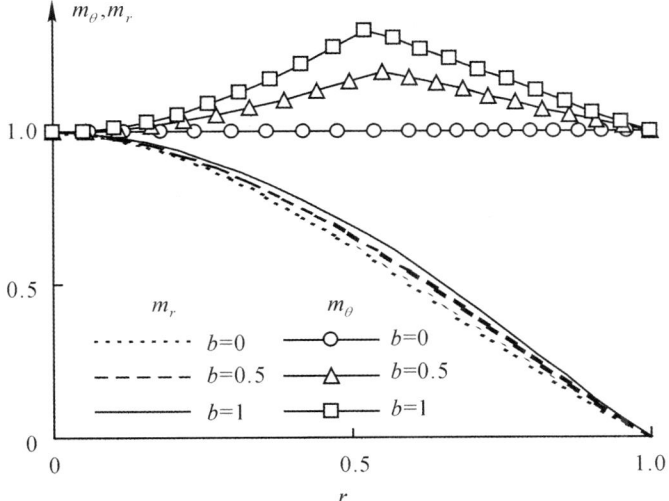

Fig. 10.15. Moment fields during the second phase of motion

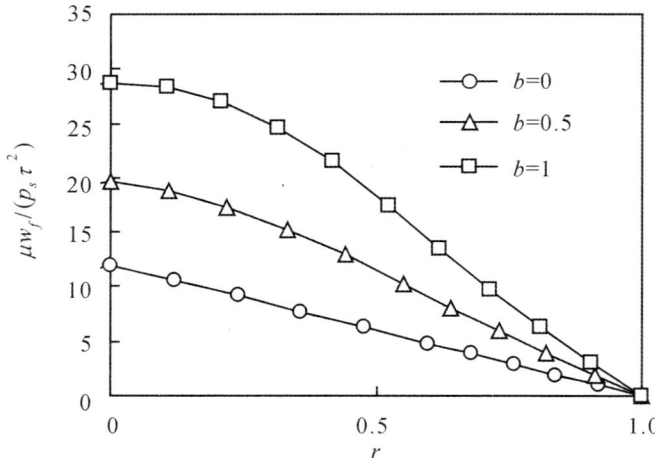

Fig. 10.16. Permanently deformed transverse displacements ($r_p = 1$, $\alpha = \alpha_d$)

Figs.10.16 to 10.18 illustrate the permanently deformed transverse displacements, displacement and velocity responses of the plate center with respect to three criteria when $r_p = 1$, $\alpha = \alpha_d$.

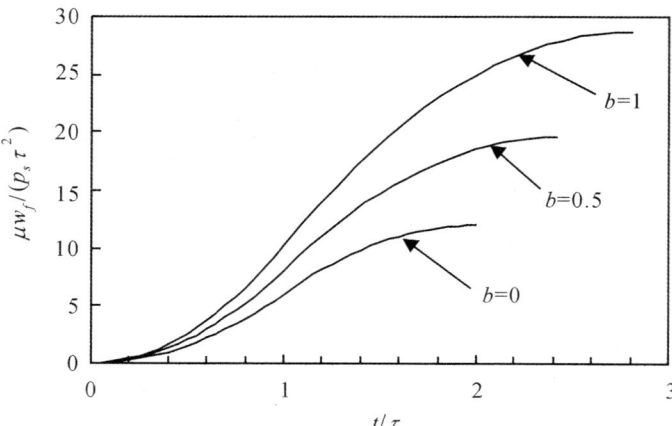

Fig. 10.17. Displacement responses at the plate center ($r_p = 1$, $\alpha = \alpha_d$)

Figs.10.19 to 10.21 show the moment and velocity profiles for the first phase of motion and the permanently deformed transverse displacements for a plate subjected to a concentrated impulsive load with $r_p = 0.01$. The moment fields are singular at the plate center because the shear force at the center is infinite. The statically admissible impulsive load p_d is close to the

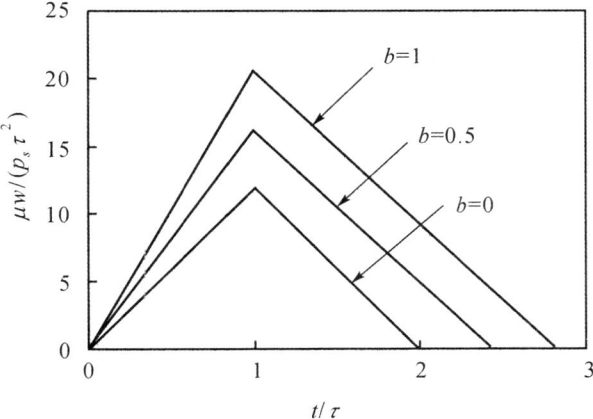

Fig. 10.18. Velocity responses at the plate center ($r_p = 1$, $\alpha = \alpha_d$)

static plastic limit load p_s. Thus we need to apply the assumption shown in Fig.10.13 (b) for a fully rigid-plastic circular plate subjected to concentrated load. Although the moment and velocity fields are quite different with different yield criteria, the static plastic limit load $P_T = \pi r_p^2 P_a^2$ for the concentrated load is equal to $P_T = 2\pi M_0$ regardless of the different parameter b (Ma et al., 1998).

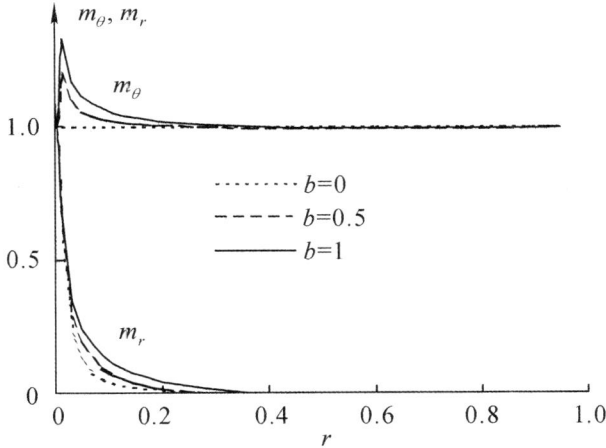

Fig. 10.19. Moment profiles of the first phase of motion ($\alpha = \alpha_d$, $r_p = 0.01$)

The profile of deformed transverse displacement converges to the profile for the case of fully uniformly distributed loading. During the second phase of motion it is independent of loading radius r_p. Unified yield criterion, except

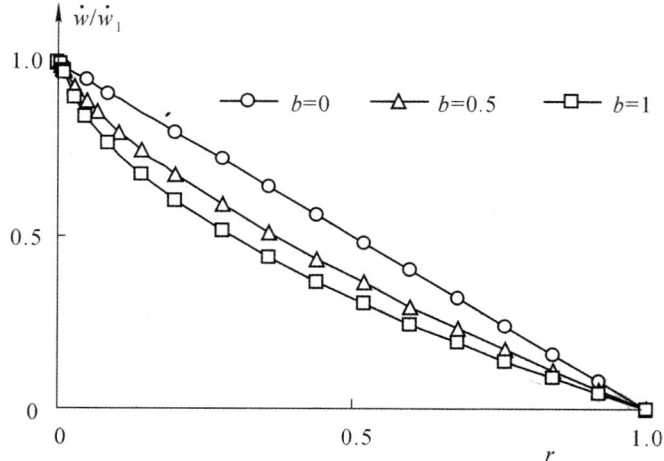

Fig. 10.20. Velocity profiles of the first phase of motion ($\alpha = \alpha_d$, $r_p = 0.01$)

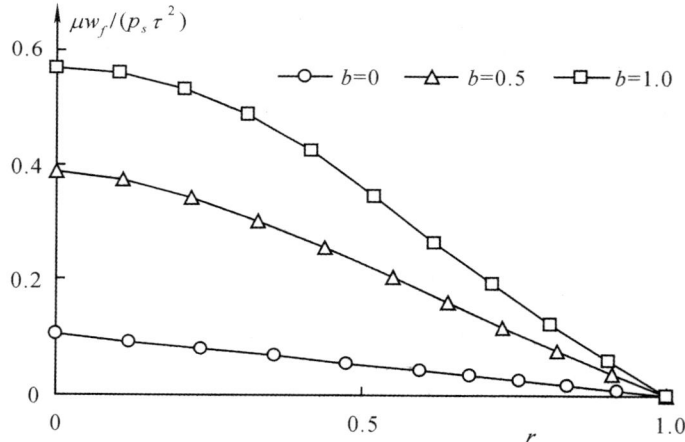

Fig. 10.21. Permanently deformed transverse displacement ($\alpha = \alpha_d$, $r_p = 0.01$)

for the special case of $b = 0$ or the Tresca criterion, leads to smooth velocity distribution at the plate center. The velocity profile varies with the different loading radius r_p. The Tresca criterion, as a special case of the unified yield criterion, leads to linear velocity profiles. The velocity fields experience singularity at the plate center regardless of r_p.

The maximum statically admissible loading factor α_d versus loading radius r_p is plotted in Fig.10.22. It shows the two loading action regions responding to the two statically admissible shown in Figs.10.11(a) and 10.11(b) with respect to the three yield criteria. The relationship of the response delaying time factor η (Fig.10.23) has an outline very close to that in Fig.10.22. However, it completely overlaps those from the Tresca criterion. It leads to $\alpha_d = 2$ and $\eta = 2$ for the case of fully uniformly distributed impulsive loading when $b = 0$, which is the same as the results with respect to the Tresca criterion reported by Hopkins and Prager (1954).

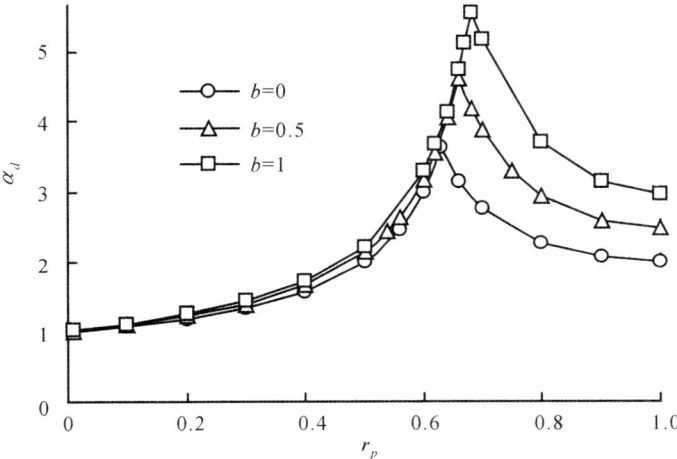

Fig. 10.22. Curves of α_d to r_p

Fig.10.24 shows that the effect of different yield criteria on the dynamic solution is higher than that in the static plastic limit state.

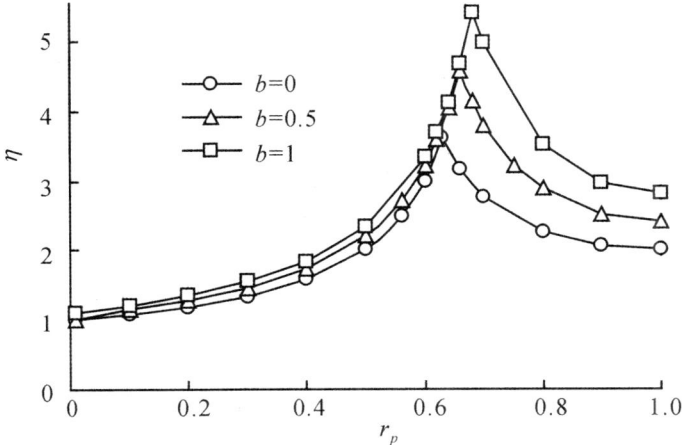

Fig. 10.23. η versus r_p

Fig.10.25 shows that the Tresca criterion gives the maximum permanent transverse displacement, which has been proved greater than the reported experimental results (Jones, 1968).

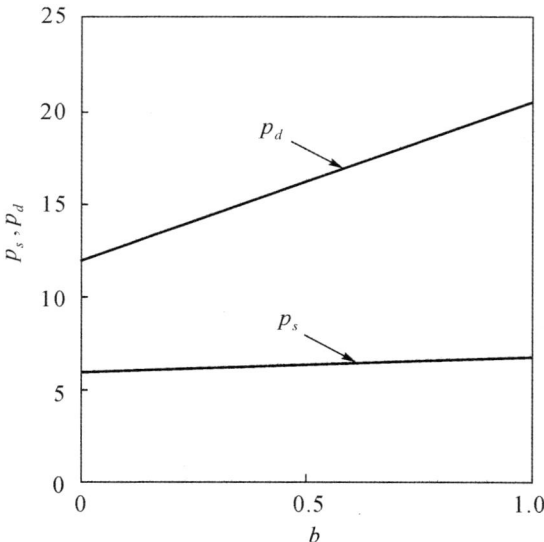

Fig. 10.24. p_d and p_s versus unified strength theory parameters b $(r_p = 1)$

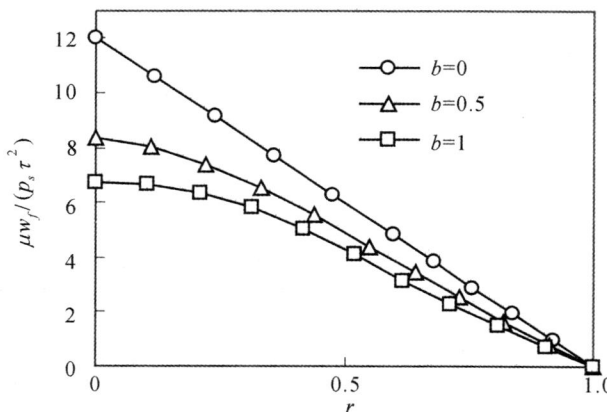

Fig. 10.25. Permanently deformed transverse displacement $(r_p = 1, p_0 = 12)$

10.6 Summary

The unified strength theory is applied to derive the dynamic responses for a simply supported circular plate subjected to moderate partial uniformly distributed impulsive loading. For dynamic plastic limit loadings, the moment and velocity fields of the plate are obtained based on the kinematical and static admissibility. A series of limit solutions based on the single-shear failure criteria (Tresca criterion and Mohr-Coulomb criterion), the twin-shear yield criterion (Yu, 1961; 1983), the Mohr-Coulomb strength theory and the twin-shear strength theory (Yu et al., 1985) are encompassed in the current unified solutions with specific parameters α and b.

(1) Since the current solutions satisfy both static admissibility and kinematical admissibility, they are the exact solutions of the response for a rigid-perfectly-plastic circular plate.

(2) Parameter and the unified strength theory parameter b have great influence on dynamic moment fields, displacement, and velocity responses. The influences on dynamic limit loading are higher than those on static limit loading. In particular when $\alpha = 1$, the percentage difference for static limit loadings with $b = 0$ and $b = 1$ is 14%, while it is 70% for dynamic limit loadings.

(3) The unified solutions are suitable for both SD and non-SD materials. When $\alpha = 1$, the unified strength theory degenerates to the unified yield criterion (UYC). The solutions with regard to UYC reported by Ma et al.(1999) are the special cases of the current unified solutions with $\alpha = 1$.

(4) Both the Tresca criterion and the Mohr-Coulomb strength theory ignore the effect of the intermediate principal stress, and thus the corresponding results deviate from experimental data. The strength theory with $b > 0$ may reflect the strength behavior of a broad range of materials and the load-bearing capacity more properly. It can be applied for proper reduction in the weight of structures, which may be of great significance for the optimization of aerospace structure.

References

Casey J, Sulivan TD (1985) Pressure dependency, strength differential effect, and plastic volume expansion in metals. Int. J. Plasticity, 1:39-61

Chait R (1972) Factors influencing the strength differential of high strength steels. Metall. Trans., 3: 365-371

Drucker DC (1973) Plasticity theory, strength differential (SD) phenomenon, and volume expansion in metals and plastics. Metall. Trans., 4:667-673

Florence AL (1977) Response of circular plates to central pulse loading. Int. J. Solids Struct., 13:1091-1102

Hopkins HG, Prager W (1954) On the dynamics of plastic circular plates. J. Appl. Mathe. Phy. (ZAMP), 5:317-330

Jones N (1968) Impulsive loading of a simple supported circular rigid plastic plate. J. Appl. Mech., 35:59-65

Jones N (1971) A theoretical study of the dynamic plastic behavior of beam and plate with finite-deflections. Int. J. Solids Struct., 7:1007-1029

Jones N, de Oliveira JG (1980) Dynamic plastic response of circular plates with transverse shear and rotatory inertia. J. Appl. Mech., 47(1):27-34

Ma GW, Hao H, Iwasaki S, Miyamoto Y, Deto H (1998) Plastic behavior of circular plate under soft impact. In: Yu MH and Fan SC (eds.) Strength Theory: Applications, Developments and Prospect for the 21st Century. Science Press, New York, Beijing, 957-962

Ma GW, Iwasaki S, Miyamoto Y, Deto H (1999) Dynamic plastic behavior of circular plate using unified yield criterion. Int. J. Solids Struct., 36:3257-3275

Shen WQ, Jones N (1993) Dynamic response and failure of fully clamped circular plates under impulsive loading. Int. J. Impact Eng., 13:259-278

Shen WQ (1995) Dynamic plastic response of thin circular plates struck transversely by non-blunt masses. Int. J. Solids Struct., 32(14):2009-2021

Symonds PS, Wierzbicki T (1979) Membrane mode solutions for impulsively loaded circular plates. J. Appl. Mech., 46(1):58-64

Stronge WJ, Yu TX (1993) Dynamic models for structural plasticity. Springer, London

Wang AJ, Hopkins HG (1954) On the plastic deformation of built-in circular plates under impulsive load. J. Mech. Phys. Solids, 3:22-37

Wang YB, Yu MH, Xiao Y, Li LS (2005) Dynamic plastic behavior of circular plate based on unified strength theory. Int. J. Impact Engineering, 31:25-40

Wei XY, Yu MH, Wang YY, Xu SQ (2001) Unified plastic limit of clamped circular plate with strength differential effect in tension and compression. Chinese Quart. Mechanics, 22(1):78-83

Yu MH (1961a) General behavior of isotropic yield function. Res. Report of Xi'an Jiaotong University, Xi'an, China (in Chinese)

Yu MH (1961b) Plastic potential and flow rules associated singular yield criterion. Res. Report of Xi'an Jiaotong University, Xi'an, China (in Chinese)

Yu MH (1983) Twin shear stress yield criterion. Int. J. Mech. Sci., 25(1):71-74

Yu MH (2002) Advances in strength theories for materials under complex stress state in the 20th century. Applied Mechanics Reviews, 55(3):169-218

Yu MH (2004) Unified strength theory and its applications. Springer, Berlin

Yu MH, He LN (1991) A new model and theory on yield and failure of materials under complex stress problems. in: Mechanical Behavior of Materials-6, Pergamon Press, Oxford, 3:841-846

Yu MH, He LN, Song LY (1985) Twin shear stress theory and its generalization. Scientia Sinica (Science in China), English ed., Series A, 28(12):1174-1183

11

Limit Angular Speed of Rotating Disc and Cylinder

11.1 Introduction

A rotating disc and cylinder, as shown in Fig.11.1, are often used as vane wheel and rotating axle of a propeller in many branches of engineering. When disc and cylinder rotate at an angular speed ω about their central axis, the stress and displacement caused by the centrifugal force are axisymmetric, i.e., σ_r, σ_θ and radial displacement u_r are related to radius r only. The rotating disc is in the generalized plane stress state, and rotating cylinder is in the generalized plane strain state.

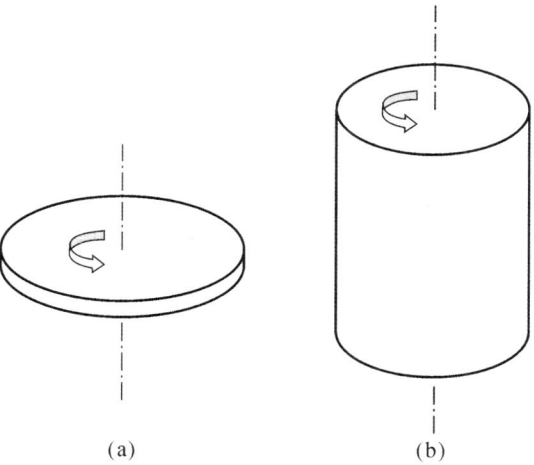

(a)   (b)

Fig. 11.1. Rotating disc and cylinder

Plastic limit analyses of disc and cylinder have been made by many researchers (Heyman, 1958; Save and Massonnet, 1972; Chakrabarty, 1987).

The Tresca yield criterion, which is a single-shear yield criterion, is usually used for elasto-plastic analysis of a rotating disc. It is understood that the analytic solution for the elasto-plastic field of the disc cannot be derived from the Huber-von Mises criterion because of the nonlinearity in its mathematical expression. The Tresca yield criterion, however, ignores the effect of the intermediate principal stresses. The unified yield criterion has also been applied to the plastic analysis of rotating disc and cylinder by Ma et al. (1995; 2001). A series of plastic limit speeds of rotating disc and cylinder were derived.

A rotating disc is a kind of rotational symmetrical plate that is loaded with inplane body force due to angular speed. The stress distributions and plastic limit angular velocity in terms of the Tresca criterion were reported by Lenard and Haddow (1972), Güven (1992; 1994), Gamer (1983), and Chakrabarty (1987).

Plastic limit analysis of rotating solid or annular discs with variable thickness based on Yu unified yield criterion (UYC) will be presented in this chapter. Stress distributions of the discs in plastic limit state corresponding to different yield curves are derived. Upper and lower bounds of the plastic limit solutions are derived by selecting proper values of parameter b in the unified yield criterion. The limit angular speed as well as the stress distributions with respect to three criteria, namely the Tresca criterion, the Huber-von Mises criterion (closed form solution), and the twin-shear yield criterion are derived and compared. The influences of different yield criteria as well as the thickness of the plastic limit solution of a rotating disc will be demonstrated and discussed.

11.2 Elastic Limit of Discs

When considering a circular disc with a uniform thickness rotating with a gradually increasing angular speed around its central axis, the thickness of the disc is assumed small enough for the disc to be considered in plane stress state. The stress state of each point satisfies $\sigma_1 = \sigma_\theta$, $\sigma_2 = \sigma_r$, $\sigma_3 = \sigma_z = 0$. The radial equilibrium of an element of the rotating disc has the form of (Timoshenko and Goodier, 1951)

$$\frac{d\sigma_r}{dr} + \frac{\sigma_r - \sigma_\theta}{r} + \rho\omega^2 r = 0, \tag{11.1}$$

where ω is the angular speed and ρ is the density of the material of the disc.

Denoting the radial displacement with u, the relevant stress-strain equations can be written as

$$\varepsilon_r = \frac{du}{dr} = \frac{1}{E}(\sigma_r - \nu\sigma_\theta), \quad \varepsilon_\theta = \frac{u}{r} = \frac{1}{E}(\sigma_\theta - \nu\sigma_r). \tag{11.2}$$

With reference to Eq.(11.1), Eq.(11.2) leads to the compatibility equation of

$$\frac{\mathrm{d}}{\mathrm{d}r}(\sigma_r + \sigma_\theta) = -(1+\nu)\rho\omega^2 r, \tag{11.3}$$

which is readily integrated to give $(\sigma_r + \sigma_\theta)$. The equilibrium equation then furnishes

$$\sigma_r = A - \frac{3+\nu}{8}\rho\omega^2 r^2, \tag{11.4a}$$

$$\sigma_\theta = A - \frac{1+3\nu}{8}\rho\omega^2 r^2, \tag{11.4b}$$

where A is a parameter depending on ω only. The boundary condition $\sigma_r = 0$ at $r = a$ gives

$$A = (3+\nu)\rho\omega^2 a^2/8. \tag{11.5}$$

The stress distribution for the solid disc becomes

$$\sigma_r = \frac{1}{8}\rho\omega^2(3+\nu)(a^2 - r^2), \tag{11.6a}$$

$$\sigma_\theta = \frac{1}{8}\rho\omega^2[(3+\nu)a^2 - (1+3\nu)r^2]. \tag{11.6b}$$

We can see from the above equations that both the stresses are tensile and $\sigma_\theta \geqslant \sigma_r$. $\sigma_\theta = \sigma$ holds only at $r = 0$ where the stress has the largest magnitude. Yielding will therefore start at the center of the disc where $\sigma_r = \sigma_\theta = \sigma_y$. The yield loci of the unified yield criterion in plane stress state are shown in Fig.11.2. From Fig.11.2, the Tresca criterion, the Huber-von Mises criterion, the twin-shear yield criterion and the unified yield criterion intersect at point A. Thus the elastic limit rotating speed ω_e based on these criteria is the same as

$$\omega_e = \frac{1}{2}\sqrt{\frac{8\sigma_y}{(3+\nu)\rho}}. \tag{11.7}$$

11.3 Elasto-plastic Analysis of Discs

If the speed of rotation is further increased, the disc will consist of an inner plastic zone surrounded by an outer elastic zone as shown in Fig.11.3. Within the plastic region, which is assumed to extend to a radius d, the stresses satisfy the equilibrium equation of Eq.(11.1) and the unified yield criterion, which can be expressed as follows (*see* Chapter 3):

$$\sigma_1 - \frac{\alpha b}{1+b}\sigma_2 = \sigma_t, \qquad \sigma_2 - \frac{\alpha b}{1+b}\sigma_1 = \sigma_t,$$

$$\frac{1}{1+b}\sigma_1 + \frac{b}{1+b}\sigma_2 = \sigma_t, \qquad \frac{1}{1+b}\sigma_2 + \frac{b}{1+b}\sigma_1 = \sigma_t,$$

$$\sigma_1 - \frac{\alpha}{1+b}\sigma_2 = -\sigma_t, \qquad \sigma_2 - \frac{\alpha}{1+b}\sigma_1 = -\sigma_t,$$

$$\frac{1}{1+b}\sigma_1 - \alpha\sigma_2 = -\sigma_t, \qquad \frac{1}{1+b}\sigma_2 - \alpha\sigma_1 = -\sigma_t,$$

$$-\frac{\alpha}{1+b}(b\sigma_1 + \sigma_2) = \sigma_t, \qquad -\frac{\alpha}{1+b}(b\sigma_2 + \sigma_1) = \sigma_t,$$

$$\frac{b}{1+b}\sigma_1 - \alpha\sigma_2 = \sigma_t, \qquad \frac{b}{1+b}\sigma_2 - \alpha\sigma_1 = \sigma_t.$$

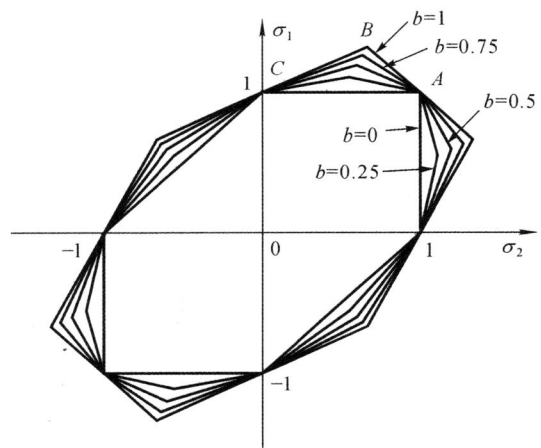

Fig. 11.2. Yield loci of the unified yield criterion in plane stress state

When the rotating disc is in an elasto-plastic state, stresses satisfy $\sigma_\theta \geqslant \sigma_r \geqslant \sigma_z = 0$, the stress state in the plastic region corresponds to either the side of AB or BC in Fig.11.2, and there are two possible cases of plasticity:

a) when the radius of the plastic zone is small as shown in Fig.11.3, the stress state of the entire plastic zone lies on the side AB (in Fig.11.2), and stresses in the elastic region satisfy Eq.(11.1);

b) when the radius of the plastic zone is larger than a specific value of r_0, as shown in Fig.11.4, the stress state of the plastic zone lies on the sides AB and BC, where G corresponds to point B in Fig.11.2. Stresses in the elastic region also satisfy equilibrium Eq.(11.1) while, in the plastic zone, the yield criterion is satisfied.

In the special case that point E and point G intersect, the first case and the second case are the same. That is the demarcating state of the two cases.

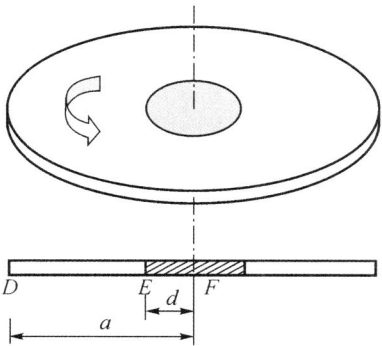

Fig. 11.3. Radius of plastic zone is smaller

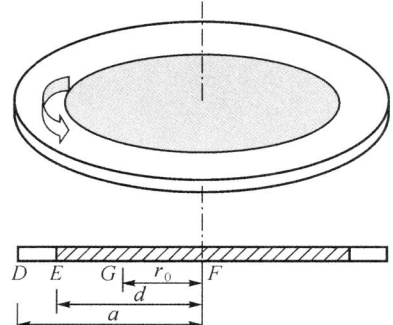

Fig. 11.4. Radius of plastic zone is bigger

Equations for yield conditions of the segments AB, BC are

$$AB: \quad \frac{1}{1+b}\sigma_r + \frac{1}{1+b}\sigma_\theta = \sigma_y, \tag{11.8a}$$

$$BC: \quad \sigma_\theta - \frac{1}{1+b}\sigma_r = \sigma_y. \tag{11.8b}$$

Boundary and continuity conditions corresponding to the first case are:
1. $\sigma_r \mid_{r=a} = 0$ at point D;
2. $\sigma_r \mid_{r=d}$ and $\sigma_\theta \mid_{r=d}$ are continuous at point E;
3. $\sigma_r \mid_{r=0}$ is a definite value at point F.
Boundary and continuity conditions corresponding to the second case are:

4. $\sigma_r \mid_{r=a} = 0$ at point D;
5. $\sigma_r \mid_{r=d}$ and $\sigma_\theta \mid_{r=d}$ are continuous at point E;
6. $\sigma_r \mid_{r=0}$ is a definite value at point F;
7. $\sigma_r \mid_{r=r_0}$ and $\sigma_\theta \mid_{r=r_0}$ are continuous at point G;
8. $\sigma_\theta \mid_{r=0} = 2\sigma_r \mid_{r=0}$ at point G.

11.4 Elasto-plastic Stress Field of Rotating Disc

For the first case shown in Fig.11.3, with reference to the plastic region from Eq.(11.1), the limit condition of Eq.(11.8) and boundary conditions 1∼3, the stresses in the elastic region are derived as

$$
\begin{cases}
\sigma_r = \sigma_y - \dfrac{\rho\omega^2}{3+b}r^2, \\[4mm]
\sigma_\theta = \sigma_y + \dfrac{b\rho\omega^2}{3+b}r^2,
\end{cases}
\qquad \text{when } 0 \leqslant r \leqslant d, \qquad (11.9)
$$

$$
\begin{cases}
\sigma_r = \sigma_y - \rho\omega^3\left(C_1 d^2 + C_2\dfrac{d^4}{r^2} + C_3 r^2\right), \\[4mm]
\sigma_\theta = \sigma_y - \rho\omega^2\left(C_1 d^2 - C_2\dfrac{d^4}{r^2} + C_4 r^2\right),
\end{cases}
\qquad \text{when } d \leqslant r \leqslant a, \qquad (11.10)
$$

where

$$
C_1 = \frac{1-b}{6+2b} - \frac{1+\nu}{4}, \qquad C_2 = \frac{1+b}{6+2b} - \frac{1-\nu}{8},
$$

$$
C_3 = \frac{3+\nu}{8}, \qquad C_4 = \frac{1+3\nu}{8}.
$$

For specific parameter b and Poisson's ratio ν, C_1, C_2, C_3 and C_4 can be determined. The stresses in the elastic region are then derived from Eq.(11.10), and the rotating speed ω in Eqs.(11.9) and (11.10) satisfies

$$
\frac{pa^2}{\omega^2\sigma_y} = C_2\left(\frac{d}{a}\right)^4 + C_1\left(\frac{d}{a}\right)^2 + C_3. \qquad (11.11)
$$

Similarly for the second case shown in Fig.11.4, with reference to equilibrium equation (Eq.(11.1)), the limit condition in Eqs.(11.8a) and (11.8b), and boundary conditions 4∼8, the stresses are deduced from

$$\begin{cases} \dfrac{\sigma_r}{\sigma_y} = 1 - \dfrac{1}{2+b}\left(\dfrac{r}{r_0}\right)^2, \\[4mm] \dfrac{\sigma_\theta}{\sigma_y} = 1 + \dfrac{b}{2+b}\left(\dfrac{r}{r_0}\right)^2, \end{cases} \qquad \text{when } 0 \leqslant r \leqslant r_0, \qquad (11.12)$$

$$\begin{cases} \dfrac{\sigma_r}{\sigma_y} = C_5 - C_6\left(\dfrac{r_0}{r}\right)^{\frac{1}{1+b}} - C_7\left(\dfrac{r}{r_0}\right)^2, \\[4mm] \dfrac{\sigma_\theta}{\sigma_y} = C_5 - \dfrac{bC_6}{1+b}\left(\dfrac{r_0}{r}\right)^{\frac{1}{1+b}} - \dfrac{bC_7}{1+b}\left(\dfrac{r}{r_0}\right)^2, \end{cases} \qquad \text{when } r_0 \leqslant r \leqslant d, \quad (11.13)$$

$$\begin{cases} \dfrac{\sigma_r}{\sigma_y} = C_5 - C_8\left(\dfrac{r}{r_0}\right)^2 - C_9\left(\dfrac{r_0}{d}\right)^{\frac{1}{1+b}} + C_{12}\left(\dfrac{d}{r_0}\right)^2 \\[2mm] \qquad - C_{10}\left(\dfrac{d}{r}\right)^2\left(\dfrac{r_0}{d}\right)^{\frac{1}{1+b}} + C_{11}\left(\dfrac{d}{r_0}\right)^2\left(\dfrac{d}{r}\right)^2, \\[4mm] \dfrac{\sigma_\theta}{\sigma_y} = C_5 - C_{13}\left(\dfrac{r}{r_0}\right)^2 - C_9\left(\dfrac{r_0}{d}\right)^{\frac{1}{1+b}} + C_{12}\left(\dfrac{d}{r_0}\right)^2 \\[2mm] \qquad + C_{10}\left(\dfrac{d}{r}\right)^2\left(\dfrac{r_0}{d}\right)^{\frac{1}{1+b}} - C_{11}\left(\dfrac{d}{r_0}\right)^2\left(\dfrac{d}{r}\right)^2, \end{cases} \quad \text{when } d \leqslant r \leqslant a.$$

$$(11.14)$$

When $d \leqslant r \leqslant a$, the limit rotating speed ω can be obtained as

$$\omega = \sqrt{\dfrac{3+b}{2+b}\dfrac{\sigma_y}{\rho}}\,\dfrac{1}{r_0}, \qquad (11.15)$$

where r_0 satisfies

$$f(r_0) = \dfrac{3+b}{2+b}C_{11}\left(\dfrac{d}{a}\right)^4 - C_{14}\left(\dfrac{r_0}{d}\right)^{\frac{1}{1+b}}\left(\dfrac{d}{a}\right)^2\left(\dfrac{r_0}{a}\right)^2 + C_{15}\left(\dfrac{r_0}{a}\right)^2$$
$$+ \dfrac{2+b}{3+b}C_1\left(\dfrac{d}{a}\right)^2 - C_{16}\left(\dfrac{r_0}{d}\right)^{\frac{1}{1+b}}\left(\dfrac{r_0}{a}\right)^2 - C_3, \qquad (11.16)$$

and

$$C_5 = 1 + b, \quad C_6 = \dfrac{2b(1+b)}{3+2b}, \quad C_7 = \dfrac{(1+b)(3+b)}{(3+2b)(2+b)}, \quad C_8 = \dfrac{3+\nu}{8}\dfrac{2+b}{3+b},$$

$$C_9 = \dfrac{b(1+2b)}{3+2b}, \quad C_{10} = \dfrac{b}{3+2b}, \quad C_{11} = \left(\dfrac{1-\nu}{8} - \dfrac{1}{6+4b}\right)\dfrac{2+b}{3+b},$$

$$C_{12} = \left(\dfrac{1+\nu}{4} - \dfrac{1+2b}{6+4b}\right)\dfrac{3+b}{2+b}, \quad C_{15} = \dfrac{(1+b)(2+b)}{3+b}, \quad C_{16} = \dfrac{b(1+2b)(2+b)}{3(3+2b)(3+b)}.$$

11.5 Solution Procedure and Results

The solution of the stress field contains the following two steps. Firstly the radius of plastic zone d_0 for the special state that demarcates the two different cases is determined. In this state points E and G overlap and $d_0 = d = r_0$. Assuming $f_0(d_0) = f(r)\ |_{r=r_0}$, d_0 satisfies

$$f_0(d_0) = \left(\frac{1-\nu}{8} - \frac{1+b}{6+2b}\right)\left(\frac{d_0}{a}\right)^4 + \left[\frac{1+\nu}{4} + \frac{3(1+b)}{2(3+b)}\right]\left(\frac{d_0}{a}\right)^2 - \frac{3+\nu}{8} = 0.$$

(11.17)

In the second step, for the case that $d \leqslant d_0$, the limit rotating speed ω and σ_r, σ_θ can be deduced from Eqs.(11.9) and (11.10). For $d_0 \leqslant d \leqslant a$, r_0 can be determined from Eq.(11.14). Substituting r_0 into Eqs.(11.8) and (11.9), σ_r, σ_θ, and the elasto-plastic limit rotating speed ω can be derived from Eqs.(11.12) and (11.13).

The procedure for solving the limit rotating speed and the stress fields is illustrated in Fig.11.5.

The deduction process can converge quickly to carry out the elasto-plastic limit rotating speed and centrifugal stress field. Figs.11.6 to 11.9 show the stress field of different radii of the plastic zone. From these figures the results with $b = 0$ conform to that based on the Tresca yield criterion. The result with $b = 0.5$ approximates to the results in terms of the von Mises yield criterion. And the result with $b = 1$ is the same as that obtained from the twin-shear yield criterion.

Fig.11.10 shows the relation of the angular speed to the plastic zone radius. Fig.11.11 demonstrates the relation of the plastic limit angular speed ω_p of circular disc to the parameter b. In the elasto-plastic state, the stress distribution over the disc depends on the unified yield criterion parameter b. When $b = 0$ (Tresca criterion), the derived stress is the smallest, and when $b = 1$ (corresponding to the twin-shear stress yield criterion), the stress is the largest. σ_θ is consistently equal to σ_y in the plastic zone of the derived elasto-plastic stress field when $b = 0$. When $b \neq 1$, σ_θ in the plastic part of elasto-plastic stress field derived from the unified yield criterion and the Huber-von Mises criterion is bigger than σ_y. The larger the plastic zone, the stronger the effect of the parameter b on the stress fields.

From Figs.11.10 and 11.11, a series of plastic solutions can be obtained with the unified yield criterion. They can be adopted for all metallic non-SD materials. The solution based on the unified yield criterion with $b = 0$ is the same as the result from the Tresca criterion (Save and Massonnet, 1972; 1997; Chakrabarty, 1987).

The plastic solutions of rotating disc in terms of the twin-shear yield criterion were reported by Li (1988), Huang and Zeng (1989). It is also a special case for the current unified solution.

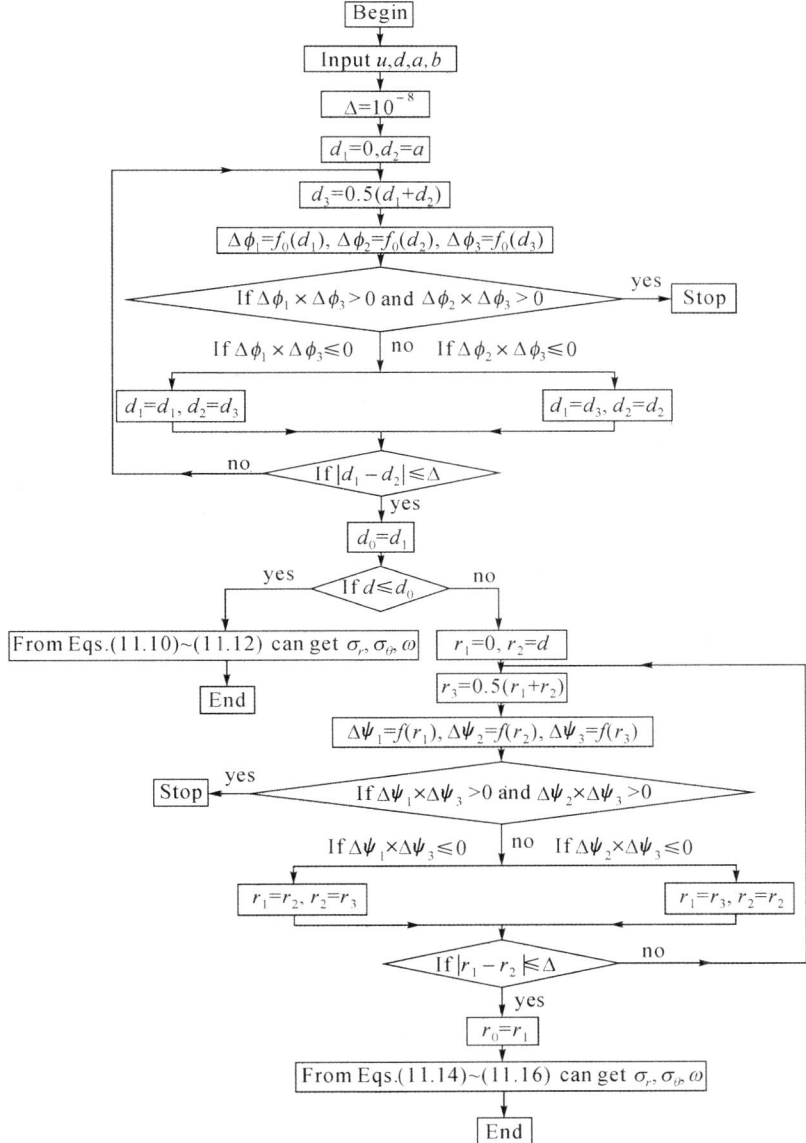

Fig. 11.5. Solution procedure of elasto-plastic analysis of circular disc

11.6 Unified Solution of Plastic Limit Analysis of Rotating Cylinder

Elastic stresses of a rotating cylinder can be written as

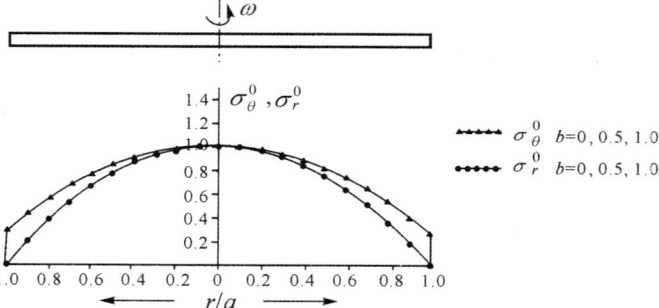

Fig. 11.6. Stress fields under elastic limit state $d = 0$

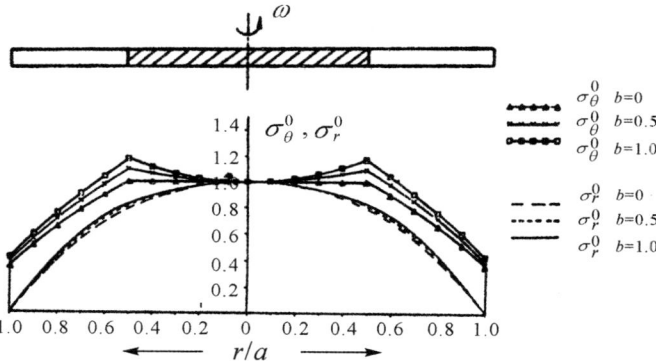

Fig. 11.7. Stress field for plastic zone with $d = 0.5a$

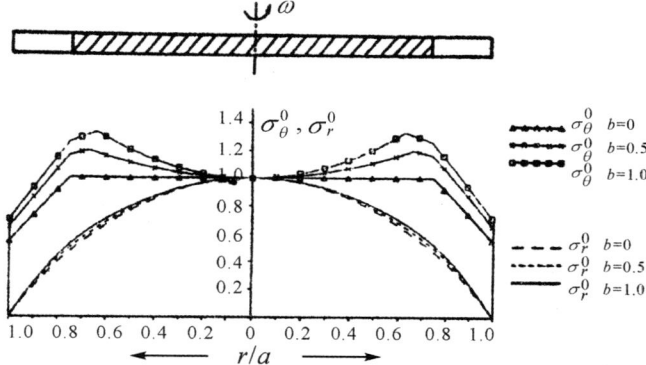

Fig. 11.8. Stress fields for plastic zone with $d = 0.75a$

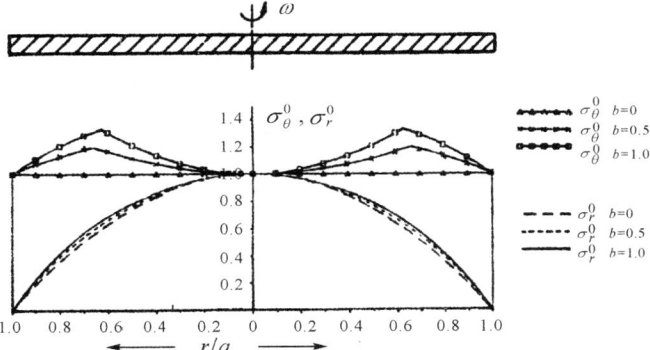

Fig. 11.9. Stress fields for plastic zone with $d = a$

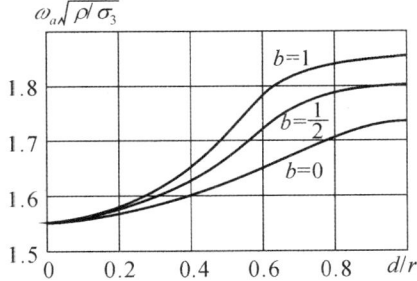

Fig. 11.10. Relation of angular speed to the radius of plastic zone

$$\begin{cases} \sigma_r = \sigma_\theta = \dfrac{1}{2}\rho\omega^2(a^2 - r^2), \\[2mm] \sigma_z = \dfrac{1}{2}\rho\omega^2(\dfrac{1}{2}a^2 - r^2). \end{cases} \tag{11.18}$$

From Eq.(11.18), $\sigma_1 = \sigma_2 = \sigma_r = \sigma_\theta$, $\sigma_3 = \sigma_z$. The unified yield criterion is independent of the parameter b and the yield condition has the form of

$$\sigma_\theta - \sigma_z = \sigma_y. \tag{11.19}$$

Substituting Eq.(11.19) into Eq.(11.18), we obtain

$$\omega_p = \omega_e = \frac{2}{a}\sqrt{\frac{\sigma_y}{\rho}}. \tag{11.20}$$

Eq.(11.20) implies that when the elastic limit rotating speed is equal to the plastic limit rotating speed for a circular cylinder, each point in the cylinder yields simultaneously. The limit velocities in view of the Tresca criterion, the

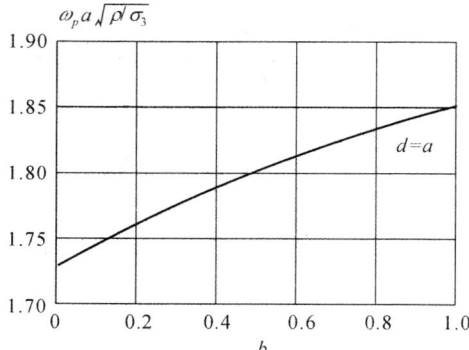

Fig. 11.11. Plastic limit angular speed ω_p versus the unified strength theory parameter b

Huber-von Mises criterion and the twin shear stress yield criterion are the same.

Substituting ω_e into Eq.(11.18), the similar centrifugal stress field is derived,

$$
\begin{cases}
\sigma_r = \sigma_\theta = 2\sigma_y \left(1 - \dfrac{r^2}{a^2} \right), \\[2mm]
\sigma_z = 2\sigma_y \left(\dfrac{1}{2} - \dfrac{r^2}{a^2} \right).
\end{cases} \tag{11.21}
$$

11.7 Limit Analysis of a Solid Disc with Variable Thickness

For convenient formulation, dimensionless variables, $r = R/a$, $\sigma_r = \tilde{\sigma}_r/\sigma_0$, $\sigma_\theta = \tilde{\sigma}_\theta/\sigma_0$ and $\Omega^2 = \rho\omega^2 a^2/\sigma_0$ are defined, where a is the radius of a rotating disc; R is radius variable in a range between 0 and a; $\tilde{\sigma}_r$, $\tilde{\sigma}_\theta$, σ_0 are the radial, circumferential and yield stresses respectively; ρ is the mass density; ω is the plastic limit angular speed of the rotating disc. The two dimensionless principal stresses σ_r and σ_θ approximately satisfy the plane stress condition when the ratio of the disc diameter to the disc thickness is very high.

The equilibrium equation of a circular rotating disc with variable thickness is

$$
d(hr\sigma_r)/dr - h\sigma_\theta = -h\Omega^2 r^2. \tag{11.22}
$$

For a solid rotating disc, radial stress σ_r at the centre $(r = 0)$ is equal to 1, which corresponds to the yield point A on the yield curves as shown in Fig.11.2; and $\sigma_r = 0$ at the outer edge $(r = 1)$ corresponds to the yield point C on the yield curve. Stresses at any other points on the disc lie on the yield lines AB and BC according to the normality requirement of plasticity (Chakrabarty, 1987). The unified yield criterion can be expressed in terms of σ_r and σ_θ as

$$\sigma_\theta = a_i \sigma_r + b_i \quad (i = 1, ..., 12), \tag{11.23}$$

where a_i and b_i are constants corresponding to the twelve line segments of the yield loci in Fig.11.2.

If the two yield constants of line AB are denoted as a_1, b_1, there are $a_1 = -b$ and $b_1 = 1 + b$. Similarly, if the constants of line BC are denoted as a_2 and b_2, there are $a_2 = b/(1+b)$ and $b_2 = 1$. The radial stress corresponding to the yield point B is dependent on the parameter b, and it has the expression of $d_1 = (1 + b)/(2 + b)$.

Substituting Eq.(11.23) into Eq.(11.22) and then integrating Eq.(11.22), σ_r on the two lines can be derived,

$$\sigma_r = e^{-\int f_i(r)\mathrm{d}r} \left[\int e^{\int f_i(r)\mathrm{d}r} g_i(r)\mathrm{d}r + c_i \right] \quad (i = 1, ..., 12), \tag{11.24a}$$

where

$$f_i(r) = \frac{(hr)'}{hr} - \frac{a_i}{r}, \quad g_i(r) = \frac{b_i - \Omega^2 r^2}{r} \quad (i = 1, 2), \tag{11.24b}$$

and $'$ denotes differentiation with respect to r. The integration constants c_i $(i = 1, 2)$ in Eq.(11.24) can be determined by continuous and boundary conditions.

If the thickness of the disc is expressed as

$$h = \sum_{j=1}^{\infty} h_j r^{j-1}, \tag{11.25}$$

the radial stress can then be derived by substituting Eqs.(11.24b) and (11.25) into Eq.(11.24a),

$$\sigma_r = h^{-1} \left[b_i \sum_{j=1}^{\infty} h_j \frac{r^{j-1}}{j - a_i} - \Omega^2 \sum_{j=1}^{\infty} h_j \frac{r^{j+1}}{j + 2 - a_i} \right] + c_i r^{-1+a_i} h^{-1} \quad (i = 1, 2), \tag{11.26}$$

Eq.(11.26) indicates that the radial stress σ_r covers two regions with respect to the two yield lines AB and BC. Assuming that r_0 is the radius of a ring where the stresses correspond to the point B in Fig.11.2, the continuity and boundary conditions can be expressed as: (1) $\sigma_r = 1$ at $r = 0$; (2)

$\sigma_r = d_1$ at $r = r_0$; (3) $\sigma_r = 0$ at $r = 1$. Applying these conditions, the two integration constants c_1 and c_2 are obtained

$$c_1 = 0, \; c_2 = -\left[b_2 \sum_{j=1}^{\infty} \frac{h_j}{j - a_2} - \Omega^2 \sum_{j=1}^{\infty} \frac{h_j}{j + 2 - a_2} \right], \qquad (11.27)$$

and the plastic limit angular speed Ω is derived

$$\Omega^2 = \frac{b_1 \sum_{j=1}^{\infty} \frac{h_j}{j - a_1} r_0^{j-1} - d_1 \sum_{j=1}^{\infty} h_j r_0^{j-1}}{\sum_{j=1}^{\infty} \frac{h_j}{j + 2 - a_1} r_0^{j+1}}, \qquad (11.28)$$

where the radius r_0 is the demarcating radius that divides the disc into two regions. r_0 has the expression of

$$d_1 \sum_{j=1}^{\infty} h_j r_0^{j-1} = -\left[b_2 \sum_{j=1}^{\infty} \frac{h_j}{j - a_2} - \Omega^2 \sum_{j=1}^{\infty} \frac{h_j}{j + 2 - a_2} \right] r_0^{-1+a_2}$$

$$+ \left[b_2 \sum_{j=1}^{\infty} \frac{h_j r_0^{j-1}}{j - a_2} - \Omega^2 \sum_{j=1}^{\infty} \frac{h_j r_0^{j+1}}{j + 2 - a_2} \right]. \qquad (11.29)$$

For a specific formulation of thickness $h(r)$, r_0 in Eq.(11.29) can be solved by half interval search method in the interval of $(0, 1)$. Substituting the calculated r_0 into Eqs.(11.27), (11.28) and (11.26), the plastic limit angular speed as well as the stress distribution is determined.

It should be mentioned that the dimensionless radial stress at any point of the disc cannot exceed 1 based on the assumption of $\sigma_\theta \geqslant \sigma_r$ (Chakrabarty, 1987) or the stability of the rotating disc. The derived limit stresses in the disc is subjected to examination to see that the yield condition of Fig.11.2 is satisfied. In some cases the radial stress calculated from Eq.(11.26) may violate the yield condition in the central area of the disc with the dimensionless radial stress exceeding the maximum value 1 ($\sigma_r > 1$) (Fig.11.12). To avoid the violation and ensure the stability of the rotating disc, stress redistribution to release the overfloated stress must be performed. In the region where the stress state violates the yield condition, there is $\partial \sigma_r / \partial r$ at the location where the maximum radial stress appears. Assuming that the maximum radial stress occurs at $r = r_m$, there exists $\partial \sigma_r / \partial r = 0$ at $r = r_m$. Differentiating Eq.(11.26) at $r = r_m$, the following equation is derived:

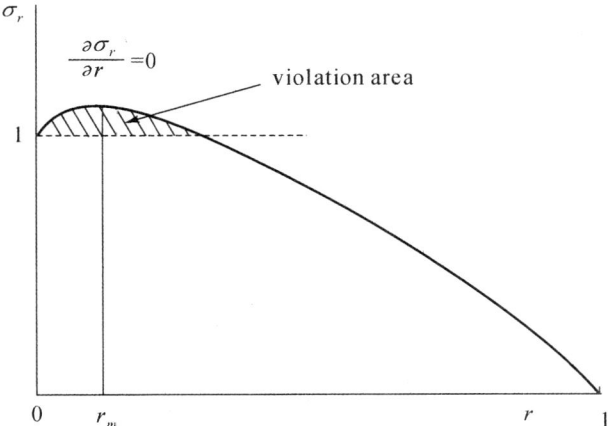

Fig. 11.12. Stress violation in the central area

$$\sigma_r \sum_{j=2}^{\infty} h_j (j-1) r_m^{j-2} = b_1 \sum_{j=2}^{\infty} h_j (j-1) \frac{r_m^{j-2}}{j-a_1}$$
$$- \Omega^2 \sum_{j=1}^{\infty} h_j (j+1) \frac{r_m^{j}}{j+2-a_1}. \tag{11.30}$$

The unknown r_m in Eq.(11.30) can be derived by iterative method.

To implement stress redistribution for the central area, it is reasonable to assume that stresses in this area correspond to the yield point A, and stresses in the outer area correspond to the yield lines AB and BC. The boundary and continuity conditions are changed to be (1) $\sigma_r = 1$ at $r = r_1$; (2) $\partial \sigma_r / \partial r = 0$ at $r = r_1$; (3) $\sigma_r = d_1$ at $r = r_2$; (4) $\sigma_r = 0$ at $r = 1$, where r_1 and r_2 are the demarcating radii corresponding to the yield points A and B in Fig.11.2. With reference to the above continuity and boundary conditions, the constants in Eq.(11.26) are derived,

$$c_1 = \left(\sum_{j=1}^{\infty} h_j r_1^{j-1} - b_1 \sum_{j=1}^{\infty} h_j \frac{r_1^{j-1}}{j-a_1} + \Omega^2 \sum_{j=1}^{\infty} h_j \frac{r_1^{j+1}}{j+2-a_1} \right) r_1^{1-a_1}, \tag{11.31}$$

$$c_2 = - \left[b_2 \sum_{j=1}^{\infty} \frac{h_j}{j-a_2} - \Omega^2 \sum_{j=1}^{\infty} \frac{h_j}{j+2-a_2} \right], \tag{11.32}$$

and r_1 and r_2 satisfy the equations

$$\sum_{j=2}^{\infty} h_j(j-1)r_1^{j-2} = b_1 \sum_{j=2}^{\infty} h_j \frac{(j-1)r_1^{j-2}}{j-a_1} - \Omega^2 \sum_{j=1}^{\infty} h_j \frac{(j+1)r_1^j}{j+2-a_1} \qquad (11.33)$$
$$+ c_1(-1+a_1)r_1^{-2+a_1},$$

$$d_1 \sum_{j=1}^{\infty} h_j r_2^{j-1} = b_1 \sum_{j=1}^{\infty} h_j \frac{r_2^{j-1}}{j-a_1} - \Omega^2 \sum_{j=1}^{\infty} h_j \frac{r_2^{j+1}}{j+2-a_1} + c_1 r_2^{-1+a_1}. \quad (11.34)$$

The plastic limit angular speed then has the expression

$$\Omega^2 = \frac{d_1 \sum_{j=1}^{\infty} h_j r_2^{j-1} - b_2 \sum_{j=1}^{\infty} h_j \frac{r_2^{j-1}}{j-a_2} + b_2 r_2^{-1+a_2} \sum_{j=1}^{\infty} \frac{h_j}{j-a_2}}{r_2^{-1+a_2} \sum_{j=1}^{\infty} \frac{h_j}{j+2-a_2} - \sum_{j=1}^{\infty} \frac{h_j r_2^{j+1}}{j+2-a_2}} \qquad (11.35)$$

Eqs.(11.33), (11.34) and (11.35) can be solved with the iterative method. r_1 can be obtained by half interval search method in the interval of $(0, r_0)$ from Eqs.(11.31) and (11.33); r_2 is then searched similarly in the interval $(r_1, 1)$ from Eq.(11.34). Ω^2 can be calculated directly from Eq.(11.35) when r_2 is determined. The calculated value of Ω^2, r_1, and r_2 are adopted to calculate new values again. The iteration cycle continues until a convergence solution with sufficient accuracy is achieved. To start the iteration process, the calculated limit angular speed before stress redistribution is adopted as the initial value of Ω^2.

11.8 Limit Analysis of an Annular Disc with Variable Thickness

For an annular disc at a constant angular speed ω, the stress fields of the entire disc satisfy the inequality $\sigma_\theta \geqslant \sigma_r \geqslant 0$, implying that the stress components σ_r and σ_θ are located on the two segments CB and BA in Fig.11.2. The boundary condition of the inner edge becomes $\sigma_r = 0$ at $r = \beta$, where β is the radius of the inner edge. It should be noted that since a dimensionless variable of radius is used, β is actually the ratio of the inner radius to the outer radius of the annular disc. Stresses at the both edges lie at the yield point C in Fig.11.2. The varying trajectory of stress components σ_r and σ_θ along the radial direction has two possible cases according to the unified yield criterion, namely (1) $C \to B \to C$ when the ratio of inner to outer radius β is higher than a specific or critical radius (Fig.11.13(a)); (2) $C \to B \to A \to B \to C$ when β is lower than the critical radius (Fig.11.13(b)). The critical radius demarcating the two cases corresponds to β_0 in Figs.11.13(a) and 11.13(b).

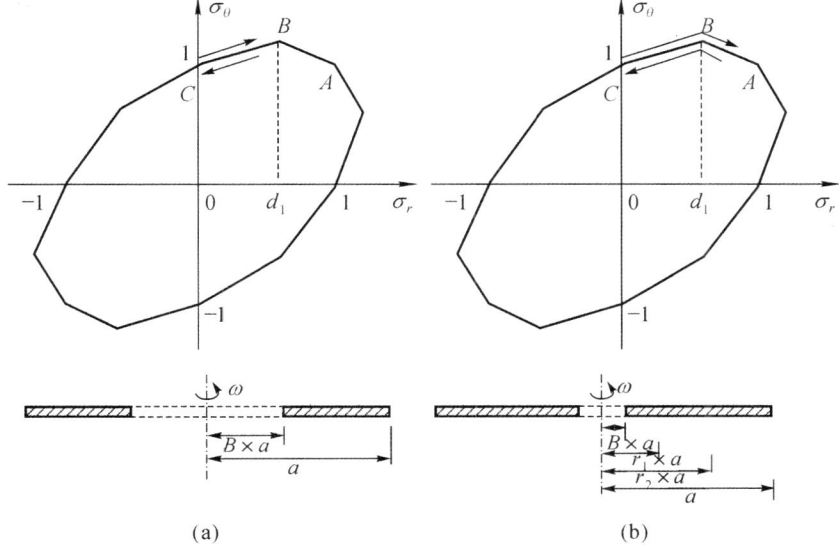

Fig. 11.13. Case (1) with $\beta \geqslant \beta_0$

11.8.1 Case (1) $(1 > \beta \geqslant \beta_0)$

When the ratio of inner radius to outer radius is higher than the critical ratio β_0, the stresses on the entire disc lie on the yield line CB in Fig.11.13 (a). The radial stress can be obtained by direct integration of Eq.(11.22). Using the boundary conditions at the outer and inner edges, the integration constant is derived,

$$c_1 = -\left[b_2 \sum_{j=1}^{\infty} \frac{h_j}{j - a_2} - \Omega^2 \sum_{j=1}^{\infty} \frac{h_j}{j + 2 - a_2} \right], \qquad (11.36)$$

and the angular speed is obtained

$$\Omega^2 = \frac{b_2 \left[\sum_{j=1}^{\infty} \dfrac{h_j}{j - a_2} - \beta^{1-a_2} \sum_{j=1}^{\infty} \dfrac{h_j \beta^{j-1}}{j - a_2} \right]}{\sum_{j=1}^{\infty} \dfrac{h_j}{j+2-a_2} - \beta^{1-a_2} \sum_{j=1}^{\infty} \dfrac{h_j \beta^{j+1}}{j+2-a_2}}. \qquad (11.37)$$

Substituting Eq.(11.36) and Eq.(11.37) into Eq.(11.26), the stress field of Case (1) is then derived. The radial stress σ_r should have a maximum value corresponding to a yield point on CB where the radial stresses return to the yield point C. As the radius ratio β decreases, the radial stress σ_r extends towards the yield line BA, which leads to the second case. For the critical case of $\beta = \beta_0$, the maximum radial stress is exactly at the yield point B in Fig.11.2, and there exists

$$\frac{\partial \sigma_r}{\partial r} = 0 \text{ and } \sigma_r = d_1 \text{ at } r = r_0, \tag{11.38}$$

or

$$d_1 \sum_{j=2}^{\infty} h_j(j-1)r_0^{j-2} = c_1(-1+a_2)r_0^{-2+a_2}$$

$$+ \left[b_2 \sum_{j=2}^{\infty} \frac{h_j(j-1)r_0^{j-2}}{j - a_2} - \Omega^2 \sum_{j=1}^{\infty} \frac{h_j(j+1)r_0^{j}}{j + 2 - a_2} \right], \tag{11.39}$$

and

$$d_1 \sum_{j=1}^{\infty} h_j r_0^{j-1} = c_1 r_0^{-1+a_2} + \left[b_2 \sum_{j=1}^{\infty} \frac{h_j r_0^{j-1}}{j - a_2} - \Omega^2 \sum_{j=1}^{\infty} \frac{h_j r_0^{j+1}}{j + 2 - a_2} \right], \tag{11.40}$$

where c_1 and Ω^2 are given by Eq.(11.36) and Eq.(11.37) respectively; r_0 denotes the radius where the radial stress σ_r reaches its maximum value d_1 in the critical case.

β_0 and r_0 satisfying both Eq.(11.39) and Eq.(11.40) can be solved with a numerical iterative method by specifying an initial value of β_0 (for example 0.3). r_0 can be calculated from Eq.(11.39) with a half interval search method. β_0 is then recalculated from Eq.(11.39) using the updated r_0. The iterative process continues until satisfactory convergence is reached.

11.8.2 Case (2) $(0 < \beta \leqslant \beta_0)$

When the radius ratio $\beta \leqslant \beta_0$, the stresses σ_r and σ_θ will extend to the line BA in Fig.11.13 (b). σ_r can be derived by integrating Eq.(11.22) with reference to the yield condition of Eq.(11.23),

$$
\sigma_r =
\begin{cases}
\left[b_2 \sum\limits_{j=1}^{\infty} h_j \dfrac{r^{j-1}}{j-a_2} - \Omega^2 \sum\limits_{j=1}^{\infty} h_j \dfrac{r^{j+1}}{j+2-a_2} \right] \Big/ h + c_1 r^{-1+a_2}/h, \\
\qquad\qquad \beta \leqslant r \leqslant r_1, \\[2mm]
\left[b_1 \sum\limits_{j=1}^{\infty} h_j \dfrac{r^{j-1}}{j-a_1} - \Omega^2 \sum\limits_{j=1}^{\infty} h_j \dfrac{r^{j+1}}{j+2-a_1} \right] \Big/ h + c_2 r^{-1+a_1}/h, \\[2mm]
\qquad\qquad r_1 \leqslant r \leqslant r_2, \\[2mm]
\left[b_2 \sum\limits_{j=1}^{\infty} h_j \dfrac{r^{j-1}}{j-a_2} - \Omega^2 \sum\limits_{j=1}^{\infty} h_j \dfrac{r^{j+1}}{j+2-a_2} \right] \Big/ h + c_3 r^{-1+a_2}/h, \\[2mm]
\qquad\qquad r_2 \leqslant r \leqslant 1,
\end{cases}
\tag{11.41}
$$

where r_1 and r_2 are the demarcating radii where the stresses fall on the point B in Fig.11.13 (b). c_1, c_2 and c_3 are integration constants. Six unknown variables of c_1, c_2, c_3, Ω^2, r_1, and r_2 in Eq.(11.41) can be numerically calculated from six boundary and continuity conditions, namely, (1) σ_r $(r=\beta)=0$; (2) $\sigma_r(r=r_1, \beta \leqslant r \leqslant r_1)=d_1$; (3) σ_r $(r=r_1, r_1 \leqslant r \leqslant r_2)=d_1$; (4) σ_r $(r=r_2, r_1 \leqslant r \leqslant r_2)=d_1$; (5) σ_r $(r=r_2, r_2 \leqslant r \leqslant 1)=d_1$; (6) σ_r $(r=1)=0$. According to these conditions, Eq.(11.41) leads to

$$
c_1 = -\beta^{1-a_2} \left[b_2 \sum_{j=1}^{\infty} h_j \frac{\beta^{j-1}}{j-a_2} - \Omega^2 \sum_{j=1}^{\infty} h_j \frac{\beta^{j+1}}{j+2-a_2} \right],
\tag{11.42}
$$

$$
c_2 = r_1^{1-a_1} \left[d_1 \sum_{j=1}^{\infty} h_j r_1^{j-1} - b_1 \sum_{j=1}^{\infty} h_j \frac{r_1^{j-1}}{j-a_1} + \Omega^2 \sum_{j=1}^{\infty} h_j \frac{r_1^{j+1}}{j+2-a_1} \right],
\tag{11.43}
$$

$$
c_3 = - \left[b_2 \sum_{j=1}^{\infty} \frac{h_j}{j-a_2} - \Omega^2 \sum_{j=1}^{\infty} \frac{h_j}{j+2-a_2} \right],
\tag{11.44}
$$

and

$$
\Omega^2 = \frac{d_1 \sum\limits_{j=1}^{\infty} h_j r_2^{j-1} + b_2 r_2^{-1+a_2} \sum\limits_{j=1}^{\infty} \frac{h_j}{j-a_2} - b_2 \sum\limits_{j=1}^{\infty} \frac{h_j r_2^{j-1}}{j-a_2}}{r_2^{-1+a_2} \sum\limits_{j=1}^{\infty} \frac{h_j}{j+2-a_2} - \sum\limits_{j=1}^{\infty} \frac{h_j r_2^{j+1}}{j+2-a_2}},
\tag{11.45}
$$

$$\left[-b_2 \sum_{j=1}^{\infty} \frac{h_j \beta^{j-1}}{j-a_2} + \Omega^2 \sum_{j=1}^{\infty} \frac{h_j \beta^{j+1}}{j+2-a_2} \right] \left(\frac{\beta}{r_1} \right)^{1-a_2}$$

$$+ \left[b_2 \sum_{j=1}^{\infty} \frac{h_j r_1^{j-1}}{j-a_2} - \Omega^2 \sum_{j=1}^{\infty} \frac{h_j r_1^{j+1}}{j+2-a_2} \right] = d_1 \sum_{j=1}^{\infty} h_j r_1^{j-1}, \tag{11.46}$$

$$\left[d_1 \sum_{j=1}^{\infty} h_j r_1^{j-1} - b_1 \sum_{j=1}^{\infty} \frac{h_j r_1^{j-1}}{j-a_1} + \Omega^2 \sum_{j=1}^{\infty} \frac{h_j r_1^{j+1}}{j+2-a_1} \right] \left(\frac{r_1}{r_2} \right)^{1-a_1}$$

$$+ \left[b_1 \sum_{j=1}^{\infty} \frac{h_j r_2^{j-1}}{j-a_1} - \Omega^2 \sum_{j=1}^{\infty} \frac{h_j r_2^{j+1}}{j+2-a_1} \right] = d_1 \sum_{j=1}^{\infty} h_j r_2^{j-1}. \tag{11.47}$$

Substituting Eq.(11.45) into Eqs.(11.46) and (11.47), two equations with respect to r_1 and r_2 are obtained, r_1 is in the range of $(0, r_0)$, and r_2 is in the range of $(r_1, 1)$. The two demarcating radii can be calculated by half interval search method from Eqs.(11.46) and (11.47) until the required accuracy is reached. The plastic limit angular speed and stresses can then be calculated with the calculated values of r_1 and r_2 from Eqs.(11.41)$\sim$(11.45).

Similar to the results derived for the solid disc, the stresses may also violate yield criterion as the combination of stresses exceeds the yield point A in Fig.11.13 (b). The location for the maximum radial stress $\sigma_r(r = r_m)$ can be determined from the following equation:

$$\sigma_r \sum_{j=2}^{\infty} h_j (j-1) r_m^{j-2} = b_1 \sum_{j=2}^{\infty} h_j (j-1) \frac{r_m^{j-2}}{j-a_1}$$

$$- \Omega^2 \sum_{j=1}^{\infty} h_j (j+1) \frac{r_m^j}{j+2-a_1} + c_2(-1+a_1) r_m^{-2+a_1}. \tag{11.48}$$

Stress redistribution is necessary if the maximum radial stress σ_r at $r = r_m$ exceeds 1. The entire disc can then be divided into five parts by four demarcating radii denoted as r_1, r_2, r_3 and r_4. The corresponding boundary and continuity conditions are (1) $\sigma_r = 0$ at $r = \beta$; (2) $\sigma_r(r = r_1, \beta \leqslant r \leqslant r_1) = d_1$; (3) $\sigma_r(r = r_1, r_1 \leqslant r \leqslant r_2) = d_1$; (4) $\sigma_r(r = r_2, r_1 \leqslant r \leqslant r_2) = 1$; (5) $\sigma_r(r = r_3, r_3 \leqslant r \leqslant r_4) = 1$; (6) $\sigma_r(r = r_4, r_3 \leqslant r \leqslant r_4) = d_1$; (7) $\sigma_r(r = r_4, r_4 \leqslant r \leqslant 1) = d_1$; (8) $\sigma_r(r = 1) = 0$; (9) $\partial \sigma_r/\partial r(r = r_2, r_1 \leqslant r \leqslant r_2) = 0$ or $\partial \sigma_r/\partial r(r = r_3, r_3 \leqslant r \leqslant r_4) = 0$. The nine conditions lead to nine equations which determine the nine constants of c_1, c_2, c_3, c_4, r_1, r_2, r_3, r_4, and Ω^2. Since the derivative of σ_r vanishes at either $r = r_2$ or $r = r_3$, two possible Ω^2 can be derived. The above conditions (1)$\sim$(4) and (9) can be used to determine c_1, c_2, r_1, r_2, and Ω^2; and the conditions (5)$\sim$(8) and (9) can be used to determine c_3, c_4, r_3, r_4, and Ω^2. The smaller Ω^2 derived from the two cases is the actual plastic limit angular speed, and the corresponding case leads to the actual stress redistribution pattern. Once Ω^2 is determined, the other eight constants can be determined according to the conditions (1)$\sim$(8).

11.9 Special Case of $b = 0$

The unified yield criterion becomes the Tresca criterion when the parameter b is equal to zero. The following context demonstrates that the solutions based on the Tresca criterion are the special cases of the present solutions.

For a rotating solid disc, the stresses σ_r, σ_θ, and Ω^2 with $b = 0$ can be calculated from Eqs.(11.6)~(11.9),

$$\sigma_r = h^{-1} \left[\sum_{j=1}^{\infty} \frac{h_j r^{j-1}}{j} - \frac{\sum_{j=1}^{\infty} \frac{h_j}{j}}{\sum_{j=1}^{\infty} \frac{h_j}{j+2}} \sum_{j=1}^{\infty} \frac{h_j r^{j+1}}{j+2} \right],$$

$$\sigma_\theta = 1 \text{ and } \Omega^2 = \frac{\sum_{j=1}^{\infty} \frac{h_j}{j}}{\sum_{j=1}^{\infty} \frac{h_j}{j+2}}.$$

$$(11.49)$$

Eq.(11.49) will converge to the following expression if the disc has a uniform thickness:

$$\sigma_r = 1 - r^2, \quad \sigma_\theta = 1 \text{ and } \Omega^2 = 3, \tag{11.50}$$

which are identical to the solutions based on the Tresca criterion (Lenard and Haddow, 1972; Güven, 1992; Gamer, 1983).

For an annular disc, the two cases in Fig.11.13(a) and Fig.11.13(b) result in the same solution when the unified yield criteria parameter b is equal to 0, which is the exact solution based on the Tresca criterion. The stresses are

$$\begin{cases} \sigma_r = h^{-1} \left[\sum_{j=1}^{\infty} \frac{h_j r^{j-1}}{j} - \Omega^2 \sum_{j=1}^{\infty} \frac{h_j r^{j+1}}{j+2} \right] \\ \\ \quad - r^{-1} h^{-1} \left[\sum_{j=1}^{\infty} \frac{h_j}{j} - \Omega^2 \sum_{j=1}^{\infty} \frac{h_j}{j+2} \right], \\ \\ \sigma_\theta = 1, \end{cases} \tag{11.51}$$

and the corresponding plastic limit angular speed becomes

$$\Omega^2 = \frac{\sum_{j=1}^{\infty} \frac{h_j}{j} - \beta \sum_{j=1}^{\infty} \frac{h_j \beta^{j-1}}{j}}{\sum_{j=1}^{\infty} \frac{h_j}{j+2} - \beta \sum_{j=1}^{\infty} \frac{h_j \beta^{j+1}}{j+2}}. \tag{11.52}$$

For the particular case of uniform thickness, the above equations give

$$\sigma_r = (1 - r^{-1}) - \frac{r^2 - r^{-1}}{\beta^2 + \beta + 1}, \quad \sigma_\theta = 1, \qquad (11.53)$$

and

$$\Omega^2 = \frac{3}{\beta^2 + \beta + 1}, \qquad (11.54)$$

which are again identical to the results based on the Tresca criterion (Lenard and Haddow, 1972; Chakrabarty, 1987). However, it should be mentioned that the solution based the Tresca criterion may lead to meaningless plastic strain at the disc center if the associated flow rule is applied (Gamer, 1984). To derive a reasonable displacement distribution of a rotating disc, a non-associated flow rule or special processes for the singularity should be considered (Gven, 1994; Gamer, 1984; Berman and Pai, 1972).

11.10   Results and Discussion

Fig.11.14 shows the stress distributions of a solid disc with uniform thickness in terms of three special yield criteria. It can be seen that both the radial stress σ_r and circumferential stress σ_θ are the smallest when b is zero, or when the Tresca criterion is used. The twin-shear yield criterion leads to the largest stresses with the maximum σ_θ at the center ($r = r_0$). The stress distributions with respect to the Mises criterion can be approximated with UYC of $b = 0.5$, which go between those from the other two yield criteria.

From Fig.11.14, the effect of different yield criteria on radial stress σ_r is minimal, while it is remarkable on σ_θ. Fig.11.15 illustrates the stress violation to the yield criterion in the central area of the disc if the thickness is reduced along the disc radius ($h = 3 - 2r$). The redistributed stress fields are plotted in Fig.11.16, where the inner area ($r < r_1$) corresponds to the yield point A in Fig.11.2. σ_r and σ_θ are constant in this area.

If the thickness increases with increasing radius r, yield criteria will not be violated and the stress distribution is shown in Fig.11.17. Compared with Fig.11.16, the location of the maximum σ_θ in Fig.11.17 is closer to the disc center, indicating that the stress distributions strongly depend on the disc thickness variation.

Figs.11.18(a) and 11.18(b) show the effects of different yield criteria on the plastic limit angular speed of a rotating solid disc. It can be seen that Ω^2 increases from 9.8% to 14.0% with b varying from 0 to 1 when the thickness function is a decreasing function of radius(Fig.11.18(a)). Ω^2 increases from 14.0% to 17.5% when the thickness function is an increasing function of

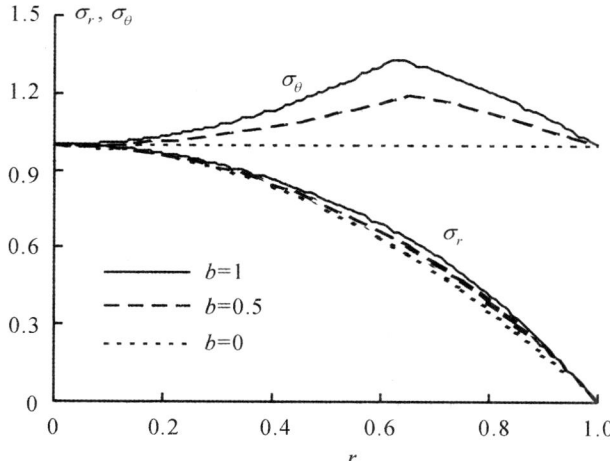

Fig. 11.14. Stress distribution of solid disc $(h = 1)$

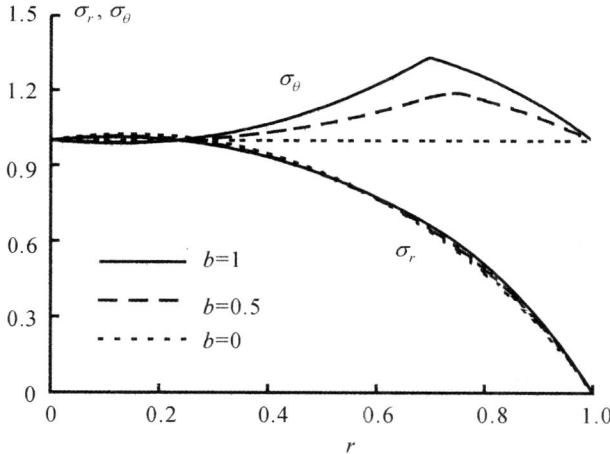

Fig. 11.15. Stress distributions of solid disc with violation of yield condition in central area $(h = 3 - 2r)$

radius(Fig.11.18(b)). It indicates that Ω^2 also depends on the thickness variation function of the disc. The distribution of stresses derived in the present study is helpful in understanding the response characteristics of a rotating disc. Economical design can be achieved for mechanical machines according to the current limit angular velocities.

For a rotating annular disc, the stress distributions in the plastic limit state are quite different from those of a rotating solid disc. Figs.11.19 and 11.20 illustrate the stress distributions of a uniform thickness disc when $\beta =$

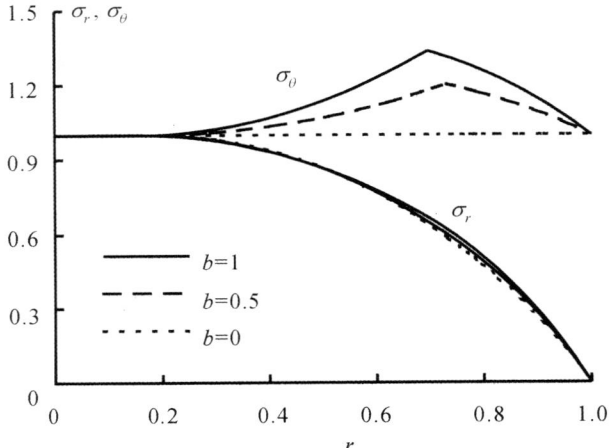

Fig. 11.16. Stress distribution of solid disc after redistribution of stress $(h = 3 - 2r)$

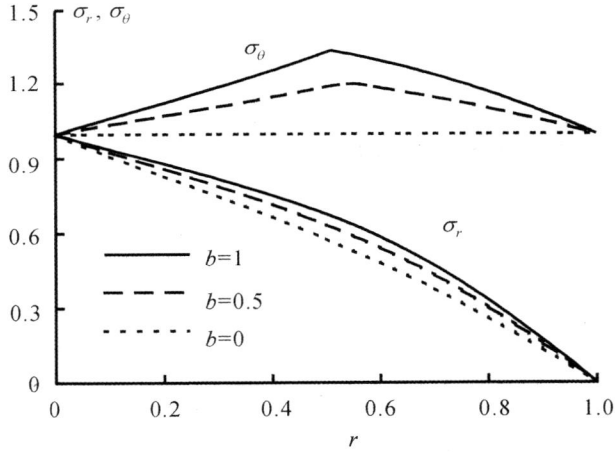

Fig. 11.17. Stress distribution of solid disc $(h = 1 + 2r)$

0.1 (Case 2) and $\beta = 0.5$ (Case 1) respectively. The circumferential stress has two peaks in Case 2; while it has one in Case 1. The circumferential stress varies along the radius for both cases.

The influence of different yield criteria becomes more and more prominent with an increasing inner radius. Figs.11.21 and 11.22 show the stress distributions when the disc has a small hole at the center for the two opposite types of thickness function.

It is seen that near the small hole the stresses change rapidly. There is no stress violation when the disc thickness increases with the radius. When

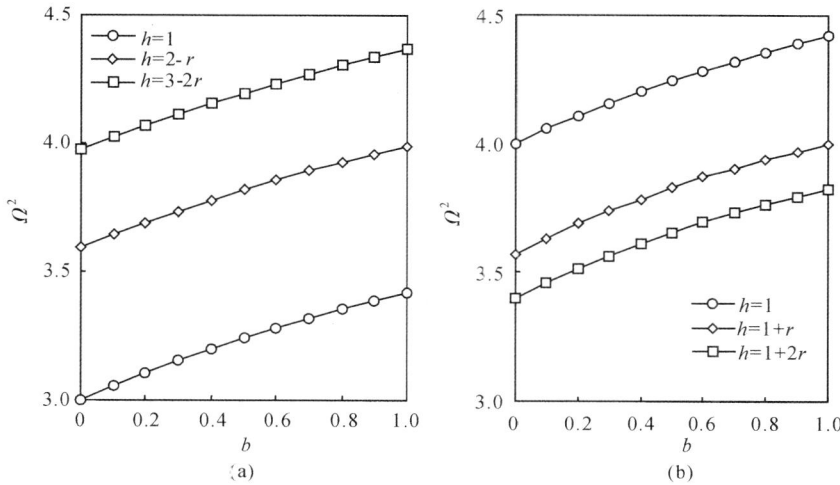

Fig. 11.18. Limit angular speed of solid disc

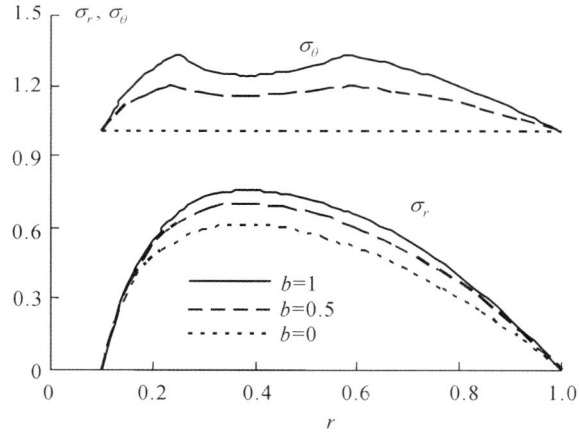

Fig. 11.19. Stress distribution of annular disc ($h = 1$, $\beta = 0.1$)

the disc thickness decreases with the radius, on the other hand, the violation to yield criterion occurs in the middle area in the disc as seen in Fig.11.22 instead of the center area for the case of the solid disc. It indicates that a small hole at the disc center will change the distribution of stress drastically. Constant distribution of σ_θ based on the Tresca criterion may not predict the stress distributions of an annular disc with variable thickness.

Fig.11.23 shows the plastic limit angular speed with respect to the three yield criteria and various thickness functions for the cases of $\beta = 0.1$ and $\beta = 0.5$. It shows that Ω^2 increases with the increase of unified yield criterion parameter b.

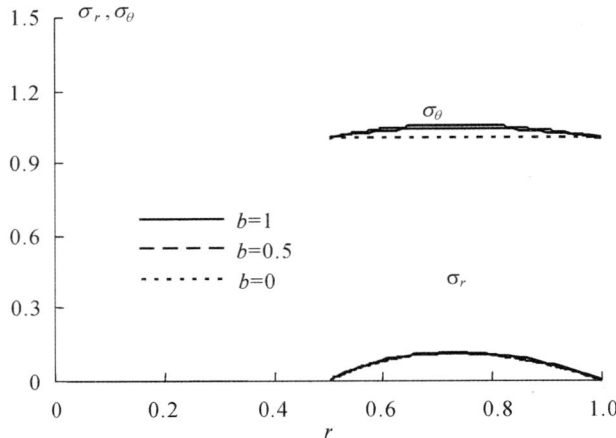

Fig. 11.20. Stress distribution of annular disc ($h = 1$, $\beta = 0.5$)

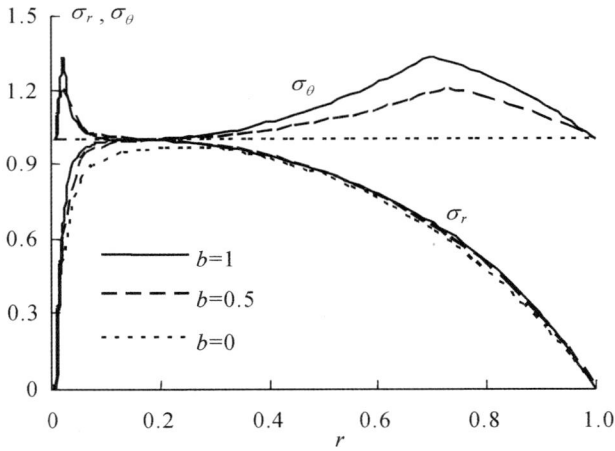

Fig. 11.21. Stress distribution with yield condition violation ($h = 3 - 2r$, $\beta = 0.1$)

The effect of the ratio of inner radius to outer radius on the plastic limit angular speed is illustrated in Fig.11.24(a) for uniform thickness disc and Fig.11.24(b) for disc with linearly varying thickness. It is seen that with the increase of ratio β, the effect of yield criteria becomes less significant. For a thin rotating ring, Ω^2 is equal to 1 and independent of yield criterion. When β approaches 0, the limit angular speed approaches that for a solid disc. It implies that a small hole at the disc center does not affect the plastic limit angular speed significantly, instead it affects stress distributions drastically.

It should be mentioned that the plane stress condition in a rotating disc with variable thickness is not exactly satisfied. However, if the disc thickness

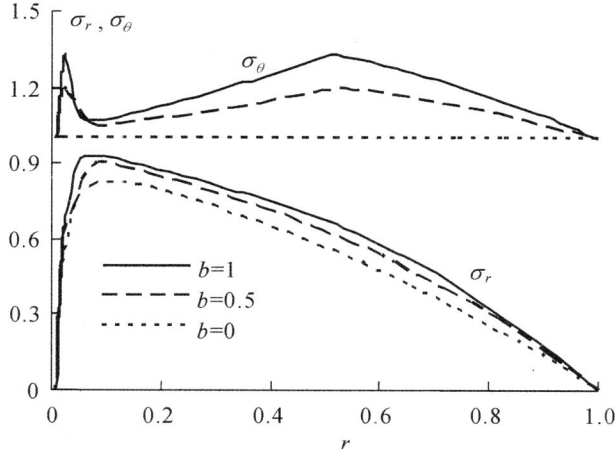

Fig. 11.22. Stress distribution of annular disc ($h = 1 + 2r$, h=1, $\beta = 0.1$)

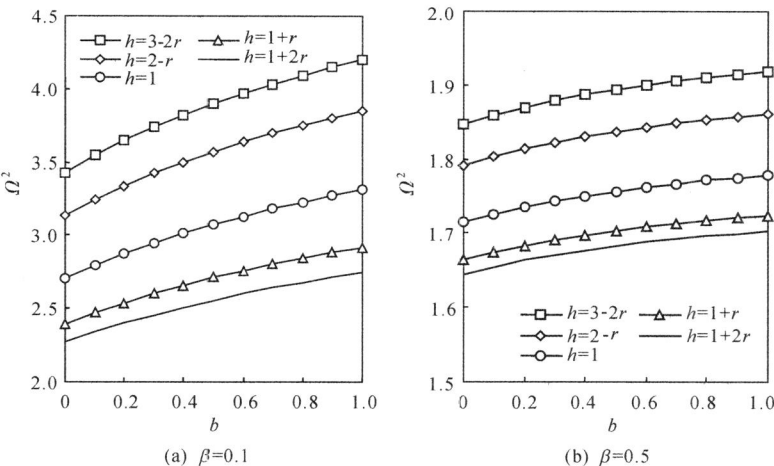

Fig. 11.23. Limit angular speed of annular disc with different unified strength theory parameter b

is small enough compared to the diameter of a solid disc or the difference between the outer and inner radii of an annular disc is small enough, the plane stress assumption is approximately valid. When the inner radius of an annular disc is very close to the outer radius, e.g., $\beta > 0.8$, the limit angular speed is almost independent of the yield criterion and the thickness variation function as illustrated in Fig.11.24. Thus the present solution based on plane stress assumption is more suited to a thin rotating disc in fully plastic state.

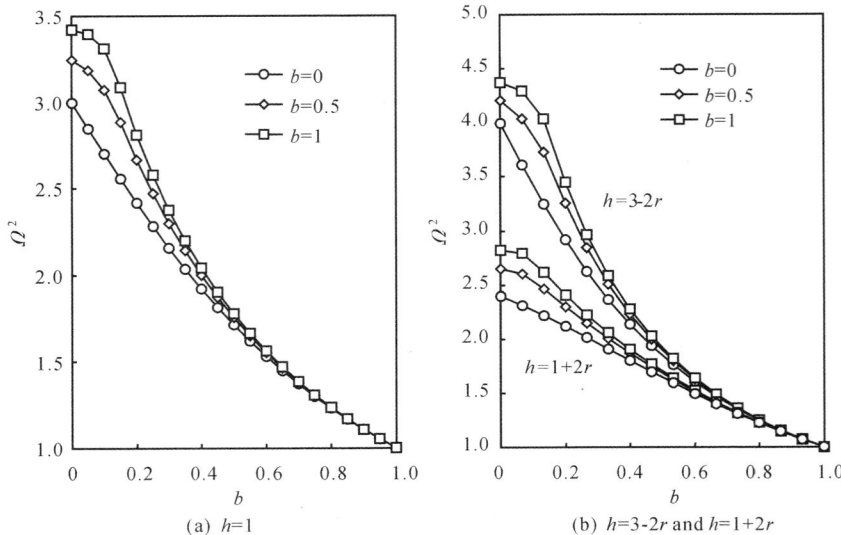

Fig. 11.24. Limit angular speed verses ratio of inner radius to outer radius

11.11 Summary

A series of elastic limits and plastic solutions for rotating discs and cylinders based on the unified yield criterion are presented. The unified solution encompasses the existing solutions based on the Tresca criterion and the Huber-von Mises criterion as special cases. The current unified solutions for the elasto-plastic stress field and elasto-plastic rotating speed of a rotating disc are useful in engineering applications. The different yield criterion parameter b has no effect on the elastic limit rotating speed of a rotating disc and cylinder, while it affects the plastic limit speed.

The percentage difference between the maximum plastic limit rotating speed derived from the twin-shear stress criterion ($b=1$) and that using the Tresca criterion ($b=0$) is 14%. Therefore it is important to choose proper yield criterion in order to obtain an economical and optimal design.

The unified yield criterion is applicable for metallic non-SD materials. It is a special case of the Yu unified strength theory (Yu, 1991; 2004).

11.12 Problems

Problem 11.1 Try your hand at an application of the unified yield criterion for non-SD materials.

Problem 11.2 Why does the solution obtained by using the unified yield criterion contain all the solutions of the Tresca yield criterion, the von

Mises yield criterion, the twin-shear yield criterion and other possible yield criteria adopted for those materials with the same yield stress in tension and in compression?

Problem 11.3 Write a paper concerning the plastic analysis of rotating annular discs using the unified yield criterion.

Problem 11.4 Try your hand at an application of the unified yield criterion for SD materials.

Problem 11.5 Try to obtain the unified solution of plastic analysis of a rotating hollow cylinder.

Problem 11.6 Can you introduce a unified solution for a plastic rotating disc using the unified strength theory? The ratio of tensile strength σ_t to compressive strength σ_c is $\alpha = \sigma_t/\sigma_c = 0.8$.

Problem 11.7 A high-strength alloy has the strength ratio in tension and compression $\alpha = 0.9$. Find the unified solution of a rotating disc made of this alloy.

Problem 11.8 Compare the plastic solutions of rotating discs using the unified yield criterion and the unified strength theory with $\alpha = 0.8$.

Problem 11.9 The plastic limit rotating speeds of solid cylinders is described in literature. Compare their results with the unified solution.

Problem 11.10 The plastic limit rotating speeds of hollow cylinders is described in literature. Compare their results with the unified solution.

References

Berman I, Pai DH (1972) A theory of anisotropic steady state creep. Int. J. Mech. Sci., 14(11):755-763

Chakrabarty J (1987) Theory of plasticity. McGraw-Hill, New York

Gamer U (1983) Tresca's yield condition and the rotating disk. J. Appl. Mech., 50:676-678

Gamer U (1984) Elastic-plastic deformation of the rotating solid disk. Ingenieur-Archiv; 54:345-354

Güven U (1992) Elastic-plastic stresses in a rotating annular disk of variable thickness and variable density. Int. J. Mech. Sci., 34(2):133-138

Güven U (1994) The fully plastic rotating solid disk of variable thickness. J. Appl. Math. Mech. (ZAMM), 74(1):61-65

Heyman J (1958) Plastic design of rotating discs. Proc. Inst. Mech. Engng., 172:531-538

Hu TC, Schield RT (1961) Minimum volume of disc. J. Appl. Math. Phys. (ZAMP) 12(5):414-421

Huang WB, Zeng GP (1989) Solving some plastic problems by using the Twin shear stress criterion, Acta Mechanica Sinica, 21(2):249-256 (English abstract)

Lenard L, Haddow JB (1972) Plastic collapse speeds for rotating cylinders. Int J. Mech. Sci., 14:285-292

Li YM (1988) Elasto-plastic limit analysis with a new yield criterion (twin-shear yield criterion), J. of Mech. Strength, 10(3):70-74 (in Chinese)

Ma GW, Hao H, Miyamoto Y (2001) Limit angular velocity of circular disc using unified yield criterion. Int. J. Mech. Sci., 43:1137-1153

Ma GW, Yu MH, Iwasaki S, Miyamoto Y, Deto H (1995) Unified elasto-plastic solution to rotating disc and cylinder. Journal of Structural Engineering, JSCE, 41A:79-85

Save MA, Massonnet CE (1972) Plastic analysis and design of plates, shells and disks, North-Holland Publ. Co., Amsterdam

Save MA, Massonnet CE, Saxce G de (1997) Plastic limit analysis of plates, shells and disks. Elsevier, Amsterdam

Timoshenko SP, Goodier JN (1951) Theory of elasticity. McGraw-Hill, New York

Yu MH (1992) A New System of Strength Theory. Xi'an Jiaotong University Press, Xi'an, China (in Chinese)

Yu MH (2004) Unified strength theory and its application, Springer, Berlin

Yu MH, He LN (1991) A new model and theory on yield and failure of materials under the complex stress state. In: Mechanical Behavior of Materials-6, Pergamon PressOxford, 641-645

Yu MH, He LN, Song LY (1985) Twin shear stress theory and its generalization, Scientia Sinica (Science in China), English ed. Series A28(11):1174-1183; Chinese ed., Series A, 28(12):1113-1120

12

Projectile Penetration into Semi-infinite Target

12.1 Introduction

A lot of research work has been conducted on impact and penetration analysis. The penetration studies include various lab and field tests, analytical derivations and numerical simulations. Early works were mainly experimental studies. In the last three decades, analytical and numerical tools have been used increasingly as a substitute for costly experiments. The critical issue in an analytical penetration model is to formulate properly the resultant penetration resistance force applied on the missile by the target medium. The most well-known resistance function is based on the so-called dynamic cavity expansion theory. The theory was pioneered by Bishop et al. (1945), who developed the equations for the quasi-static expansion of cylindrical and spherical cavities and estimated forces on conical nose punches pushed slowly into metal targets. Later Hill (1950) and Hopkins (1960) derived and discussed the dynamic and spherically symmetric cavity-expansion equations for an incompressible target material.

The cavity expansion theory was further developed by Luk and Forrestal (1987), Forrestal and Tzou (1997), and Mastilovic and Krajcinovic (1999) to model the penetration of projectiles through soil, porous rock, ceramic and concrete targets. An overview on projectile penetration into geological targets was given by Heuze (1990). Li QM (2005) summarized the recent progress in the penetration mechanics of a hard missile and extended Forrestal's concrete penetration model to missiles of general nose shapes.

By comparison with the analytical results derived from the cylindrical and spherical cavity expansion theories, it is found that the cylindrical assumption gives closer results to test data for low and medium velocity impacts. Tresca criterion and Huber-von Mises criterion were often used for penetration problems of metallic targets, and the Mohr-Coulomb strength theory was used for penetration problems of geomaterials (Longcope and Forrestal, 1983). The selection of failure criteria is of great importance (Zukas et al.,

1982). To reflect the strength difference effect and the effect of failure criteria, the unified strength theory (Yu, 1992; 2004) has been adopted in penetration analysis for both metallic and geological targets (Li and Yu, 2000; Li, 2001; Wei and Yu, 2002; Wei, 2002; Wang et al., 2004; 2005).

The present chapter firstly presents the spatial axisymmetric form of the unified strength theory as the failure condition of target materials in Section 12.2. The governing differential equations for concrete targets are summarized in Section 12.3. The cylindrical cavity expansion model is then applied to incompressible and compressible materials in Sections 12.4. Explicit forms of the pressure on the cavity expansion surface and the cavity expansion velocity are derived in Section 12.5. Section 12.6 gives the resistance force on different nose shapes of the projectile, which is simplified as a rigid body. The penetration depth of the projectile is obtained and compared with test results available in the published literature in Section 12.7.

12.2 Spatial Axisymmetric Form of Unified Strength Theory

There are four stress components σ_r, σ_θ, σ_z and τ_{rz} in a spatial axisymmetric problem. The other components, namely, $\tau_{r\theta}$ and $\tau_{\theta z}$, are zero. According to the spatial axisymmetric unified characteristics line theory (Yu et al., 2001), the stress σ_2 can be expressed as

$$\sigma_2 = \sigma_3 + m\left(\frac{\sigma_1 + \sigma_3}{2} - \sigma_3\right), \tag{12.1}$$

where m is a parameter and $0 \leqslant m \leqslant 2$. When $m=0$ and $m=2$, Eq.(12.1) is the Haar-von Karman complete plastic condition. If we define

$$P = \frac{\sigma_1 + \sigma_3}{2}, \ R = \frac{\sigma_1 - \sigma_3}{2}, \tag{12.2}$$

then

$$\sigma_1 = P + R, \ \sigma_2 = P + (m-1)R, \ \sigma_3 = P - R. \tag{12.3}$$

The non-zero stress components of an axisymmetric problem can be expressed as

$$\sigma_r = P + R\cos 2\theta, \ \sigma_z = P - R\cos 2\theta,$$
$$\tau_{rz} = R\sin 2\theta, \ \sigma_\theta = P + (m-1)R, \tag{12.4}$$

where θ is the angle between the directions of the maximum principal stress and axis r.

Since $m-1 \leqslant \sin \varphi_0$, that is $\sigma_2 \leqslant P - R \sin \varphi_0$, the unified strength theory (UST) has an expression with respect to the internal friction angle φ_0 and cohesion C_0,

$$F = \sigma_1 - \frac{1 - \sin \varphi_0}{(1+b)(1 + \sin \varphi_0)}(b\sigma_2 + \sigma_3) = \frac{2C_0 \cos \varphi_0}{1 + \sin \varphi_0},$$

$$\text{when} \quad \sigma_2 \leqslant \frac{1}{2}(\sigma_1 + \sigma_3) + \frac{\sin \varphi}{2}(\sigma_1 - \sigma_3), \tag{12.5a}$$

$$F' = \frac{1}{1+b}(\sigma_1 + b\sigma_2) - \frac{1 - \sin \varphi}{1 + \sin \varphi}\sigma_3 = \frac{2C_0 \cos \varphi_0}{1 + \sin \varphi_0},$$

$$\text{when} \quad \sigma_2 \geqslant \frac{1}{2}(\sigma_1 + \sigma_3) + \frac{\sin \varphi}{2}(\sigma_1 - \sigma_3). \tag{12.5b}$$

Then, there is

$$R = -\frac{2(1+b)\sin \varphi_0}{2(1+b) + mb(\sin \varphi_0 - 1)}P + \frac{2(1+b)C_0 \cos \varphi_0}{2(1+b) + mb(\sin \varphi_0 - 1)}. \tag{12.6}$$

The above equation can be rewritten as (Yu et al., 1997; 2001)

$$R = -P \sin \varphi_{\text{uni}} + C_{\text{uni}} \cos \varphi_{\text{uni}}, \tag{12.7}$$

where the unified strength parameters C_{uni} and φ_{uni} were proposed and derived by Yu et al. in 1997 and 2001. These two parameters are referred as the unified cohesion and unified internal friction angle corresponding to the UST respectively. Their relations to the material constants C_0 and φ_0 can be written as (Yu et al., 1997; 2001)

$$\sin \varphi_{\text{uni}} = \frac{2(1+b)\sin \varphi_0}{2(1+b) + mb(\sin \varphi_0 - 1)}, \tag{12.8}$$

$$C_{\text{uni}} = \frac{2(1+b)C_0 \cos \varphi_0}{2(1+b) + mb(\sin \varphi_0 - 1)} \cdot \frac{1}{\cos \varphi_{\text{uni}}}. \tag{12.9}$$

Denoting compressive stress as P, Eq.(12.7) can be expressed as (Yu et al., 1997; 2001)

$$R = P \sin \varphi_{\text{uni}} + C_{\text{uni}} \cos \varphi_{\text{uni}}. \tag{12.10}$$

12.3 Fundamental Equations for Concrete Targets

12.3.1 Conservation Equations

In cylindrical coordinates, the conservation equations of momentum and mass for the target materials can be expressed as

$$\frac{\partial v}{\partial r} + \frac{v}{r} = -\frac{1}{\rho}\frac{d\rho}{dt}, \qquad (12.11)$$

$$\frac{\partial \sigma_r}{\partial r} + \frac{\sigma_r - \sigma_\theta}{r} = -\rho\frac{dv}{dt}, \qquad (12.12)$$

where v is the radial velocity of a particle in the target material and v is positive if it is in the outward direction.

12.3.2 Relation between Pressure and Bulk Strain

If the material is compressive, the relation between pressure and bulk strain can be expressed as

$$P = K\eta = K(1 - \frac{\rho_0}{\rho}), \qquad (12.13)$$

where η is the bulk strain, K is the bulk modulus, P is the hydrostatic pressure and can be written as

$$P = \frac{1}{3}(\sigma_r + \sigma_\theta + \sigma_z). \qquad (12.14)$$

For the problem of cavity expansions, the relation among stresses is

$$\sigma_z = \nu\,(\sigma_r + \sigma_\theta) \quad \text{in elastic zone,} \qquad (12.15a)$$

$$\sigma_z = \frac{1}{2}\,(\sigma_r + \sigma_\theta) \quad \text{in plastic zone.} \qquad (12.15b)$$

Eqs.(12.15a) and (12.15b) are applied respectively for the elastic zone and plastic zone when material is compressible, while only Eq. (12.15b) is used when material is incompressible.

12.3.3 Failure Criterion Expressed by σ_r and σ_θ

The UST is used as the failure condition for the target material in this chapter. According to the axisymmetric stress state of the target material impacted and penetrated by a long rod, Eq.(12.10) has another form of

$$\sigma_r - \sigma_\theta = A_{\text{uni}}\sigma_r + B_{\text{uni}}, \qquad (12.16)$$

where

$$A_{\text{uni}} = \frac{2\sin\varphi_{\text{uni}}}{1 + \sin\varphi_{\text{uni}}}, \quad B_{\text{uni}} = \frac{2C_{\text{uni}}\cos\varphi_{\text{uni}}}{1 + \sin\varphi_{\text{uni}}}.$$

12.3.4 Interface Conditions

The target medium can be divided into four zones during the cavity expansion, i.e., a plastic zone, a radial cracked zone, an elastic zone and an undisturbed zone. At the two interfaces between the plastic and radial cracked zones, radial cracked and elastic zones, the Hugoniot jump conditions are valid. According to the conservation of mass and momentum across the interface, there are

$$[\rho(v - c_J)] = 0, \tag{12.17}$$

$$[\sigma_r + \rho v(v - c_J)] = 0, \tag{12.18}$$

where the expression $[G] = G^+ - G^-$ stands for the magnitude of the discontinuity of the square-bracketed variable across the wave front (interface) that propagates with an interface velocity of c_J. The above equations can also be rewritten as

$$\rho_1(v_1 - c_J) = \rho_2(v_2 - c_J), \tag{12.19}$$

$$\sigma_2 - \sigma_1 = \rho_1(c_J - v_1)(v_2 - v_1). \tag{12.20}$$

12.4 Cylindrical Cavity Expansion Analysis

A cylindrical symmetric cavity expands with velocity v_r from an initial radius of zero when the target is impacted and penetrated by a long rod (Fig.12.1). In Fig.12.2, c is the interface velocity between the plastic and radial cracked zones; c_1 is the interface velocity between the radial cracked and elastic zones; c_d is the elastic dilatation velocity. The stress in the plastic zone ($v_r t \leqslant r \leqslant ct$) has reached the yield surface of the unified strength theory. Because geomaterials are always very weak in tension, radial cracks adjacent to the plastic zone are often observed in a penetration process for targets made of geomaterials. The formation and the magnitude of the area for a cracked zone or a damaged zone depend on the circumferential tensile stress. If the circumferential stress exceeds the tensile strength of the target material, a radial cracked zone forms. The range of the radial cracked zone can be represented by $ct < r \leqslant c_1 t$. The elastic zone is in the range of $c_1 t < r \leqslant c_d t$; and the undisturbed zone is in the range of $r > c_d t$. Defining a dimensionless variable of $\xi = r/ct$, the four different zones can be categorized in Fig.12.3.

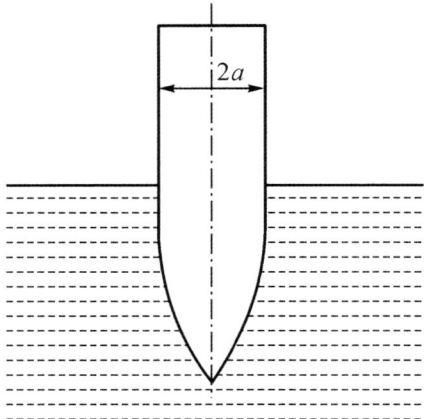

Fig. 12.1. Penetration by a long rod

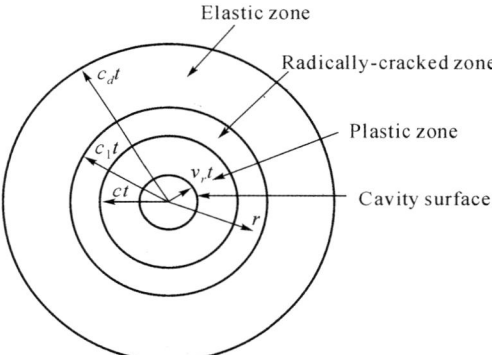

Fig. 12.2. Different zones of target material

12.4.1 Elastic Zone ($c_1 t \leqslant r \leqslant c_d t,\ \beta_1/\beta \leqslant \xi \leqslant 1/\alpha$)

In the elastic zone, the target materials satisfy the linear stress-strain relations. According to the generalized Hooke's law,

$$\sigma_r = -\frac{E}{(1-2\nu)(1+\nu)}\left[(1-\nu)\frac{\partial u}{\partial r} + \nu\frac{u}{r}\right], \qquad (12.21)$$

$$\sigma_\theta = -\frac{E}{(1-2\nu)(1+\nu)}\left[\nu\frac{\partial u}{\partial r} + (1-\nu)\frac{u}{r}\right], \qquad (12.22)$$

where E and ν are the modulus of elasticity and Poisson's ratio; u is the radial displacement. The normal stresses are positive in compression for convenient formulation.

The conservation equation of momentum can be expressed as

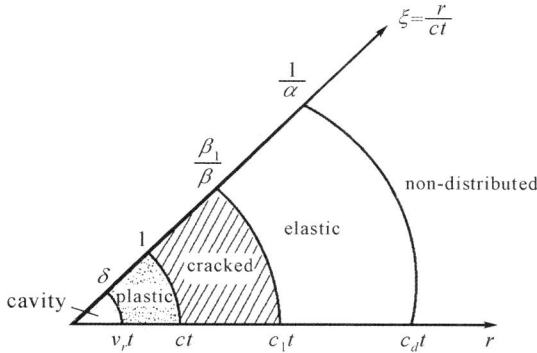

Fig. 12.3. Dimensionless expression of the four zones

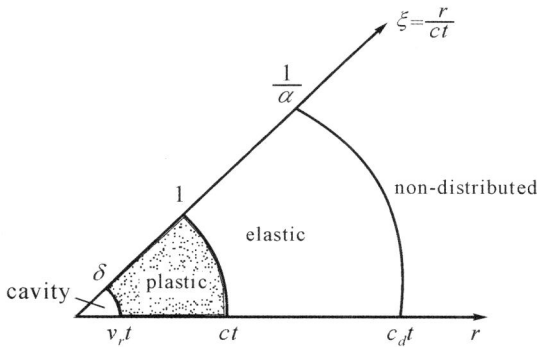

Fig. 12.4. Dimensionless expression of the three zones

$$\frac{\partial \sigma_r}{\partial r} + \frac{\sigma_r - \sigma_\theta}{r} = -\rho \left(\frac{\partial \nu}{\partial t} + \nu \frac{\partial \nu}{\partial r} \right). \tag{12.23}$$

Substituting Eqs.(12.21) and (12.22) into Eq.(12.23), we obtain

$$\frac{\partial^2 u}{\partial r^2} + \frac{1}{r} \frac{\partial u}{\partial r} - \frac{u}{r^2} = \frac{1}{c_d^2} \frac{\mathrm{d}^2 u}{\mathrm{d} t^2}, \tag{12.24}$$

where c_d is the elastic wave velocity in the semi-infinite medium and

$$c_d = \sqrt{\frac{E(1 - \nu_0)}{(1 + \nu_0)(1 - 2\nu_0)\rho_0}}. \tag{12.25}$$

Defining

$$u = \frac{\bar{u}}{ct}, \tag{12.26}$$

Eq.(12.24) can be rewritten as

$$(1 - \alpha^2 \xi^2)\frac{d^2\bar{u}}{d\xi^2} + \frac{1}{\xi}\frac{d\bar{u}}{d\xi} - \frac{1}{\xi^2}\bar{u} = 0, \tag{12.27}$$

where $\alpha = c/c_d$.

Defining

$$z = \alpha\xi, \bar{u} = z\phi, F = \frac{d\phi}{dz}, \tag{12.28}$$

then

$$\begin{cases} \dfrac{d\bar{u}}{d\xi} = \alpha\phi + \alpha z F, \\[2mm] \dfrac{d^2\bar{u}}{d\xi^2} = 2\alpha^2 F + \alpha^2 z\dfrac{dF}{dz}. \end{cases} \tag{12.29}$$

Eq.(12.27) can be simplified into a first-order differential equation with reference to Eq.(12.29). Integrating Eq.(12.27), we obtain

$$\bar{u} = A\alpha\xi + \left[\frac{\alpha\xi}{2}\ln\frac{1 + \sqrt{1 - \alpha^2\xi^2}}{\alpha\xi} - \frac{\sqrt{1 - \alpha^2\xi^2}}{2\alpha\xi}\right] \cdot B, \tag{12.30}$$

where A and B are integration constants that can be determined by considering the following boundary conditions:

$$\bar{u}(\xi = \frac{1}{\alpha}) = 0, \tag{12.31}$$

$$\sigma_\theta(\xi = \frac{\beta_1}{\beta}) = -\sigma_t, \tag{12.32}$$

where Eq.(12.31) indicates that the radial displacement is zero at the interface of the elastic and undistributed zones; Eq.(12.32) indicates that the circumferential stress reaches the tensile strength at the interface of the elastic and radial crack zones.

With reference to Eq.(12.30) and Eq.(12.31),

$$A = 0. \tag{12.33}$$

The displacement distribution in the elastic zone is then derived as

$$\bar{u} = \left[\frac{\alpha\xi}{2}\ln\frac{1 + \sqrt{1 - \alpha^2\xi^2}}{\alpha\xi} - \frac{\sqrt{1 - \alpha^2\xi^2}}{2\alpha\xi}\right] \cdot B. \tag{12.34}$$

From Eq.(12.34),

$$\frac{\bar{u}}{\xi} = \left[\frac{\alpha}{2} \ln \frac{1 + \sqrt{1 - \alpha^2 \xi^2}}{\alpha \xi} - \frac{\sqrt{1 - \alpha^2 \xi^2}}{2\alpha \xi^2} \right] \cdot B, \qquad (12.35)$$

$$\frac{\partial \bar{u}}{\partial \xi} = \left[\frac{\alpha}{2} \ln \frac{1 + \sqrt{1 - \alpha^2 \xi^2}}{\alpha \xi} + \frac{\sqrt{1 - \alpha^2 \xi^2}}{2\alpha \xi^2} \right] \cdot B. \qquad (12.36)$$

Defining

$$\bar{\sigma}_r = \frac{\sigma_r}{K}, \ \bar{\sigma}_\theta = \frac{\sigma_\theta}{K}, \ \bar{\sigma}_t = \frac{\sigma_t}{K}, \qquad (12.37)$$

the dimensionless circumferential stress and radial stress in the elastic zone can be derived from Eqs.(12.21) and (12.22),

$$\bar{\sigma}_\theta = -\frac{3}{1 + \nu} \left[\frac{\alpha}{2} \ln \frac{1 + \sqrt{1 - \alpha^2 \xi^2}}{\alpha \xi} - \frac{(1 - 2\nu)\sqrt{1 - \alpha^2 \xi^2}}{2\alpha \xi^2} \right] \cdot B, \qquad (12.38)$$

$$\bar{\sigma}_r = -\frac{3}{1 + \nu} \left[\frac{\alpha}{2} \ln \frac{1 + \sqrt{1 - \alpha^2 \xi^2}}{\alpha \xi} + \frac{(1 - 2\nu)\sqrt{1 - \alpha^2 \xi^2}}{2\alpha \xi^2} \right] \cdot B. \qquad (12.39)$$

The integration constant B can be obtained from Eq.(12.32) with reference to Eq.(12.38),

$$B = \frac{1 + \nu}{3} \sigma_t \left[\frac{\alpha}{2} \ln \frac{1 + \sqrt{1 - \alpha^2 (\beta_1/\beta)^2}}{\alpha (\beta_1/\beta)} - \frac{(1 - 2\nu)\sqrt{1 - \alpha^2 (\beta_1/\beta)^2}}{2\alpha (\beta_1/\beta)^2} \right]^{-1} . \qquad (12.40)$$

Defining a dimensionless radial velocity as

$$\bar{\nu}(\xi) = \frac{\bar{\nu}}{c}, \qquad (12.41)$$

Eq.(12.34) yields

$$\bar{\nu}(\xi) = \left[-\frac{\sqrt{1 - \alpha^2 \xi^2}}{\alpha \xi} \right] \cdot B. \qquad (12.42)$$

At the interface of elastic-radial cracked zones, i.e., $\xi = \beta_1/\beta$, the radial stress, the circumferential stress and the velocity in the elastic zone are

$$\bar{\sigma}_{r1} = -\frac{3}{1+\nu} \left[\frac{\alpha}{2} \ln \frac{1 + \sqrt{1 - \alpha^2 (\beta_1/\beta)^2}}{\alpha (\beta_1/\beta)} + \frac{(1 - 2\nu)\sqrt{1 - \alpha^2 (\beta_1/\beta)^2}}{2\alpha (\beta_1/\beta)^2} \right] \cdot B,$$
$$(12.43)$$

$$\bar{\sigma}_{\theta 1} = -\frac{3}{1+\nu} \left[\frac{\alpha}{2} \ln \frac{1 + \sqrt{1 - \alpha^2 (\beta_1/\beta)^2}}{\alpha (\beta_1/\beta)} - \frac{(1 - 2\nu)\sqrt{1 - \alpha^2 (\beta_1/\beta)^2}}{2\alpha (\beta_1/\beta)^2} \right] \cdot B,$$
$$(12.44)$$

$$\bar{\nu}_1(\xi) = \left[-\frac{\sqrt{1 - \alpha^2 (\beta_1/\beta)^2}}{\alpha (\beta_1/\beta)} \right] \cdot B.$$
$$(12.45)$$

12.4.2 Interface of Elastic-cracked Zones ($r = c_1 t,\ \xi = \beta_1/\beta$)

Defining at the interface the dimensionless radial stress and radial velocity in the cracked zone as $\bar{\sigma}_{r2}$ and $\bar{\nu}_2$, respectively, with reference to the Hugoniot jump condition,

$$\rho_1 \left(\bar{\nu}_1 - \frac{\beta_1}{\beta} \right) = \rho_2 \left(\bar{\nu}_2 - \frac{\beta_1}{\beta} \right),$$
$$(12.46)$$

$$\bar{\sigma}_{r2} = \bar{\sigma}_{r1} + \frac{\rho_1}{\rho_0} \beta^2 \left(\frac{\beta_1}{\beta} - \bar{\nu}_1 \right) (\bar{\nu}_2 - \bar{\nu}_1),$$
$$(12.47)$$

where ρ_1 and ρ_2 are the density of materials in the elastic and crack zones respectively.

According to the pressure-bulk strain relation, in the elastic zone,

$$\frac{(1 + \nu)\sigma_r + (2 - \nu)\sigma_\theta}{3} = K \left(1 - \frac{\rho_0}{\rho} \right).$$
$$(12.48)$$

Since the circumferential stress is zero in the cracked zone, the above equation can be rewritten as

$$\frac{1 + \nu}{3} \sigma_r = K \left(1 - \frac{\rho_0}{\rho} \right).$$
$$(12.49)$$

At the interface, the circumferential stress in the elastic zone reaches the tensile strength, which implies

$$\bar{\sigma}_{\theta 1} = -\bar{\sigma}_t.$$
$$(12.50)$$

Substituting the above equation into Eq.(12.48), the radial stress in the elastic zone near to the interface can be expressed as

$$\bar{\sigma}_{r1} = \frac{2-\nu}{1+\nu}\bar{\sigma}_t + \frac{3}{1+\nu}\left(1 - \frac{\rho_0}{\rho_1}\right). \tag{12.51}$$

From Eq.(12.49), the radial stress in the cracked zone near to the interface can be derived as

$$\bar{\sigma}_{r2} = \frac{3}{1+\nu}\left(1 - \frac{\rho_0}{\rho_2}\right). \tag{12.52}$$

Substituting Eq.(12.51) into Eq.(12.47),

$$\bar{\sigma}_{r2} = \bar{\sigma}_{r1} + \frac{3(\beta_1 - \beta\bar{\nu}_1)(\beta\bar{\nu}_2 - \beta\bar{\nu}_1)}{3 - (1+\nu)(\bar{\sigma}_{r1} - \bar{\sigma}_t)}. \tag{12.53}$$

Putting the expression of ρ_1 and ρ_2 from Eqs.(12.51) and (12.52) into Eq.(12.46),

$$\beta\bar{\nu}_2 = \beta_1 + \frac{3 - \bar{\sigma}_{r2}(1+\nu)}{3 - (\bar{\sigma}_{r1} - \bar{\sigma}_t)(1-\nu)}(\beta\bar{\nu}_1 - \beta_1). \tag{12.54}$$

12.4.3 Radial Cracked Zone ($ct \leqslant r \leqslant c_1 t$, $1 \leqslant \xi \leqslant \beta_1/\beta$)

When the circumferential stress reaches the tensile strength, the radial cracks occur. Once the radial cracks occur, σ_θ diminishes immediately to zero. The conservation equations of mass and momentum can be expressed as

$$\frac{\partial \nu}{\partial r} + \frac{\nu}{r} = -\frac{1}{\rho}\frac{d\rho}{dt}, \tag{12.55}$$

$$\frac{\partial \sigma_r}{\partial r} + \frac{\sigma_r}{r} = -\rho\left(\frac{\partial \nu}{\partial t} + \nu\frac{\partial \nu}{\partial r}\right). \tag{12.56}$$

According to the pressure-bulk strain relation,

$$P = \frac{1+\nu}{3}\sigma_r = K\left(1 - \frac{\rho_0}{\rho}\right) = K\eta. \tag{12.57}$$

Putting Eq.(12.57) into Eqs.(12.55) and (12.56) respectively,

$$\frac{\partial \nu}{\partial r} + \frac{\nu}{r} = -\frac{1+\nu}{3K(1-\eta)}\frac{d\sigma_r}{dt}, \tag{12.58}$$

$$\frac{\partial \sigma_r}{\partial r} + \frac{\sigma_r}{r} = -\frac{\rho_0}{1-\eta}\left(\frac{\partial \nu}{\partial t} + \nu\frac{\partial \nu}{\partial r}\right). \tag{12.59}$$

Because the bulk strain is very small, there is $1 - \eta \approx 1$. The above equations can be rewritten as

$$\begin{cases} \dfrac{d\bar{\nu}}{d\xi} + \dfrac{\bar{\nu}}{\xi} = \dfrac{1+\nu}{3}\xi\dfrac{d\bar{\sigma}_r}{d\xi}, \\[3mm] \dfrac{d\bar{\sigma}_r}{d\xi} + \dfrac{\bar{\sigma}_r}{\xi} = \beta^2\xi\dfrac{d\bar{\nu}}{d\xi}. \end{cases} \tag{12.60}$$

Integrating Eq.(12.60), we obtain

$$\bar{\sigma}_r(\xi) = -\dfrac{D_1}{\xi} + D_2, \tag{12.61}$$

$$\bar{\nu}(\xi) = -\dfrac{D_2}{\beta^2\xi} + \dfrac{1+\nu}{3}D_1, \tag{12.62}$$

where D_1 and D_2 are the integration constants that can be determined with application of boundary conditions.

At the interface of the cracked and plastic zones ($r = ct$, $\xi = 1$), defining radial stress and radial velocity in the cracked zone $\bar{\sigma}_{r3}$ and $\bar{\nu}_3$, respectively, the boundary conditions can be expressed as

$$r = ct(\xi = 1), \quad \begin{cases} \bar{\sigma}_r(\xi = 1) = \bar{\sigma}_{r3} = \dfrac{\bar{B}_t}{1 - A_t}, \\[3mm] \bar{\nu}(\xi = 1) = \bar{\nu}_3, \end{cases} \tag{12.63}$$

$$r = c_1 t(\xi = \beta_1/\beta), \quad \begin{cases} \bar{\sigma}_r(\xi = \beta_1/\beta) = \bar{\sigma}_{r2}, \\[3mm] \bar{\nu}(\xi = \beta_1/\beta) = \bar{\nu}_2. \end{cases} \tag{12.64}$$

From Eqs.(12.61) and (12.63),

$$-D_1 + D_2 = \dfrac{\bar{B}_t}{1 - A_t}. \tag{12.65}$$

The radial stress and the radial velocity in the cracked zone adjacent to the interface of the cracked and elastic zones ($r = c_1 t$, $\xi = \beta_1/\beta$) can be expressed as

$$\bar{\sigma}_{r2} = -\dfrac{D_1}{\beta_1/\beta} + D_2, \tag{12.66}$$

$$\bar{\nu}_2 = -\dfrac{D_2}{\beta_1\beta} + \dfrac{1+\nu}{3}D_1, \tag{12.67}$$

the integration constants D_1, D_2 can then be expressed as

$$D_1 = \dfrac{3\beta_1(\bar{\sigma}_{r2} + \beta_1\beta\bar{\nu}_2)}{\beta[\beta_1^2(1 + \nu) - 3]}, \tag{12.68}$$

$$D_2 = \dfrac{\beta_1^2(1 + \nu)\bar{\sigma}_{r2} + 3\beta_1\beta\bar{\nu}_2}{\beta_1^2(1 + \nu) - 3}. \tag{12.69}$$

Putting Eqs.(12.68) and (12.69) into Eq.(12.65),

$$\frac{\bar{B}_t}{1 - A_t} = -\frac{3\beta_1(\bar{\sigma}_{r2} + \beta_1\beta\bar{\nu}_2)}{\beta[\beta_1^2(1+\nu)-3]} + \frac{\beta_1^2(1+\nu)\bar{\sigma}_{r2} + 3\beta_1\beta\bar{\nu}_2}{\beta_1^2(1+\nu)-3}. \tag{12.70}$$

The velocity in the cracked zone adjacent to the cracked-elastic zones interface can be written as

$$\bar{\nu}_3 = -\frac{1}{\beta^2}\frac{\beta_1^2(1+\nu)\bar{\sigma}_{r2} + 3\beta\beta_1\bar{\nu}_2}{\beta^2(1+\nu)-3} + \frac{\beta_1(1+\nu)(\bar{\sigma}_{r2} + \beta\beta_1\bar{\nu}_2)}{\beta[\beta_1^2(1+\nu)-3]}. \tag{12.71}$$

12.4.4 Interface of the Plastic and Cracked Zones ($r = ct$, $\xi = 1$)

Defining $\bar{\sigma}_{r4}$ and $\bar{\nu}_4$ the radial stress and velocity, respectively, in the plastic zone adjacent to the interface, with reference to the Hugoniot jump condition,

$$\rho_4\left(\bar{\nu}_4 - 1\right) = \rho_3\left(\bar{\nu}_3 - 1\right), \tag{12.72}$$

$$\bar{\sigma}_{r4} = \bar{\sigma}_{r3} + \frac{\rho_3}{\rho_0}\beta^2\left(1 - \bar{\nu}_3\right)\left(\bar{\nu}_4 - \bar{\nu}_3\right), \tag{12.73}$$

where ρ_4 and ρ_3 are the density in the plastic and cracked zones adjacent to the interface respectively.

From Eq.(12.63),

$$\bar{\sigma}_{r3} = \frac{\bar{B}_{\text{uni}}}{1 - A_{\text{uni}}}. \tag{12.74}$$

According to the pressure-bulk strain relation, the following expression can be obtained:

$$\left(1 - \frac{\rho_0}{\rho_3}\right) = \frac{1+\nu}{3}\frac{\bar{B}_{\text{uni}}}{1 - A_{\text{uni}}}. \tag{12.75}$$

Putting Eq.(12.75) into Eq.(12.73),

$$\bar{\sigma}_{r4} = \bar{\sigma}_{r3} + \frac{3(1 - A_{\text{uni}})\beta^2\left(1 - \bar{\nu}_3\right)\left(\bar{\nu}_4 - \bar{\nu}_3\right)}{3(1 - A_{\text{uni}}) - (1+\nu)\bar{B}_{\text{uni}}}. \tag{12.76}$$

Based on the unified strength theory, the material in the plastic zone satisfies

$$\bar{\sigma}_r - \bar{\sigma}_\theta = A_{\text{uni}}\bar{\sigma}_r + \bar{B}_{\text{uni}}, \tag{12.77}$$

where

$$A_{\text{uni}} = \frac{2 \sin \varphi_{\text{uni}}}{1 + \sin \varphi_{\text{uni}}}, \quad B_{\text{uni}} = \frac{2C_{\text{uni}} \cos \varphi_{\text{uni}}}{1 + \sin \varphi_{\text{uni}}}.$$

C_{uni} and φ_{uni} are the unified internal friction angle and unified cohesion, respectively, corresponding to the unified strength theory and they have the form of

$$\sin \varphi_{\text{uni}} = \frac{2(1 + b) \sin \varphi_0}{2 + b + b \sin \varphi_0}, \tag{12.78}$$

$$C_{\text{uni}} = \frac{2(1 + b)c_0 \cos \varphi_0}{2 + b + b \sin \varphi_0} \cdot \frac{1}{\cos \varphi_{\text{uni}}}. \tag{12.79}$$

At the interface, Eq.(12.77) can be rewritten as

$$\bar{\sigma}_{r4} - \bar{\sigma}_{\theta 4} = A_{\text{uni}}\bar{\sigma}_{r4} + \bar{B}_{\text{uni}}. \tag{12.80}$$

According to the pressure-bulk strain relation,

$$\frac{1}{2}(\bar{\sigma}_{r4} + \bar{\sigma}_{\theta 4}) = 1 - \frac{\rho_0}{\rho_4}. \tag{12.81}$$

From the above equations it derives

$$\rho_4 = \frac{2\rho_0}{2 - (2 - A_{\text{uni}})\bar{\sigma}_{r4} + \bar{B}_{\text{uni}}}. \tag{12.82}$$

Putting Eq.(12.82) into Eq.(12.72), then there is

$$\bar{\nu}_4 = 1 + \frac{\rho_3(2 + \bar{B}_{\text{uni}})(\bar{\nu}_3 - 1)}{2\rho_0} - \frac{\rho_3(2 - A_{\text{uni}})\bar{\sigma}_{r4}(\bar{\nu}_3 - 1)}{2\rho_0}. \tag{12.83}$$

From Eqs.(12.83) and (12.76), we can get

$$\bar{\sigma}_{r4} = \bar{n}\bar{\sigma}_{r3} + \bar{n}\frac{\rho_3}{\rho_0}\beta^2(\bar{\nu}_3 - 1)^2 \left[1 - \frac{\rho_3(2 + \bar{B}_{\text{uni}})}{2\rho_0}\right], \tag{12.84}$$

where

$$\bar{n} = \frac{2\rho_0^2}{2\rho_0^2 - \rho_3^2\beta^2(1 - \bar{\nu}_3)^2(2 - A_{\text{uni}})}.$$

12.4.5 Plastic Zone ($v_r t \leqslant r \leqslant ct$, $\delta \leqslant \xi \leqslant 1$)

The mass and momentum equations in Eqs.(12.55) and (12.56) are still valid in the plastic zone. The boundary stress and velocity conditions in the plastic zone can be expressed as

$$r = v_r t, \quad \bar{\nu}(\xi = \delta) = \delta, \tag{12.85}$$

$$r = ct, \quad \begin{cases} \bar{\sigma}_r(\xi = 1) = \bar{\sigma}_{r4}, \\ \bar{\nu}(\xi = 1) = \bar{\nu}_4. \end{cases} \tag{12.86}$$

According to the pressure-bulk strain relation,

$$(\sigma_r + \sigma_\theta) = 2K \left(1 - \frac{\rho_0}{\rho_4} \right) = 2K\eta. \tag{12.87}$$

The material in the plastic zone satisfies the unified strength theory,

$$\sigma_r - \sigma_\theta = A_t \sigma_r + B_t. \tag{12.88}$$

From Eqs.(12.87) and (12.88),

$$\eta = \frac{(2 - A_t)\bar{\sigma}_r - \bar{B}_t}{2}. \tag{12.89}$$

Putting Eqs.(12.87) and (12.88) into the mass and momentum conservation equations, the following differential equations can be derived,

$$\frac{\partial \nu}{\partial r} + \frac{\nu}{r} = -\frac{2 - A_t}{2K(1 - \eta)} \left(\frac{\partial \sigma_r}{\partial t} + \nu \frac{\partial \sigma_r}{\partial r} \right), \tag{12.90}$$

$$\frac{\partial \sigma_r}{\partial r} + \frac{A_t \sigma_r + B_t}{r} = -\frac{\rho_0}{1 - \eta} \left(\frac{\partial \nu}{\partial t} + \nu \frac{\partial \nu}{\partial r} \right). \tag{12.91}$$

The dimensionless expressions of the above equations are

$$\frac{d\bar{\nu}}{d\xi} + \frac{\bar{\nu}}{\xi} = -\frac{2 - A_{\text{uni}}}{2(1 - \eta)} (\xi - \bar{\nu}) \frac{d\bar{\sigma}_r}{d\xi}, \tag{12.92}$$

$$\frac{d\bar{\sigma}_r}{d\xi} + \frac{A_{\text{uni}}\bar{\sigma}_r + \bar{B}_{\text{uni}}}{\xi} = \frac{\beta^2}{1 - \eta}(\xi - \bar{\nu})\frac{d\bar{\nu}}{d\xi}. \tag{12.93}$$

Eqs.(12.92) and (12.93) can be rewritten as

$$\frac{d\bar{\sigma}_r}{d\xi} = \frac{2(1 - \eta) \left[\beta\bar{\nu}(\beta\xi - \beta\bar{\nu}) + (1 - \eta)(A_{\text{uni}}\bar{\sigma}_r + \bar{B}_{\text{uni}}) \right]}{\xi[(2 - A_{\text{uni}})(\beta\xi - \beta\bar{\nu})^2 - 2(1 - \eta)^2]}, \tag{12.94}$$

$$\frac{\mathrm{d}\,(\beta\bar{\nu})}{\mathrm{d}\xi} = \frac{(1-\eta)\left[(2-A_{\mathrm{uni}})(A_{\mathrm{uni}}\bar{\sigma}_r + \bar{B}_{\mathrm{uni}})(\beta\xi - \beta\bar{\nu}) + 2\beta\bar{\nu}(1-\eta)\right]}{\xi[(2-A_{\mathrm{uni}})(\beta\xi - \beta\bar{\nu})^2 - 2(1-\eta)^2]}.$$

(12.95)

Eqs.(12.94) and (12.95) can be solved using the Runge-Kutta method. Defining $y_1 = \bar{\sigma}_r$ and $y_2 = \bar{\nu}$, the boundary conditions can be written as $y_1(0) = \bar{\sigma}_{r4}$ and $y_2(0) = \bar{\nu}_4$. When the stress and velocity in the plastic zone are deduced, Eqs.(12.94) and (12.95) can be expressed in the form

$$\begin{cases} y_1' = f(\xi, y_1, y_2), & y_1(0) = \sigma_{r4}, \\ y_2' = g(\xi, y_1, y_2), & y_2(0) = \bar{\nu}_4. \end{cases}$$

(12.96)

The integral domain borders the plastic-cracked zones interface ($\xi = 1$) and the cavity surface ($\xi = \delta$). According to the boundary conditions, when $\xi_n = 1 - n\Delta\xi$, $y_{2n} = \xi_n$. The radial stress and velocity at the cavity surface can be obtained when $\delta = \xi_n$ and $y_1(n) = \bar{\sigma}_r(\delta)$.

The detailed procedures for solving the differential equations are given as follows:

Step 1. Substituting Eqs.(12.43), (12.45), (12.53), and (12.54) into Eq. (12.70), the relation between β_1 and β is deduced.

Step 2. Assuming an initial value for β_1, β can be calculated with reference to the relations between β_1 and β deduced in Step 1.

Step 3. $\bar{\sigma}_{r1}$ and $\bar{\nu}$ are calculated from Eqs.(12.43) and (12.45) with reference to β_1 and β. Putting $\bar{\sigma}_{r1}$ and $\bar{\nu}_1$ into Eqs.(12.53), (12.54), (12.68), (12.69), and (12.71), $\bar{\sigma}_{r2}$ and $\bar{\nu}_2$, the integration constants D_1 and D_2, and $\bar{\nu}_3$ are determined. Putting the above quantities into Eqs.(12.74) and (12.75), $\bar{\sigma}_{r3}$ and $\bar{\rho}_3$ are obtained.

Step 4. Substituting $\bar{\sigma}_{r3}$, $\bar{\rho}_3$, and $\bar{\nu}_3$ into Eqs.(12.84) and (12.83), $\bar{\sigma}_{r4}$ and $\bar{\nu}_4$ are determined.

Step 5. Based on boundary conditions, the differential equation in Eq.(12.96) is solved from $\xi = 1$ to $\xi = \delta$. The stress and velocity distribution in the plastic zone is then calculated. With application of the boundary conditions, the radial stress and the expansion velocity are obtained.

When the bulk strain is zero, i.e., $\eta = 0$, and $\rho_i = \rho_0$ ($i = 1, ..., 4$), the solutions for incompressible materials can be deduced from Eqs.(12.21) to (12.96). If the interface velocities c and c_1 are the same, the radial cracked zone disappears and there are only plastic, elastic, and undisturbed zones in the materials.

12.5 Cavity Expansion Pressure and Velocity

With application of the concrete parameters given by Forrestal (1997), i.e., bulk modulus K of 6.7 GPa, compressive strength Y of 130 MPa, elastic modulus E of 11.3 GPa, Poisson's ratio ν of 0.22, tensile strength $\sigma_t = 13$

MPa, density $\rho_0 = 2260$ kg/m^3, according to the derived equations based on the cylindrical cavity expansion theory, the material is considered to be incompressible or compressible. The response of the target can be elastic-plastic or elastic-crack-plastic, respectively.

12.5.1 Incompressible Material

Fig.12.5 illustrates the relation between the radial stress at the cavity surface and the cavity expansion velocity for the elastic-plastic response of the target. From Fig.12.5, the radial stress increases with increasing cavity expansion velocity, the radial stress and the unified strength theory parameter b.

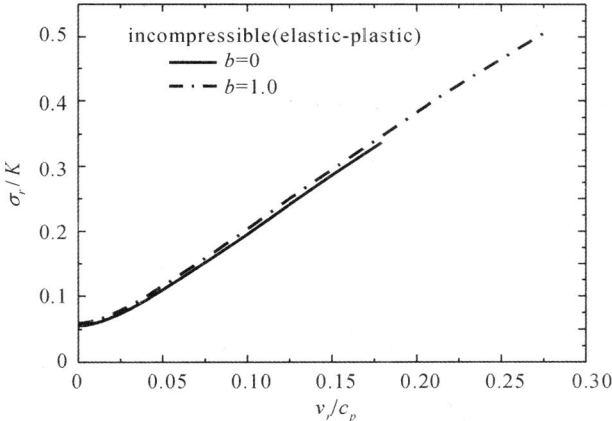

Fig. 12.5. Radial stress versus cavity expansion velocity (incompressible material, elastic-plastic response)

Figs.12.6 and 12.7 plot the curves of the plastic-crack interface velocity c and the elastic-crack interface velocity c_1 versus the cavity expansion velocity v_r for incompressible material for $b = 1.0$ and $b = 0$, respectively. It is shown that for a given velocity v_r, c_1 is higher than c. When $b = 1.0$ and $v_r/(Y/\rho_0)^{1/2} = 0.82$, the curves of c_1 and c intersect, i.e., the cracked zone vanishes at this cavity expansion velocity and there are only elastic and plastic zones in the material. When $b = 0$ and $v_r/(Y/\rho_0)^{1/2} = 0.7$, the curves of c_1 and c also intersect, the response of material shifts from elastic-crack-plastic to elastic-plastic. The current solution with $b = 0$ conforms to the result reported by Forrestal (1997), who applied the spherical cavity expansion theory and discovered that the cracked zone disappears when $v_r/(Y/\rho_0)^{1/2} = 0.71$.

Fig.12.8 illustrates schematically the relation between the radial stress in the cavity surface and the cavity expansion velocity for incompressible materials under elastic-crack-plastic response for $b = 0$ and $b = 1.0$, respectively.

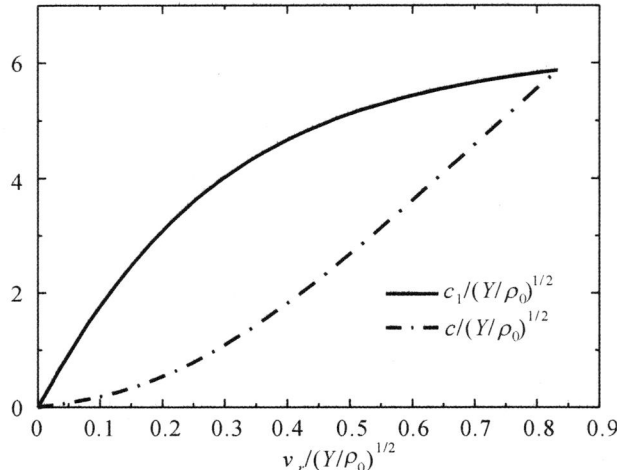

Fig. 12.6. Cavity expansion velocity versus interface velocity (incompressible material, elastic-crack-plastic response, $b = 1.0$)

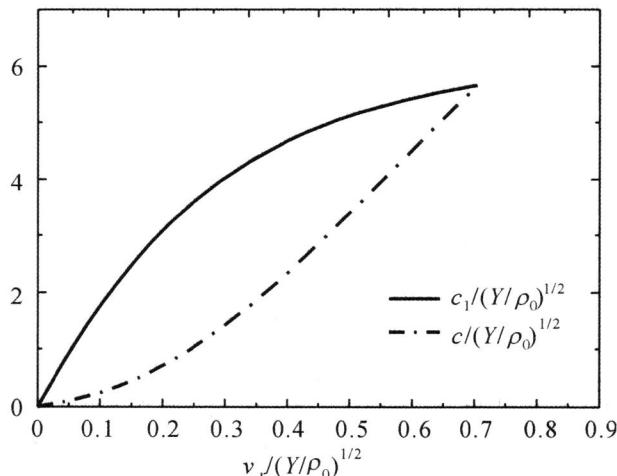

Fig. 12.7. Cavity expansion velocity versus interface velocity (incompressible material, elastic-crack-plastic response, $b = 0$)

The radial stress for $b = 1.0$ is higher than that for $b = 0$. The radial stress increases with increasing cavity expansion velocity. The quasi-static cavity expansion pressure is the radial stress at the cavity surface when $v_r = 0$.

Compared with the results given by Forrestal (1997) which are based on the spherical cavity expansion theory, the radial stress derived based on the cylindrical cavity expansion theory is smaller when the impact velocity is relatively low. However, it is higher when the impact velocity is relatively high. It agrees with the statements by Forrestal (1997).

Fig.12.9 compares the radial stresses between the elastic-plastic response and the elastic-crack-plastic response for incompressible material. From Fig. 12.9, the stress is higher for the elastic-plastic response when the velocity is lower. When the velocity increases, the stress of the elastic-crack-plastic response gradually transfers to that of elastic-plastic response. Finally, the curves of the two responses intersect.

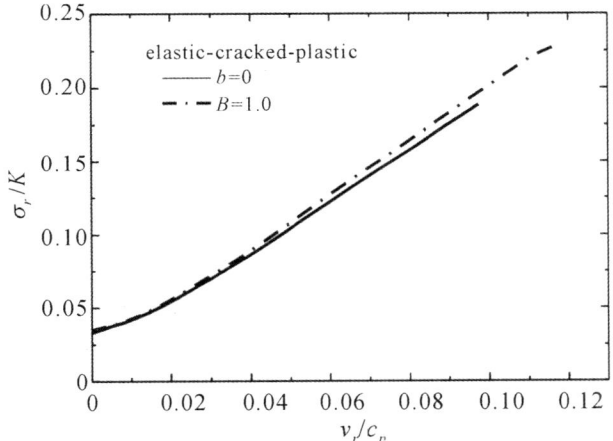

Fig. 12.8. Radial stress at cavity surface versus cavity expansion velocity (incompressible material, elastic-crack-plastic response)

Fig.12.10 shows the curves of elastic-crack interface velocity c_1 versus plastic-crack interface velocity c during the cavity expansion. The curves are different for different parameter b. Fig.12.11 plots the curves of the cavity expansion velocity v_r versus the plastic-crack interface velocity c for incompressible material. Fig.12.12 plots the curves of the radial stress versus the radius for incompressible material under the elastic-crack-plastic response. Fig.12.12 shows that the stress at the cavity surface is the highest and reduces gradually with the increasing radius.

12.5.2 Compressible Material

Figs.12.13 and 12.14 compare the cavity expansion stress of compressible materials with that of incompressible materials under elastic-plastic response for $b = 0$ (Single-shear theory) and $b = 1.0$ (Twin-shear theory) respectively. It is seen that for a given cavity expansion velocity, the cavity expansion surface pressure of incompressible materials is much higher than that of compressible materials.

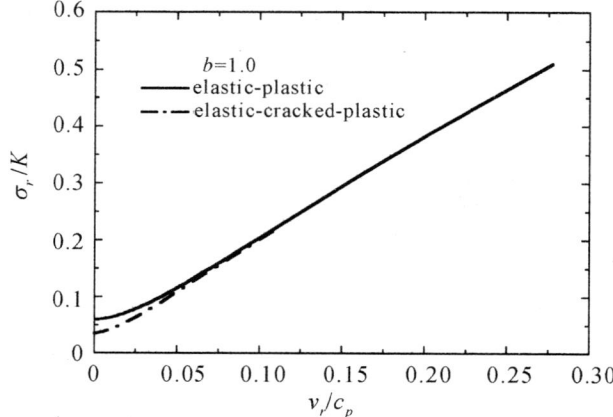

Fig. 12.9. Comparison of radial stress for elastic-plastic and elastic-crack-plastic responses (incompressible material)

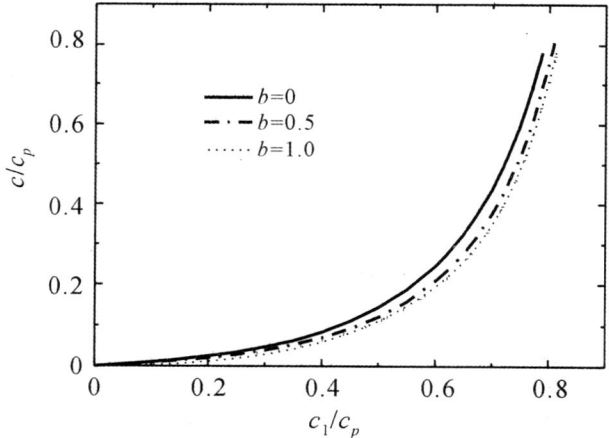

Fig. 12.10. Curves of c_1 versus c for incompressible material

Fig.12.15 illustrate the curves of cavity expansion velocity v_r versus the plastic-crack interface velocity c. From Fig.12.15, for a given v_r, c is the highest when $b = 0$, while it is the lowest when $b = 1.0$. Fig.12.16 plots the relations between the cavity expansion velocity, plastic-crack interface velocity c and plastic-crack interface velocity c_1 for compressible materials with $b = 0.5$. Similar to the incompressible materials, for a given v_r the elastic-crack interface velocity c_1 is higher than the plastic-crack interface velocity c, where $f(\xi, y_1, y_2)$ and $g(\xi, y_1, y_2)$ correspond with Eqs.(12.94) and (12.95). The curves of c_1 and c intersect at the point when $v_r/(Y/\rho_0)^{1/2} =$

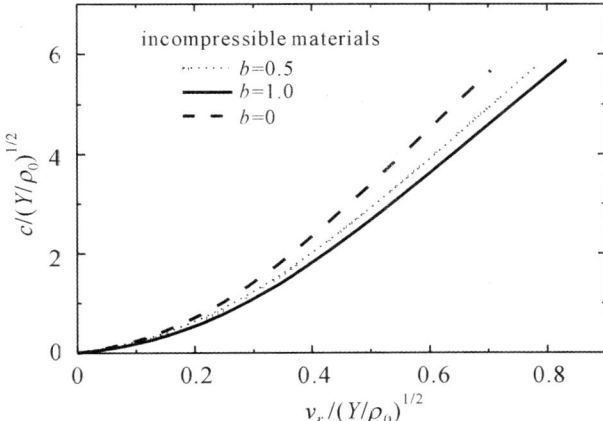

Fig. 12.11. Curves of v_r versus c for incompressible material

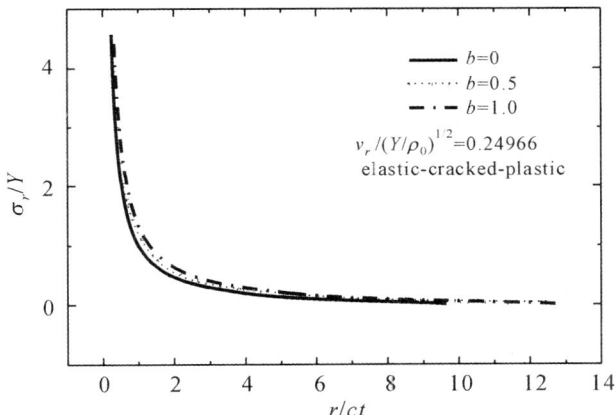

Fig. 12.12. Radial stress versus radius for incompressible material

1.05, where the cracked zone disappears and the response of the material is elastic-plastic.

Figs.12.17 and 12.18 compare the cavity expansion pressures between the elastic-plastic and elastic-crack-plastic responses for compressible materials for $b = 1.0$ (Twin-shear theory), $b = 0.5$ (Median theory) and $b = 0$ (Single-shear theory), respectively.

From Figs.12.17, 12.18, and 12.19, the cavity expansion pressure for elastic-crack-plastic response is lower when the expansion velocity is lower. With the increase of the cavity expansion velocity, the cracked zone disappears and the response of the target becomes elastic-plastic.

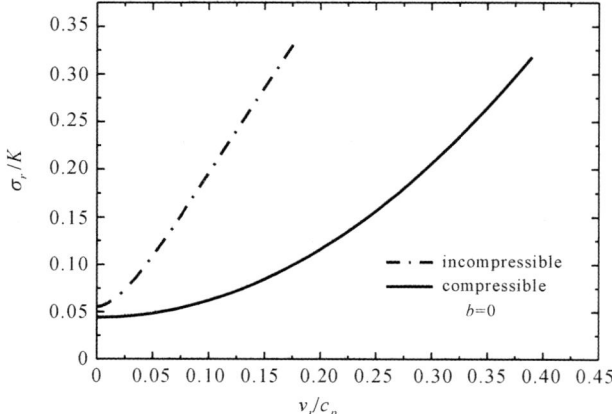

Fig. 12.13. Comparison of radial stress between incompressible and compressible materials ($b = 0$)

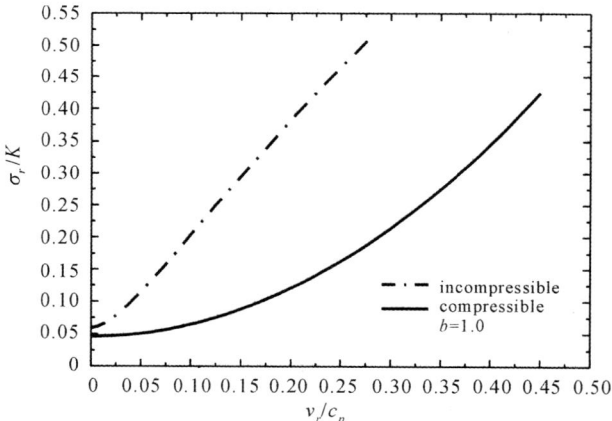

Fig. 12.14. Comparison of radial stress between incompressible and compressible materials ($b = 1.0$)

For compressible materials the curves of the cavity expansion pressure versus cavity expansion velocity can be expressed as a quadric parabola,

$$\sigma_r/K = A_1 + B_1 \left(\frac{v_r}{c_p} \right)^2, \tag{12.97}$$

where v_r is the cavity expansion velocity; A_1 is the quasi-static cavity expansion pressure.

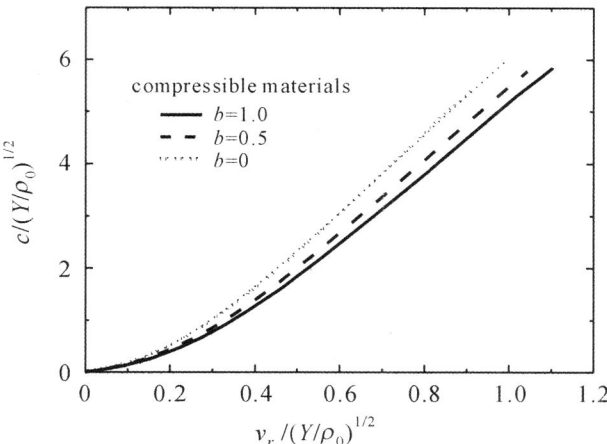

Fig. 12.15. v_r versus c for elastic-crack-plastic response (incompressible materials)

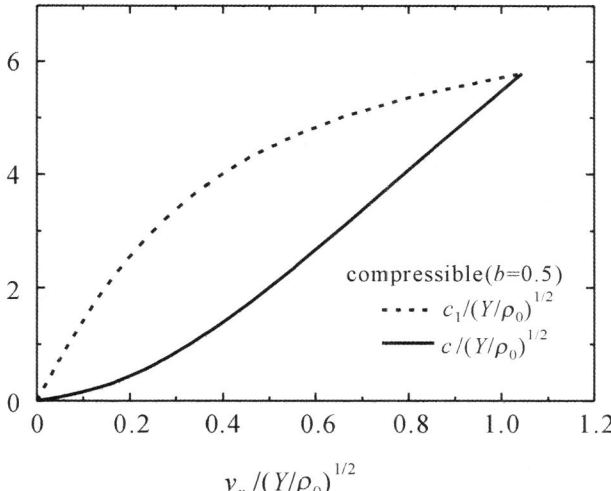

Fig. 12.16. Cavity expansion velocity versus interface velocity (compressible materials, $b = 0.5$)

The coefficients in Eq.(12.97) are listed in Table 12.1 for the elastic-plastic and elastic-crack-plastic responses, respectively, with different unified strength theory parameter b.

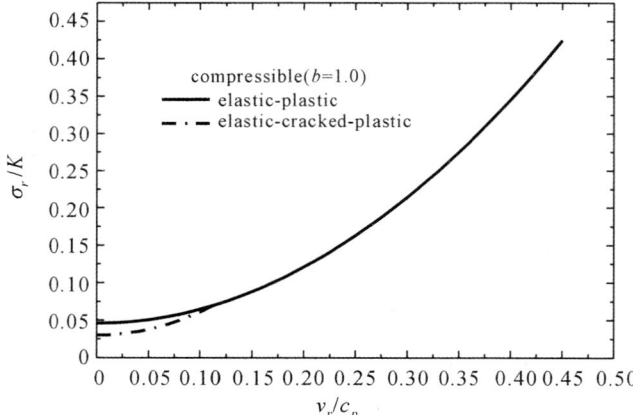

Fig. 12.17. Comparison of cavity expansion velocities between elastic-plastic and elastic-crack-plastic responses (compressible materials, $b = 1.0$)

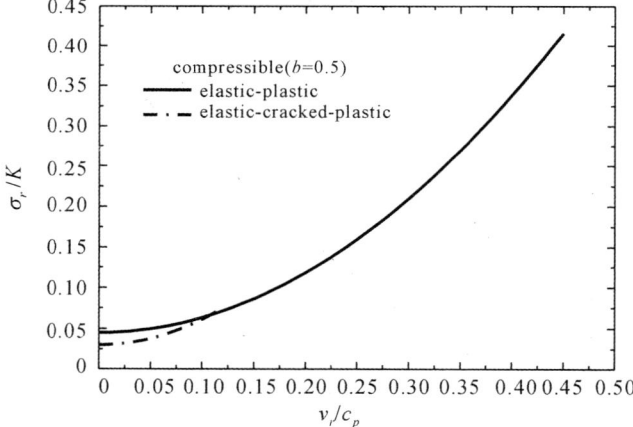

Fig. 12.18. Comparison of cavity expansion velocities between elastic-plastic and elastic-crack-plastic responses (compressible materials, $b = 0.5$)

12.6 Penetration Resistance Analysis

The capabilities of penetration and destruction of a long rod projectile are much higher than those of the old-style armor-piercing projectile since the long rods have a higher length-diameter ratio. The long rod projectiles can be divided into straight-shank type and cone-shank type. According to the shape of warhead the long rod can be categorized into a spherical, ogive, and conical warhead nose, respectively, as shown in Fig.12.20.

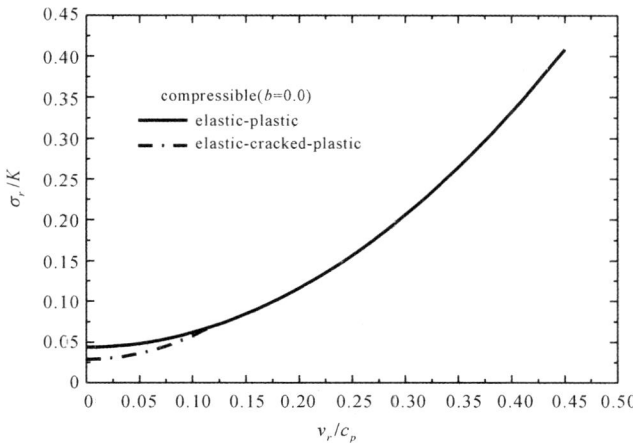

Fig. 12.19. Comparison of cavity expansion velocities between elastic-plastic and elastic-crack-plastic responses (compressible materials, $b = 0$)

Table 12.1. Curve fitting of cavity expansion pressure for compressible materials under elastic-plastic response

Response	Material strength parameter b	A_1	B_1
	1.0	0.044	1.80
Elastic-plastic	0.5	0.045	1.83
	0.0	0.046	1.87
	1.0	0.029	2.90
Elastic-crack-plastic	0.5	0.030	3.02
	0.0	0.031	3.10

Because a long rod impacts and penetrates a target with an impact velocity V_0 and a penetration velocity V_z, the coordination xOz of the target is established as shown in Fig.12.21. The origin is the impacting point of the long rod, the positive z axial is downwards vertically, and the x axial is horizontal. The resistance on the long rod includes the resistance on the warhead and that on the surface of the shank. The resistance on the shank surface is very small and can be omitted because the velocity of the impact and penetration is low (Jones et al., 1993; Bless et al., 1987).

The tractions that resist the penetration are the normal force F_n, and the tangential force F_t. The resistance is analyzed in the following context for

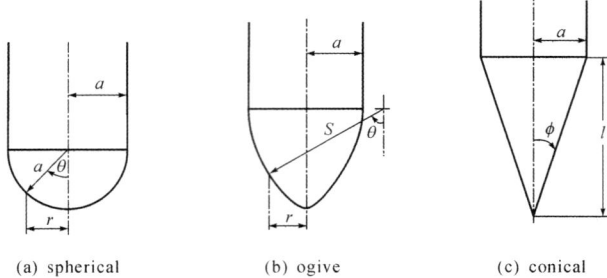

(a) spherical   (b) ogive   (c) conical

Fig. 12.20. Different head shapes of straight long rods

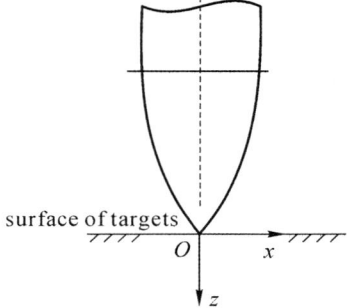

Fig. 12.21. Rod-target system

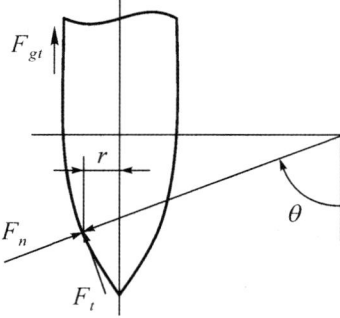

Fig. 12.22. Resistance on the long rod

the ogive-nose projectile. The resistances for other nose shape projectiles can be similarly derived.

For an ogive-nose projectile as shown in Fig.12.20(b) with radius of s, central angle of θ_0,

$$\theta_0 = \sin^{-1}\left(\frac{s-a}{s}\right). \tag{12.98}$$

The normal force on the tiny surface of the warhead is

$$\mathrm{d}F_n = 2\pi s^2 \left[\sin\theta - \left(\frac{s-a}{s}\right)\right]\sigma_r\mathrm{d}\theta. \tag{12.99}$$

The tangential force on the tiny surface of the warhead is

$$\mathrm{d}F_t = 2\pi s^2 \left[\sin\theta - \left(\frac{s-a}{s}\right)\right]\mu\sigma_r\mathrm{d}\theta, \tag{12.100}$$

where σ_r is the principal stress on the warhead, which is also the radial stress on the cavity surface using the cavity expansion theory. μ is the friction coefficient. The value of μ depends on the penetration depth, velocity and the material properties of the rod and the target. Its linear experiential formula is proposed by Bowden and Tabor (1966),

$$\mu = \begin{cases} \mu_d, & V_z \geqslant V_d, \\ \mu_s - \dfrac{\mu_s - \mu_d}{V_d}V_z, & V_z \leqslant V_d, \end{cases} \tag{12.101}$$

where V_d is the critical penetration velocity which can be obtained by trial-and-error; μ_s and μ_d are friction parameters of the targets, which are related to the penetration velocity. For the geomaterials, there are $V_d = 300$ m/s, $\mu_s = 0.5$, $\mu_d = 0.08$ (Bowden and Tabor, 1966).

The radial stress can be written as

$$\sigma_r/K = A_1 + B_1 \left(\frac{\nu_r}{c_p}\right)^2. \tag{12.102}$$

The relation between the cavity expansion velocity and the penetration velocity is

$$V_r = V_z \cos\theta, \tag{12.103}$$

where θ is the angle between the surface normal and the rod axial directions. Putting Eq.(12.103) into Eq.(12.102),

$$\sigma_r/K = A_1 + B_1 \left(\frac{V_z \cos\theta}{c_p}\right)^2. \tag{12.104}$$

The total resistance on the warhead is

$$F_z = 2\pi s^2 \int_{\theta_0}^{\frac{\pi}{2}} \left[\left(\sin\theta - \frac{s-a}{s}\right)(\cos\theta + \mu\sin\theta)\right]\sigma_r\mathrm{d}\theta. \tag{12.105}$$

Integrating Eq.(12.102) with reference to Eq.(12.105),

$$F_z = \alpha_s + \beta_s V_z^2,\tag{12.106}$$

where

$$\alpha_s = \pi a^2 K A_1[1 + 4\mu\psi^2(\pi/2 - \theta_0) - \mu(2\psi - 1)(4\psi - 1)^{1/2}],\tag{12.107}$$

$$\beta_s = \pi a^2 \rho B_1 \left[\frac{(8\psi - 1)}{24\psi^2} + \mu\psi^2(\pi/2 - \theta_0) \right. \\ \left. - \frac{\mu(2\psi - 1)(6\psi^2 + 4\psi - 1)(4\psi - 1)^{1/2}}{24\psi^2} \right].\tag{12.108}$$

The resistance on rods of other shapes can also be deduced similarly with the replacement of different geometry shape functions based on the shapes of the nose of the rod.

12.7 Analysis and Verification of Penetration Depth

Assuming a long rod is non-deformable during the penetration, with reference to the Newton's second law,

$$m_p \frac{\mathrm{d}V_z}{\mathrm{d}t} = m_p V_z \frac{\mathrm{d}V_z}{\mathrm{d}z} = -F_z = -(\alpha_s + \beta_s V_z^2),\tag{12.109}$$

where α_s and β_s are the shape parameters of the nose; m_p is the mass of the rod.

Integrating Eq.(12.109) with reference to the initial and final conditions, the penetration depth can be deduced,

$$Z_{\max} = \frac{m_p}{2\beta_s} \ln \left[1 + \frac{\beta_s V_0^2}{\alpha_s} \right].\tag{12.110}$$

The parameters for the target material in Section 12.5 are also applied for the current problem. The parameters used for the long rod are m_p=1.6 kg, s=91.5 mm, a=15.25 mm.

Based on the results of the cavity expansion pressure for compressible materials in Section 12.5, the final penetration depth can be derived for the target material impacted by an ovate straight long rod with a velocity of 300 *sim* 1100 m/s. The analytical results from the current penetration model agree very well with the test data (Forrestal et al., 1996) when the initial velocity is low.

The penetration depths are illustrated schematically in Fig.12.23 for the friction parameters $\mu = 0.1$ and $\mu = 0.2$ respectively. From Fig.12.23 the friction parameter has great influence on the penetration depth. The penetration depth for $\mu = 0.1$ is closer to the test results (Forrestal et al., 1996).

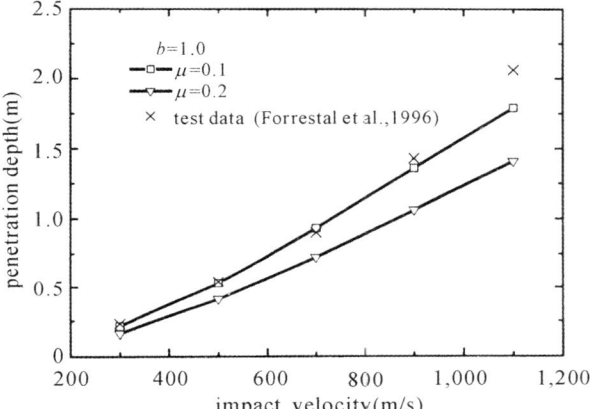

Fig. 12.23. Comparison of penetration depths

Fig.12.24 shows the influence of rod mass on the penetration depth. It is seen that the heavier the rod, the greater the penetration capability.

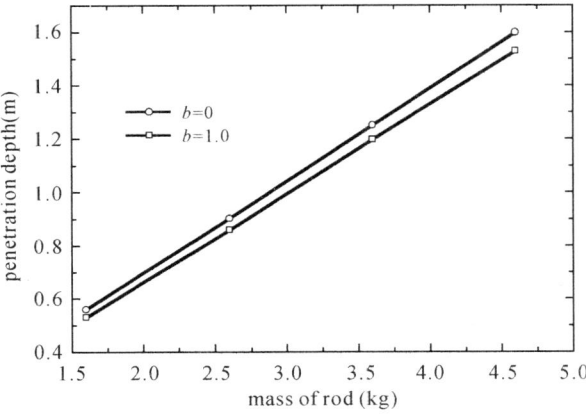

Fig. 12.24. Influence of the rod mass on the penetration depth

The penetration depths are solved for the ovate and spherical warheads with the same shank diameter as that for the ogive-nose projectile as shown

Fig.12.25. It can be seen that the ovate-warhead rod can penetrate deeper than the spherical warhead rod.

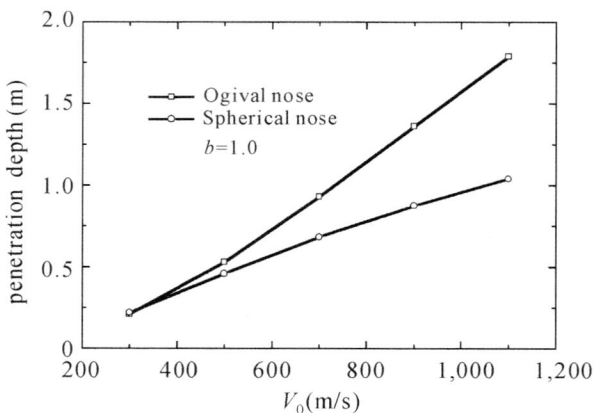

Fig. 12.25. Influence of different shapes of warhead on penetration depth

12.8 Summary

Based on the cylindrical cavity-expansion theory, the unified strength theory is applied as the failure condition for penetration analysis. The cavity expansion pressure is deduced from the elastic-plastic and elastic-crack-plastic responses for incompressible and compressible materials. Assuming the long rods are rigid during the penetration, the penetration depths of rods are obtained and are compared with the test results. The following conclusions are derived:

(1) It is convenient to use the unified strength theory for penetration problems.

(2) When the cavity expansion velocity is low, the target response is elastic-crack-plastic. When the cavity expansion velocity increases to a certain value, the cracked zone disappears and the target responds elastic-plastically. It agrees with the results reported by Forrestal (1997).

(3) The cavity expansion pressure and the penetration depth are different when the different parameter b is used, which represents a different failure criterion. The penetration depth is higher for $b = 0$ than that for $b = 1$. The current solutions agree well with test results in the published literature for low impact velocity cases (Forrestal et al., 1996). However, the predicted penetration depth is smaller than that in the test results when the impact velocity is larger than 1000 m/s.

(4) For a given shape and dimension, the higher the rod mass or the density of the rod material, the deeper the penetration. This may be the underlying reason why those high-density metals are used as warheads for modern weapons.

(5) The shape of a warhead has a significant influence on the penetration depth. When the shank diameter is the same, the ovate-warhead rod can penetrate deeper than the spherical warhead.

References

Bishop RF, Hill R, Mott NF (1945) The theory of identification and hardness tests. The Proc. Phys. Soc. 57(3):147-159

Bless SJ, Rosenberg Z, Yoon B (1987) Hypervelocity penetration of ceramics. Int. J. Impact Eng., 5:165-171

Bowden FP, Tabor D (1966) Friction, lubrication and wear—survey of work during last decade. British Journal of Applied Physics, 17(12):1521-1544

Forrestal MJ, Frew DJ, Hanchak SJ, Brar NS (1996) Penetration of grout and concrete targets with ogive-nose steel projectiles. Int. J. Impact Eng., 18(5):465-476

Forrestal MJ, Tzou DY (1997) Spherical cavity-expansion penetration model for concrete targets. Int. J. Solids Struct., 34(31-32):4127-4146

Forrestal MJ, Zhou DY, Askari E, Longcope DB (1995) Penetration into ductile metal targets with rigid spherical-nose rods. Int. J. Impact. Eng., 16(5/6):699-710

Heuze FE (1990) Overview of projectile penetration into geological materials, with emphasis on rocks. Int. J. Rock Mechanics and Mining Sciences & Geomechanics Abstracts, 27(1):1-14

Hill R (1950) The mathematical theory of plasticity. Clarendon Press, London

Hopkins HG (1960) Dynamic expansion of spherical cavities in metal. In: Sneddon IN, Hill R (eds.) Progress in Solid Mechanics, Vol.1, Chapter III, North-Holland Publ. Co., Amsterdam, New York

Jones SE, Marlow RB, House JW, et al. (1993) A one-dimensional analysis of the penetration of semi-infinite 1100-0 aluminum targets by rods. Int. J. Impact Eng., 14: 407-416

Li JC (2001) Investigation of high velocity long-rod penetration semi-infinite concrete plate. Ph.D. dissertation, Xi'an Jiaotong University (in Chinese, English abstract)

Li JC, Yu MH, et al. (2000) The dynamic investigation of semi-infinite concrete targets penetrated by long rods. Mechanics 2000, Forest Press, Beijing, 519-520.(in Chinese)

Li JC, Yu MH, Gong YN (2000) Dynamic investigation of semi-infinite concrete target penetrated by long rod. In: Proceedings of the 3rd Asian-Pacific Conference. On Aerospace Technology and Science, Beijing University of Aeronautics and Astronautics Press, Beijing, 263-269

Li QM (2005) Penetration of a hard missile. 6th Asia-Pacific Conference on Shock & Impact Loads on Structures, Dec 7-9, 2005, Perth W, Australia, 49-62

Longcope DB, Forrestal MJ (1983) Penetration of targets described by a Mohr-Coulomb failure criterion with a tension cutoff. J. Appl. Mech., 50:327-333

Luk VK, Forrestal MJ (1987) Penetration into semi-infinite reinforced-concrete targets with spherical and ogival noise projectiles. Int. J. Impact Eng., 6(4):291-301

Mastilovic S, Krajcinovic D (1999) Penetration of rigid projectiles through quasi-brittle materials. J. Appl. Mech., 66:585-592

Wang YB (2004) Research in structural impacting problems based on the unified strength theory. Ph.D. dissertation, Xi'an Jiaotong University (in Chinese, English abstract)

Wang YB, Li ZH, Wei XY, Yu MH (2005) Analysis of high-velocity tungsten rod on penetration brittle target. Chinese J. High Pressure Physics, 19(3):257-263 (in Chinese, English abstract)

Wang YB, Zhu YY, Yu MH (2004) Penetration analysis of high-velocity tungsten rod on ceramic targets using unified strength theory. Explosion and Shock Waves, 24(6):534-540 (in Chinese, English abstract)

Wei XY (2002) Investigation of long-rod penetration problems. Ph.D. dissertation, Xi'an Jiaotong University (in Chinese, English abstract)

Wei XY, Yu MH (2002) Analysis of tungsten rod on penetration ceramic targets at high velocity. Acta Armamentarii, 23(2):167-170 (in Chinese, English abstract)

Yu MH (1992) A new system of strength theory. Xi'an Jiaotong University Press, Xi'an, China (In Chinese)

Yu MH (2004) Unified Strength theory and its applications. Springer, Berlin

Yu MH, He LN (1991) A new model and theory on yield and failure of materials under complex stress state. In: Mechanical Behavior of Materials-6, Pergamon Press, Oxford, 3:841-846

Yu MH, He LN, Liu CY (1992) Generalized twin shear stress yield criterion and its generalization. Chinese Science Bulletin, 37(24):2085-2089

Yu MH, Li JC, Zhang YQ (2001) Unified characteristics line theory of spatial axisymmetric plastic problem. Science in China (Series E), English ed., 44(2):207-215; Chinese ed., 44(4):323-331

Yu MH, Ma GW, Qiang HF, et al. (2006) Generalized plasticity. Springer, Berlin

Yu MH, Yang SY, Liu CY, Liu JY (1997) Unified plane-strain slip line theory. China Civil Eng. J. 30(2):14-26 (in Chinese, English abstract)

Zukas JA, Nicholas T, Swift HF, et al. (1982) Impact dynamics. John Wiley & Sons, Inc., New York

13
Plastic Analysis of Orthogonal Circular Plate

13.1 Introduction

Many plates used in practical engineering are strengthened by stiffeners to achieve high strength and low structural weight. Stiffeners are usually placed along the orthogonal directions. Thus a plate strengthened with stiffeners will exhibit structural plane orthotropy with two orthogonal axes of symmetry (Tsai, 1968; Daniel and Ishai, 1994).

Material orthotropy may also arise from the cold forming process, so that yield stresses in different directions are different. Regardless of the source, the orthotropy can be described with a proper yield criterion that has been verified by experimental evidence. A pioneering work that investigated the plastic limit behavior of an orthotropic circular plate was carried out by Markowitz and Hu (1965), who employed a modified Tresca criterion. Save et al. (1985; 1997) summarized the plastic limit solutions for an orthotropic circular plate. Plastic limit analysis of orthotropic circular plates using the unified yield criterion was derived by Ma et al. (2002). A series of new results were given.

In the following context an orthotropic yield criterion which is the extension of the unified yield condition (UYC) will be explored. The normality law is assumed for the yield condition. The directions of the principal stresses and the orthotropy are assumed to be the radial and circumferential directions.

13.2 Orthotropic Yield Criteria

The unified yield criterion (UYC) may represent or approximate various different yield criteria, e.g., the single-shear stress criterion (Tresca criterion, 1864), octahedral shear stress criterion (Huber-von Mises criterion, 1904-1913) and twin-shear stress criterion (1961; 1983). However, it is applicable to isotropic materials only. Proper modification to the criteria should be made if it is applied to orthotropic materials or orthotropic structures.

For an anisotropic plate, if it is assumed that the UYC is still valid, the piecewise linear mathematical formula of the unified yield criterion in terms of the moment M_r and M_θ may take another form of

$$\frac{M_\theta}{M_0} = a_i \frac{M_r}{M_0} + b_i, \quad (i = 1, ..., 12), \quad (13.1)$$

where M_r and M_θ are the actual radial and circumferential moments respectively; M_0 is the yield moment that is independent of the orientations; a_i and b_i ($i=1, ..., 12$), are parameters corresponding to the 12 segments of the yield loci determined by the unified yield criterion parameter b.

For an orthotropic circular plate with the principal stress directions coinciding with the radial and circumferential directions, the unified yield criterion in generalized stresses form can be expressed as

$$\frac{M_\theta}{M_{\theta 0}} = a_i \frac{M_r}{M_{r0}} + b_i, \quad (i = 1, ..., 12), \quad (13.2)$$

where M_{r0} and $M_{\theta 0}$ are the yield moments in the radial and circumferential directions respectively.

With the normalized quantities of $\kappa = M_{\theta 0}/M_{r0}$, $m_\theta = M_\theta/M_{r0}$, and $m_r = M_r/M_{r0}$, Eq.(13.2) can be rewritten as

$$m_\theta = A_i m_r + B_i, \quad (i = 1, ..., 12), \quad (13.3)$$

where $A_i = a_i \kappa$, $B_i = b_i \kappa$ ($i=1, ..., 12$). κ is the ratio of the circumferential yield moment to the radial yield moment, called orthotropy coefficient. $\kappa = 1$ corresponds to the special case of isotropy, while $\kappa > 1$ or $\kappa < 1$ represents the circumferential or radial strengthening of the plate.

Similarly, the Huber-von Mises yield criterion can be rewritten with the normalized quantities as

$$\frac{M_\theta^2}{M_{\theta 0}^2} - \frac{M_\theta M_r}{M_{\theta 0} M_{r0}} + \frac{M_r^2}{M_{r0}^2} = 1. \quad (13.4)$$

In terms of the dimensionless variables, Eq.(13.4) becomes

$$m_\theta^2 - \kappa m_\theta m_r + \kappa^2 m_r^2 = \kappa^2. \quad (13.5)$$

It is seen that Eq.(13.5) is nonlinear with respect to the two variables m_r and m_θ, which makes it not straightforward for the plastic limit analysis of plates.

The yield loci with different orthotropic ratios of the modified unified yield criterion (UYC) defined in Eq.(13.3) are plotted in Figs.13.1(a)~(c) with respect to $b=0, 1, 0.5$. Fig.13.1(d) illustrates the modified Mises criterion defined in Eq.(13.5).

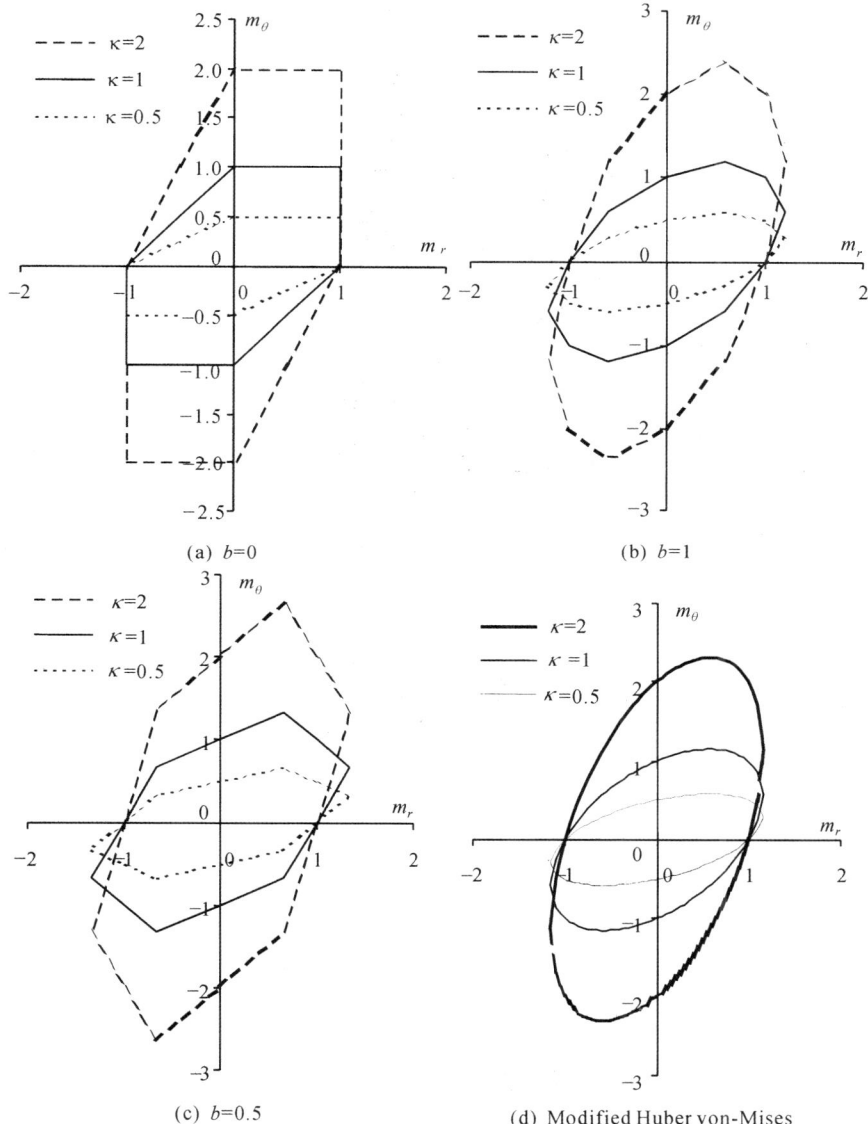

Fig. 13.1. Orthotropic yield criteria

The orthotropic yield criteria are compared in Fig.13.2 with different orthotropic ratios, i.e., $\kappa = 2$, $\kappa = 0.5$, and the isotropic case of $\kappa = 1$. It shows again that even in orthotropic conditions, the Huber-von Mises criterion can be approximately represented by the UYC with $b = 0.5$. Thus, it is possible to obtain an approximate plastic limit solution to an orthotropic circular plate in terms of the Huber-von Mises yield criterion.

The moments at the corners on the yield loci in Fig.13.2 can be derived by solving the equations for the two intersecting lines which have the forms of

$$m_{r(i-j)} = -\frac{B_i - B_j}{A_i - A_j}, m_{\theta(i-j)} = \frac{A_i B_j - A_j B_i}{A_i - A_j},$$
$$(i = 1, ..., 11, j = i + 1; i = 12, j = 1),$$
(13.6)

$$m_{r(i-j)} = -\frac{b_i - b_j}{a_i - a_j} m_{\theta(i-j)} = \frac{a_i b_j - a_j b_i}{a_i - a_j} \kappa,$$
$$(i = 1, ..., 11, j = i + 1; i = 12, j = 1),$$
(13.7)

where the line with $i = 1$ corresponds to line AB, and i increases from 1 to 12 in anti-clockwise order. Thus, $i = 12$ corresponds to line LA. The symbol $(i - j)$ denotes the corner intersected by line i and line j.

The values of the parameters a_i and b_i (i=1, ..., 12) are listed in Table 13.1. The moments at the 12 corners are listed in Table 13.2. The moment and

Table 13.1. Constants a_i and b_i in the modified unified yield criterion

	AB	BC	CD	DE	EF	FG
	$(i = 1\ )$	$(i = 2\ )$	$(i = 3)$	$(i = 4)$	$(i = 5)$	$(i = 6)$
a_i	$-b$	$\frac{b}{1+b}$	$\frac{1}{1+b}$	$1 + b$	$\frac{1+b}{b}$	$-\frac{1}{b}$
b_i	$1 + b$	1	1	$1 + b$	$\frac{1+b}{b}$	$-\frac{1+b}{b}$
	GH	HI	IJ	JK	KL	LA
	$(i = 7)$	$(i = 8)$	$(i = 9)$	$(i = 10)$	$(i = 11)$	$(i = 12)$
a_i	$-b$	$\frac{b}{1+b}$	$\frac{1}{1+b}$	$1 + b$	$\frac{1+b}{b}$	$-\frac{1}{b}$
b_i	$1 + b$	1	1	$1 + b$	$\frac{1+b}{b}$	$-\frac{1+b}{b}$

velocity fields, the plastic limit loads of orthotropic circular plates with different orthotropic ratios and boundary conditions with respect to the modified anisotropic yield criteria will be presented in the following sections.

13.3 General Solutions

For clear notification, dimensionless variables p and r, i.e., $p = Pa^2/M_{r0}$ and $r = R/a$ are defined for an orthotropic plate with uniform thickness, where

Table 13.2. Moments at corners

Node	A	B	C	D	E	F
m_θ	κ	$\frac{2(1+b)}{2+b}\kappa$	κ	$\frac{1+b}{2+b}\kappa$	0	$-\frac{1+b}{2+b}\kappa$
m_r	1	$\frac{1+b}{2+b}$	0	$-\frac{1+b}{2+b}$	-1	$-\frac{2(1+b)}{2+b}$
Node	G	H	I	J	K	L
m_θ	$-\kappa$	$-\frac{2(1+b)}{2+b}\kappa$	$-\kappa$	$-\frac{1+b}{2+b}\kappa$	0	$\frac{1+b}{2+b}\kappa$
m_r	-1	$-\frac{1+b}{2+b}$	0	$\frac{1+b}{2+b}$	1	$\frac{2(1+b)}{2+b}$

P is the uniformly distributed load, a is the outer radius of the plate and R is the radius variable in the range of 0 to a. For an axisymmetric loading, any two orthogonal directions at the central point are principal directions for stresses and orthotropy. Hence the plate is locally isotropic at the center and there exists $M_r = M_\theta$ or $m_r = m_\theta$. Any stress profile should be symmetric about that point, which corresponds to point A' in Fig.13.2.

Point A' may be located on different line segments, namely lines with $i = 11, 12, 1,$ or 2, according to the orthotropic ratio κ. If $\kappa = 1$, A' coincides with the corner point A. When $\kappa > 1$, A' falls on line KL ($i = 11$) or LA ($i = 12$), while A' falls on line AB ($i = 1$) or BC ($i = 2$) when $\kappa < 1$.

For a simply supported circular plate in a fully plastic limit state, the moment at the outer edge satisfies the yield point C in Fig.13.2. The moments in the whole plate correspond to the yield segments from A' to C in anti-clockwise order. The total number of the segments could be 4 when A' is on line KL, and could be reduced to 1 when A' is on line BC. On the other hand, the edge of a fixed supported circular plate corresponds to a yield point on line EF, which has been discussed in Chapter 3 and Chapter 4. In the present study, the moments at the edge are assumed to exactly correspond with the corner point F in Fig.13.2. Thus the total number of the segments regarding the fixed supported circular plate varies from 4 to 7 with reference to the different orthotropic ratio κ.

According to the piecewise linear yield criterion in Eq.(13.3), the radial moment can be integrated as

$$m_r = \frac{B_i}{1 - A_i} - \frac{pr^2}{2(3 - A_i)} + C_i r^{-1+A_i}, \tag{13.8}$$

where i represents each valid segment of the yield loci; c_i are integration constants, which could be determined by continuity and boundary conditions.

In a general case where the moments in the plate cover total n segments on the yield loci, e.g., if $\kappa > 1$, the yield moments are on segments $A'A, AB, BC$ ($n = 3$) or on segments $A'L, LA, AB, BC$ ($n = 4$) with the

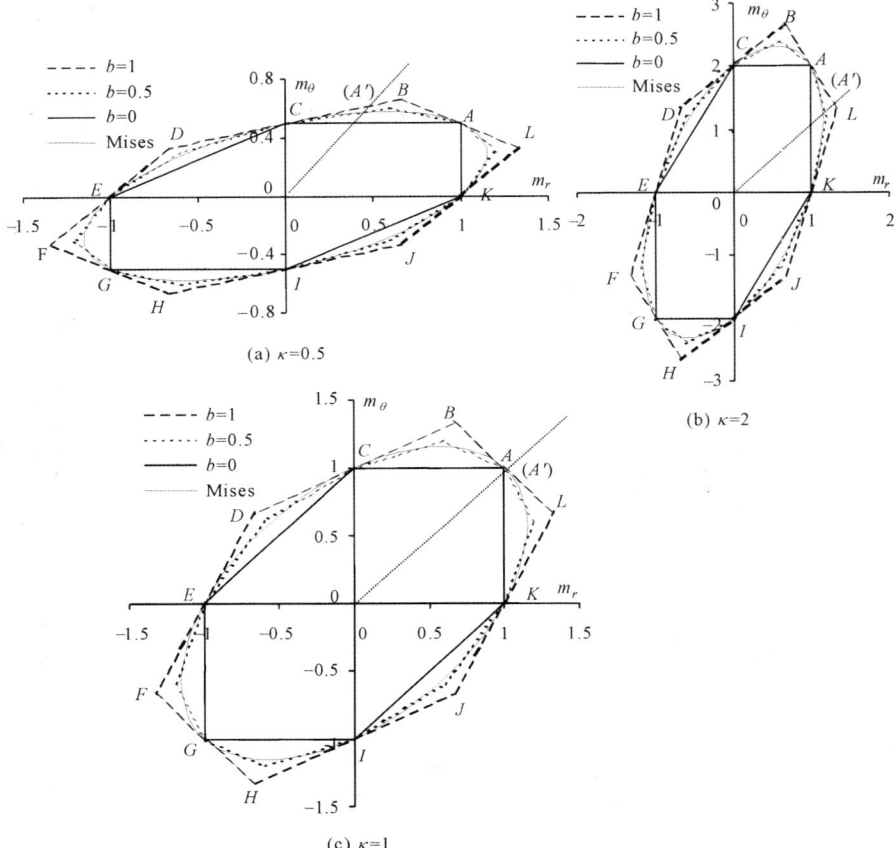

Fig. 13.2. Comparison of the modified yield criteria

moments varying from A' to C with r increasing from 0 to 1, the continuity and boundary conditions of the radial moment will yield

$$\frac{B_{(1)}}{1 - A_{(1)}} - \frac{pr_0^2}{2(3 - A_{(1)})} + C_1 r_0^{-1+A_{(1)}} = m_{r(0)}, \qquad (13.9a)$$

$$\frac{B_{(1)}}{1 - A_{(1)}} - \frac{pr_1^2}{2(3 - A_{(1)})} + C_1 r_1^{-1+A_{(1)}} = m_{r(1)}, \qquad (13.9b)$$

$$\frac{B_{(2)}}{1 - A_{(2)}} - \frac{pr_1^2}{2(3 - A_{(2)})} + C_2 r_1^{-1+A_{(2)}} = m_{r(1)}, \qquad (13.9c)$$

$$\frac{B_{(2)}}{1 - A_{(2)}} - \frac{pr_2^2}{2(3 - A_{(2)})} + C_2 r_2^{-1+A_{(2)}} = m_{r(2)}, \qquad (13.9d)$$

$$\frac{B_{(i)}}{1 - A_{(i)}} - \frac{pr_{i-1}^2}{2(3 - A_{(i)})} + C_i r_{i-1}^{-1+A_{(i)}} = m_{r(i-1)}, \tag{13.9e}$$

$$\frac{B_{(i)}}{1 - A_{(i)}} - \frac{pr_i^2}{2(3 - A_{(i)})} + C_i r_i^{-1+A_{(i)}} = m_{r(i)}, \tag{13.9f}$$

$$\frac{B_{(n)}}{1 - A_{(n)}} - \frac{pr_{n-1}^2}{2(3 - A_{(n)})} + C_n r_{n-1}^{-1+A_{(n)}} = m_{r(n-1)}, \tag{13.9g}$$

$$\frac{B_{(n)}}{1 - A_{(n)}} - \frac{pr_n^2}{2(3 - A_{(n)})} + C_n r_n^{-1+A_{(n)}} = m_{r(n)}, \tag{13.9h}$$

where the subscript i of the parameters $A_{(i)}$ and $B_{(i)}$ denotes the ith valid segment from A' on the yield loci in Fig. 13.2. The subscript i of the radial moment $m_{r(i)}$ represents the ith corner point from A', e.g., $m_{r(0)}$ corresponds to the yield moment on point A', while $m_{r(n)}$ corresponds to the yield moment at the corner point C.

Defining $\alpha_1 = r_1/r_2$, $\alpha_2 = r_2/r_3$, $\alpha_i = r_i/r_{i+1}$ and $\alpha_{n-1} = r_{n-1}$, there then exist $r_1 = \alpha_1 \alpha_2 \cdots \alpha_{n-1}$, $r_2 = \alpha_2 \alpha_3 \cdots \alpha_{n-1}$, $r_i = \alpha_i \cdots \alpha_{n-1}$ and $r_{n-1} = \alpha_{n-1}$. With reference to the continuity and boundary conditions, C_i $(i = 1, ..., n)$ in Eq.(13.9a) to Eq.(13.9h) are determined,

$$C_1 = 0, \tag{13.10a}$$

$$C_2 = \left[-\left(\frac{B_{(2)}}{1 - A_{(2)}} - m_{r(1)} \right) + \left(\frac{B_{(1)}}{1 - A_{(1)}} - m_{r(1)} \right) \frac{3 - A_{(1)}}{3 - A_{(2)}} \right] (r_1)^{1-A_{(2)}}, \tag{13.10b}$$

$$C_i = \left[-\left(\frac{B_{(i)}}{1 - A_{(i)}} - m_{r(i-1)} \right) + \left(\frac{B_{(1)}}{1 - A_{(1)}} \right. \right.$$
$$\left. \left. - m_{r(1)} \right) \frac{3 - A_{(1)}}{3 - A_{(i)}} (\alpha_1 \alpha_2 \cdots \alpha_{i-2})^{-2} \right] (r_{i-1})^{1-A_{(i)}}, \tag{13.10c}$$

$$C_n = \left[-\left(\frac{B_{(n)}}{1 - A_{(n)}} - m_{r(n-1)} \right) + \left(\frac{B_{(1)}}{1 - A_{(1)}} \right. \right.$$
$$\left. \left. - m_{r(1)} \right) \frac{3 - A_{(1)}}{3 - A_{(n)}} (\alpha_1 \alpha_2 \cdots \alpha_{n-2})^{-2} \right] (r_{n-1})^{1-A_{(n)}}, \tag{13.10d}$$

where the ratios $\kappa_1, \kappa_2, \cdots, \kappa_i, \cdots, \kappa_{n-2}$ are given by

$$\frac{B_{(2)}}{1-A_{(2)}} - m_{r(2)} - \left(\frac{B_{(1)}}{1-A_{(1)}} - m_{r(1)}\right)\frac{3-A_{(1)}}{3-A_{(2)}}\alpha_1^{-2}$$

$$+\left[-\left(\frac{B_{(2)}}{1-A_{(2)}} - m_{r(1)}\right) + \left(\frac{B_{(1)}}{1-A_{(1)}} - m_{r(1)}\right)\frac{3-A_{(1)}}{3-A_{(2)}}\right]\alpha_1^{1-A_{(2)}} = 0,$$

$$(13.11a)$$

$$\frac{B_{(i)}}{1-A_{(i)}} - m_{r(i)} - \left(\frac{B_{(1)}}{1-A_{(1)}} - m_{r(1)}\right)\frac{3-A_{(1)}}{3-A_{(i)}}(\alpha_1\alpha_2\cdots\alpha_{i-1})^{-2}$$

$$+\left[-\left(\frac{B_{(i)}}{1-A_{(i)}} - m_{r(i-1)}\right) + \left(\frac{B_{(1)}}{1-A_{(1)}}\right.\right. \qquad (13.11b)$$

$$\left.\left. - m_{r(1)}\right)\frac{3-A_{(1)}}{3-A_{(i)}}(\alpha_1\alpha_2\cdots\alpha_{i-2})^{-2}\right]\alpha_{i-1}^{1-A_{(i)}} = 0.$$

$$\frac{B_{(n)}}{1-A_{(n)}} - m_{r(n)} - \left(\frac{B_{(1)}}{1-A_{(1)}} - m_{r(1)}\right)\frac{3-A_{(1)}}{3-A_{(n)}}(\alpha_1\alpha_2\cdots\alpha_{n-1})^{-2}$$

$$+\left[-\left(\frac{B_{(n)}}{1-A_{(n)}} - m_{r(n-1)}\right) + \left(\frac{B_{(1)}}{1-A_{(1)}}\right.\right. \qquad (13.11c)$$

$$\left.\left. - m_{r(1)}\right)\frac{3-A_{(1)}}{3-A_{(n)}}(\alpha_1\alpha_2\cdots\alpha_{n-2})^{-2}\right]\alpha_{n-1}^{1-A_{(n)}} = 0.$$

The above simultaneous equations can be solved by numerical iteration method, such as a half interval search method. When α_i ($i=1, ..., n-1$) is derived, the demarcating radius r_i ($i=1, ..., n-1$) is then obtained.

The plastic limit load can be derived from Eq.(13.9b) with reference to Eq.(13.10a),

$$p = \left[\frac{B_{(1)}}{1-A_{(1)}} - m_{r(1)}\right]\frac{2(3-A_{(1)})}{r_1^2}. \qquad (13.12)$$

Substituting the integration constants C_i ($i=1, ..., n$), $r_i(i=0, ..., n)$, and the plastic limit load p into Eq.(13.8), the radial moment field m_r of the whole plate is then obtained.

The velocity field of the plate can be integrated from the compatible conditions and the associated plastic flow conditions, which are the same as that for the isotropic case,

$$\dot{w} = \dot{w}_0\left(c_{1i}r^{1-A_{(i)}} + c_{2i}\right), r_{i-1} \leqslant r \leqslant r_i, \quad (i=1, ..., n), \qquad (13.13)$$

where c_{1i} and c_{2i} ($i=1, ..., n$) are the integration constants corresponding to the n valid segments on the yield loci. r_i ($i=0, ..., n$) are the demarcating radii, where the moments are located at the corner points of the yield loci.

The continuity and boundary conditions of the velocity field in the current condition gives: (1) $\dot{w}(r = 0) = \dot{w}_0$; (2) $\dot{w}$ and $d\dot{w}/dr$ ($r = r_i, i = 1, ..., n-1$) are continuous; (3) $\dot{w}(r = 1) = 0$. With reference to these conditions, the constants c_{1i} and c_{2i} in Eq.(13.13) can be derived,

$$c_{21} = 1, \tag{13.14a}$$

$$c_{11}r_1^{1-A_{(1)}} + c_{21} = c_{12}r_1^{1-A_{(2)}} + c_{22}, \tag{13.14b}$$

$$(1 - A_{(1)})c_{11}r_1^{-A_{(1)}} = (1 - A_{(2)})c_{12}r_1^{-A_{(2)}}, \tag{13.14c}$$

$$c_{1i}r_i^{1-A_{(i)}} + c_{2i} = c_{1(i+1)}r_i^{1-A_{(i+1)}} + c_{2(i+1)}, \tag{13.14d}$$

$$(1 - A_{(i)})c_{1i}r_i^{-A_{(i)}} = (1 - A_{(i+1)})c_{1(i+1)}r_i^{-A_{(i+1)}}, \tag{13.14e}$$

$$c_{1(n-1)}r_{n-1}^{1-A_{(n-1)}} + c_{2(n-1)} = c_{1(n)}r_{n-1}^{1-A_{(n)}} + c_{2n}, \tag{13.14f}$$

$$(1 - A_{(n-1)})c_{1(n-1)}r_{n-1}^{-A_{(n-1)}} = (1 - A_{(n)})c_{1n}r_{n-1}^{-A_{(n)}}, \tag{13.14g}$$

$$c_{1n}r_n^{1-A_{(n)}} + c_{2n} = 0. \tag{13.14h}$$

The $2n$ constants c_{1i} and c_{2i} ($i=1, ..., n$) can be solved directly from the above $2n$ simultaneous linear equations.

Substituting these integration constants into Eq.(13.13), the velocity field of the circular plate is then obtained.

13.4 Simply Supported Orthotropic Circular Plate

For a simply supported circular plate with orthotropic ratio κ, there are four possible cases, namely point A' where the moment corresponding to the plate center is on segments KL, LA, AB or BC (Fig.13.2). The following sections will discuss the four cases with reference to $\kappa < 1$ and $\kappa > 1$.

13.4.1 Case I: Point A' Falls on Segment KL

When $\kappa > 1$, the circumferential yield moment $m_{\theta0}$ is less than the radial yield moment m_{r0}. Point A' may be located on either KL or LA. Considering the case when point A' coincides with point L, there are

$$\begin{cases} m_\theta = \kappa a_{11} m_r + \kappa b_{11}, \\ m_\theta = \kappa a_{12} m_r + \kappa b_{12}, \\ m_\theta = m_r. \end{cases} \tag{13.15}$$

The orthotropic ratio κ in this case is derived as

$$\kappa = \frac{b_{11} - b_{12}}{a_{12} b_{11} - a_{11} b_{12}}. \tag{13.16}$$

When $\kappa > \frac{b_{11} - b_{12}}{a_{12} b_{11} - a_{11} b_{12}}$, point A' falls on segment KL. The four segments $(n = 4)$ of $A'L, LA, AB$, and BC are valid. Because of the rotational symmetry of the moments at the plate center, the radial moment at $r = 0$ corresponding to point A' and $m_r = m_\theta$ is derived,

$$m_{r0} = \frac{B_{11}}{1 - A_{11}} \quad \text{or} \quad m_{r0} = \frac{\kappa b_{11}}{1 - \kappa a_{11}}, \tag{13.17}$$

which is determined by the unified yield criterion parameter b and the orthotropic ratio κ.

13.4.2 Case II: Point A' Falls on Segment LA

When $\frac{b_{11} - b_{12}}{a_{12} b_{11} - a_{11} b_{12}} \geqslant \kappa > 1$, point A' falls on segment LA. The valid segments are $A'A, AB$ and BC with the radial moment m_r varying from point A' to C on the yield loci when r increases from 0 to 1. The radial moment at point A' has the form of

$$m_{r0} = \frac{B_{12}}{1 - A_{12}} \quad \text{or} \quad m_{r0} = \frac{\kappa b_{12}}{1 - \kappa a_{12}}. \tag{13.18}$$

13.4.3 Case III: Point A' Falls on Segment AB

When $1 \geqslant \kappa > \frac{b_1 - b_2}{a_2 b_1 - a_1 b_2}$, the moments in the whole plate correspond to the two yield segments of $A'B$ and BC. The radial moment m_r at the plate center is

$$m_{r0} = \frac{B_1}{1 - A_1} \quad \text{or} \quad m_{r0} = \frac{\kappa b_1}{1 - \kappa a_1}. \tag{13.19}$$

13.4.4 Case IV: Point A' Falls on Segment BC

when $\kappa \leqslant \frac{b_1 - b_2}{a_2 b_1 - a_1 b_2}$, point A' locates on BC. The moments in the whole plate correspond to the yield segment BC. The valid segment is $A'C$ and the radial moment at the plate center is

$$m_{r0} = \frac{B_2}{1 - A_2} \quad \text{or} \quad m_{r0} = \frac{\kappa b_2}{1 - \kappa a_2}. \tag{13.20}$$

13.4.5 Moment, Velocity Fields and Plastic Limit Load

The moments at other corner points and the parameters $A_{(i)}$ and $B_{(i)}$ are listed in Table 13.3, in which $m_{r(i-j)}$ is given by Eqs.(13.6) and (13.7).

Table 13.3. Yield moments at corner points and the corresponding $A_{(i)}$ and $B_{(i)}$

	i	1	2	3	4
Case I	$m_{r(i)}$	$m_{r(11-12)}$	$m_{r(12-1)}$	$m_{r(1-2)}$	$m_{r(2-3)}$
($n=4$)	$A_{(i)}$	A_{11}	A_{12}	A_1	A_2
	$B_{(i)}$	B_{11}	B_{12}	B_1	B_2
Case II	$m_{r(i)}$	$m_{r(12-1)}$	$m_{r(1-2)}$	$m_{r(2-3)}$	$m_{r(3-4)}$
($n=3$)	$A_{(i)}$	A_{12}	A_1	A_2	A_3
	$B_{(i)}$	B_{12}	B_1	B_2	B_3
Case III	$m_{r(i)}$	$m_{r(1-2)}$	$m_{r(2-3)}$	$m_{r(3-4)}$	$m_{r(4-5)}$
($n=2$)	$A_{(i)}$	A_1	A_2	A_3	A_4
	$B_{(i)}$	B_1	B_2	B_3	B_4
Case IV	$m_{r(i)}$	$m_{r(2-3)}$	$m_{r(3-4)}$	$m_{r(4-5)}$	$m_{r(5-6)}$
($n=1$)	$A_{(i)}$	A_2	A_3	A_4	A_5
	$B_{(i)}$	B_2	B_3	B_4	B_5

Fig.13.3(a) to Fig.13.3(c) illustrate the moment fields in terms of the orthotropic unified yield criterion with respect to $\kappa = 2$, $\kappa = 1$, and $\kappa = 0.5$ respectively. The three figures correspond to Case II, Case III and Case IV respectively. From Fig.13.3, when $\kappa \neq 1$ or the plate has different yield moments in the radial and circumferential directions, both the moment distribution profiles and the moment state at the plate centre in terms of different parameter b ($b=0$, $b=0.5$ and $b=1$) in the modified unified yield criterion are quite different.

The yield loci with $b = 0$, which is the same as the Tresca criterion, gives the lowest yield moment at the plate center. When the Tresca criterion is applied, the circumferential moment m_θ keeps constant on the whole plate for Case III and Case IV, while it is constant only in the range of $r_1 \leqslant r \leqslant 1$ for Case II. Although it is unreasonable, the solutions with reference to the Tresca criterion are generally preferred in practical engineering.

Based on the modified unified yield criterion on the other hand, the moment distribution of an orthotropic plate with varying yield properties can also be conveniently derived.

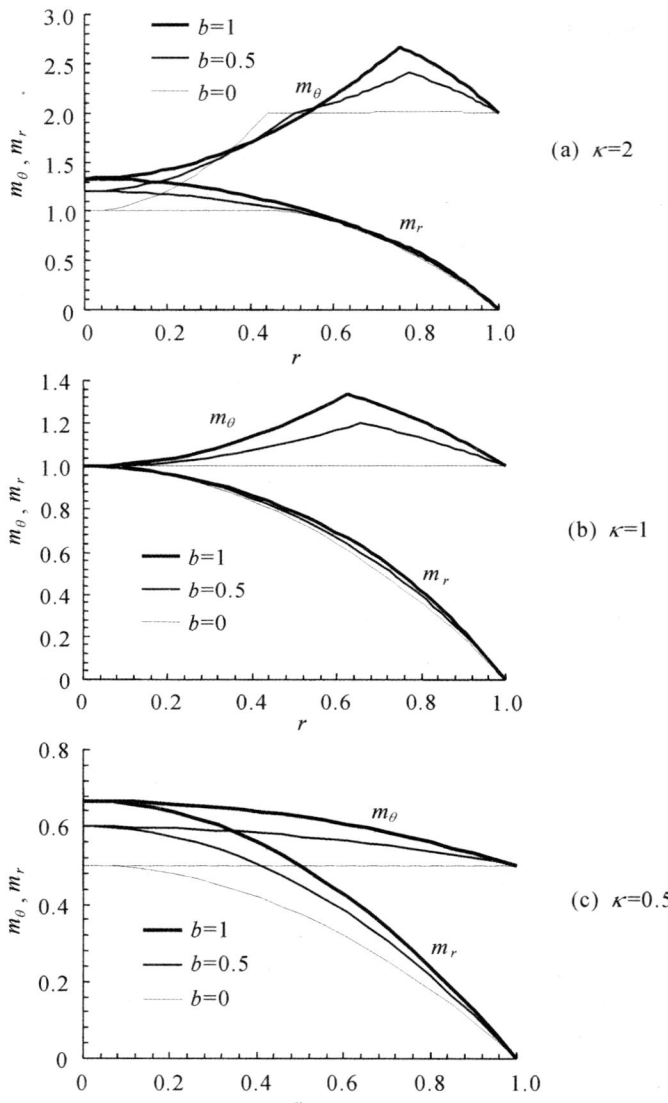

Fig. 13.3. Moment fields of orthotropic simply supported circular plate

The velocity fields from different cases are shown in Fig.13.4. The Tresca criterion ($b = 0$) gives a linear line (Fig.13.4(b) and Fig.13.4(c) or piecewise linear lines in Fig.13.4(a)) for velocity distributions with different orthotropic ratios. The criteria with $b = 0.5$ and $b = 1$, which represent approximately the modified Huber von-Mises criterion and the Yu criterion respectively, yield smooth and orthotropic ratio dependent velocity distributions.

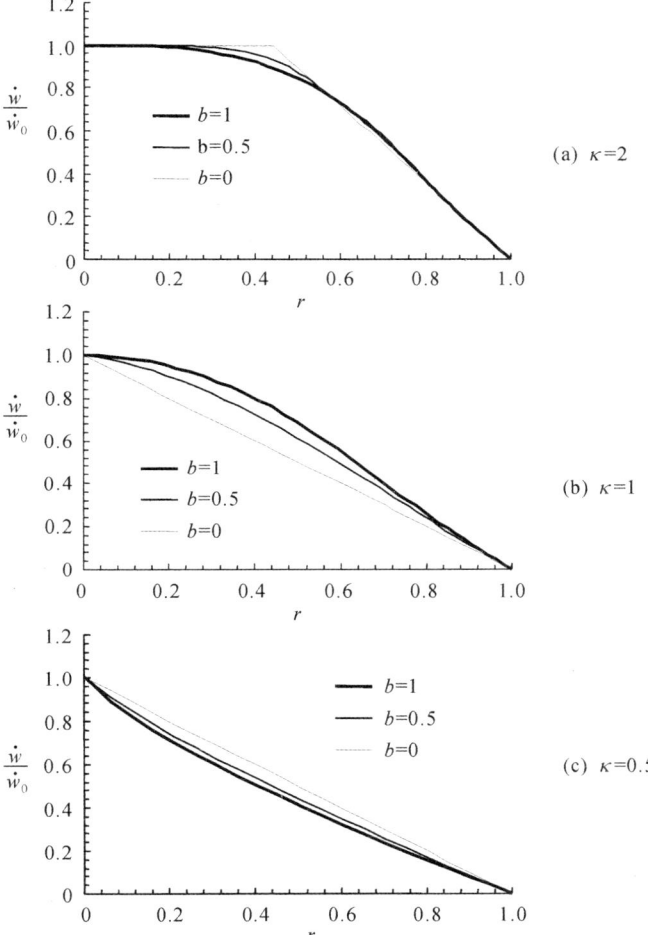

Fig. 13.4. Velocity fields of orthotropic simply supported circular plate

The unified yield criterion parameter b is the key factor in unifying different yield criteria. The curve of plastic limit load versus the parameter b is plotted in Fig.13.5(a) for three orthotropic ratios of $\kappa = 2$, $\kappa = 1$ and $\kappa = 0.5$. It is seen that for all the three cases, the plastic limit load increases monotonically with the increase of the parameter b. The plastic limit load when $\kappa = 2$ is much higher than those with $\kappa = 1$ and $\kappa = 0.5$. This indicates that a higher plastic limit load can be achieved when the plate is stiffened along the circumferential direction ($\kappa > 1$) instead of the radial direction.

Denoting a difference ratio ρ as

$$\rho = \frac{p - p_{\text{Tresca}}}{p_{\text{Tresca}}} \times 100\%. \tag{13.21}$$

It is convenient to compare the effect of different yield criteria on the plastic limit load. Fig.13.5(b) gives the difference ratio versus the parameter b. It is found that the difference ratio increases monotonically with the increase of parameter b. The difference ratio is as much as 7.1% and 12.6% when $\kappa = 2$; 8.1%, 14.0% when $\kappa = 1$, and 13.3% and 22.2% when $\kappa = 0.5$ with $b = 0.5$ and 1 respectively. This implies that the plastic limit loads based on different yield criteria are very different, which is helpful in the optimal design of an orthotropic circular plate.

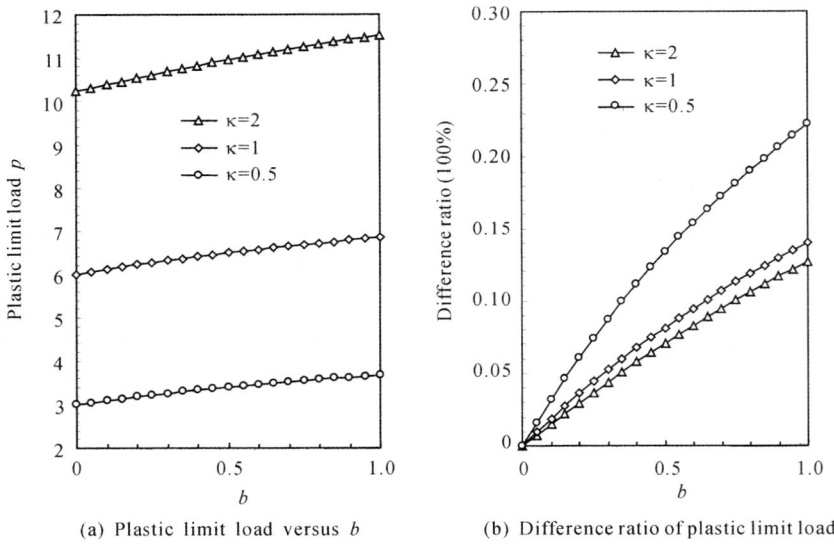

(a) Plastic limit load versus b (b) Difference ratio of plastic limit load

Fig. 13.5. Effects of the unified yield criteria parameter b

The effect of the orthotropic ratio on the plastic limit load is shown in Fig.13.6(a). It can be seen that the plastic limit load increases monotonically and sharply with the increase of the orthotropic ratio κ, which indicates that the strengthening along the circumferential direction does improve the load bearing capacity of a plate. The curve of difference ratio versus the orthotropic ratio with respect to two values of the unified yield criterion parameter, i.e., $b = 0.5$ and $b = 1$, is plotted in Fig.13.6(b). The maximum difference occurs at $\kappa = 0.56$ with percentage differences of 13.8% and 22.9% with respect to $b = 0.5$ and $b = 1$.

The percentage difference is almost constant when $\kappa > 1$ and they are about 7% and 13% corresponding to, respectively, $b = 0.5$ and $b = 1$. This indicates again that different yield criteria affect the plastic limit load significantly for a wide range of orthotropic ratios.

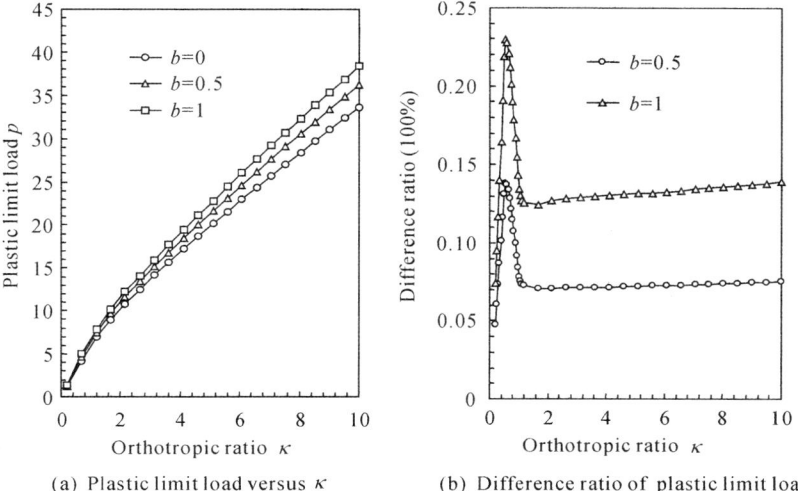

(a) Plastic limit load versus κ (b) Difference ratio of plastic limit load

Fig. 13.6. Effect of orthotropic ratio for simply supported plate

13.5 Fixed Supported Circular Plate

For the fixed supported circular plate, there also exist four possible cases, namely point A' falls on KL, LA, AB, or BC corresponding to the different orthotropic ratio κ. The moments in the whole plate correspond to the segments from A' to F in anti-clockwise order.

13.5.1 Case I: Point A' Falls on Segment KL

When $\kappa > 1$ and $\kappa > \frac{b_{11}-b_{12}}{a_{12}b_{11}-a_{11}b_{12}}$, point A' locates on the segment KL. There are in total seven valid segments ($n = 7$), namely, $A'L$, LA, AB, BC, CD, DE, and EF, on the yield surface. Because of the rotational symmetry of the moments about the plate center, the radial moment at $r = 0$ corresponding to point A' where $m_r = m_\theta$ is the same as that in Eq.(13.17).

13.5.2 Case II: Point A' Falls on Segment LA

When $\frac{b_{11}-b_{12}}{a_{12}b_{11}-a_{11}b_{12}} \geqslant \kappa > 1$, point A' falls on segment LA. The valid segments are $A'A$, AB, BC, CD, DE, and EF with the radial moment m_r varying from point A' to C on the yield loci when r increases from 0 to 1. The radial moment at A' is the same as that in Eq.(13.18).

13.5.3 Case III: Point A' Locates on Segment AB

When $1 \geqslant \kappa > \frac{b_1-b_2}{a_2b_1-a_1b_2}$, point A' falls on the yield segment AB and the moments in the whole plate correspond to the yield segments $A'B$, BC, CD,

DE, and EF. The radial moment m_r at the plate center is the same as that in Eq.(13.19).

13.5.4 Case IV: Point A' Falls on Segment BC

When point A' falls on segment BC with $\kappa \leqslant \frac{b_1 - b_2}{a_2 b_1 - a_1 b_2}$, the moments in the whole plate correspond to the yield segments of BC, CD, DE and EF. The valid segment is $A'C$ and the radial moment at plate center is the same as that in Eq.(13.20).

13.5.5 Moment Fields, Velocity Fields, and Plastic Limit Load

The moments at the corner points in different cases and the parameters $A_{(i)}$ and $B_{(i)}$ in Eqs.(13.9) to (13.14) are listed in Table 13.4.

Table 13.4. Yield moments at corner points and the corresponding $A_{(i)}$ and $B_{(i)}$

	i	1	2	3	4	5	6	7
Case I	$m_{r(i)}$	$m_{r(11-12)}$	$m_{r(12-1)}$	$m_{r(1-2)}$	$m_{r(2-3)}$	$m_{r(3-4)}$	$m_{r(4-5)}$	$m_{r(5-6)}$
$(n = 7)$	$A_{(i)}$	A_{11}	A_{12}	A_1	A_2	A_3	A_4	A_5
	$B_{(i)}$	B_{11}	B_{12}	B_1	B_2	B_3	B_4	B_5
Case II	$m_{r(i)}$	$m_{r(12-1)}$	$m_{r(1-2)}$	$m_{r(2-3)}$	$m_{r(3-4)}$	$m_{r(4-5)}$	$m_{r(5-6)}$	
$(n = 6)$	$A_{(i)}$	A_{12}	A_1	A_2	A_3	A_4	A_5	
	$B_{(i)}$	B_{12}	B_1	B_2	B_3	B_4	B_5	
Case III	$m_{r(i)}$	$m_{r(1-2)}$	$m_{r(2-3)}$	$m_{r(3-4)}$	$m_{r(4-5)}$	$m_{r(5-6)}$		
$(n = 5)$	$A_{(i)}$	A_1	A_2	A_3	A_4	A_5		
	$B_{(i)}$	B_1	B_2	B_3	B_4	B_5		
Case IV	$m_{r(i)}$	$m_{r(2-3)}$	$m_{r(3-4)}$	$m_{r(4-5)}$	$m_{r(5-6)}$			
$(n = 4)$	$A_{(i)}$	A_2	A_3	A_4	A_5			
	$B_{(i)}$	B_2	B_3	B_4	B_5			

The moment fields of the fixed circular plate with different orthotropic ratios, namely $\kappa = 2$, $\kappa = 1$ and $\kappa = 0.5$, are shown in Fig.13.7(a) to Fig.13.7(c) respectively. Fig.13.7(a) covers 6 valid yield segments, and Fig.13.7(b) and Fig.13.7(c) cover 5 and 4 yield segments respectively. It is seen that when $\kappa \neq 1$ or the plate has different yield moments in the radial and circumferential directions, the moment distribution profiles and the moment state at

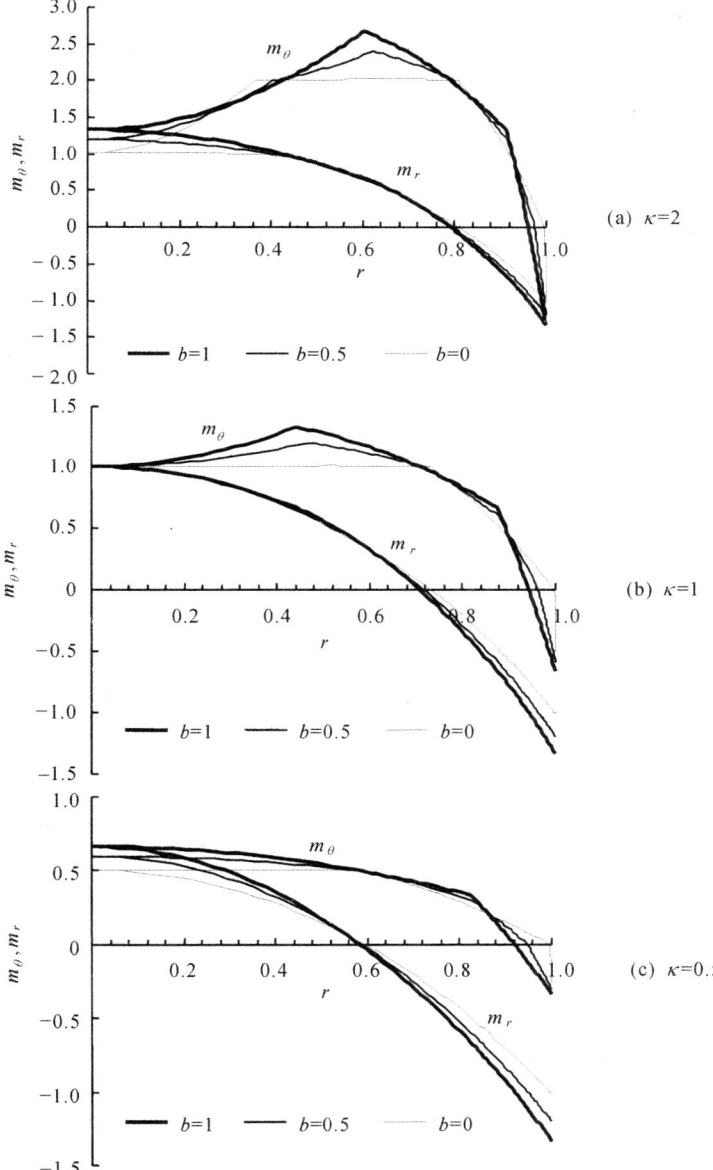

Fig. 13.7. Moment fields of orthotropic fixed supported circular plate

the plate center, with reference to different values of parameter b, i.e., $b = 0$, $b = 0.5$ and $b = 1$, of the unified yield criterion are quite different.

The velocity fields of the fixed plate for different cases are shown in Fig.13.8. The Tresca criterion ($b = 0$) gives a linear line (Fig.13.8(b) and Fig.13.8(c)) when $\kappa = 1$ and $\kappa = 0.5$. The criteria with $b = 0.5$ and $b = 1$, which represent approximately Huber-von Mises criterion and Yu criterion respectively, derive the smooth and orthotropic ratio dependent velocity distributions.

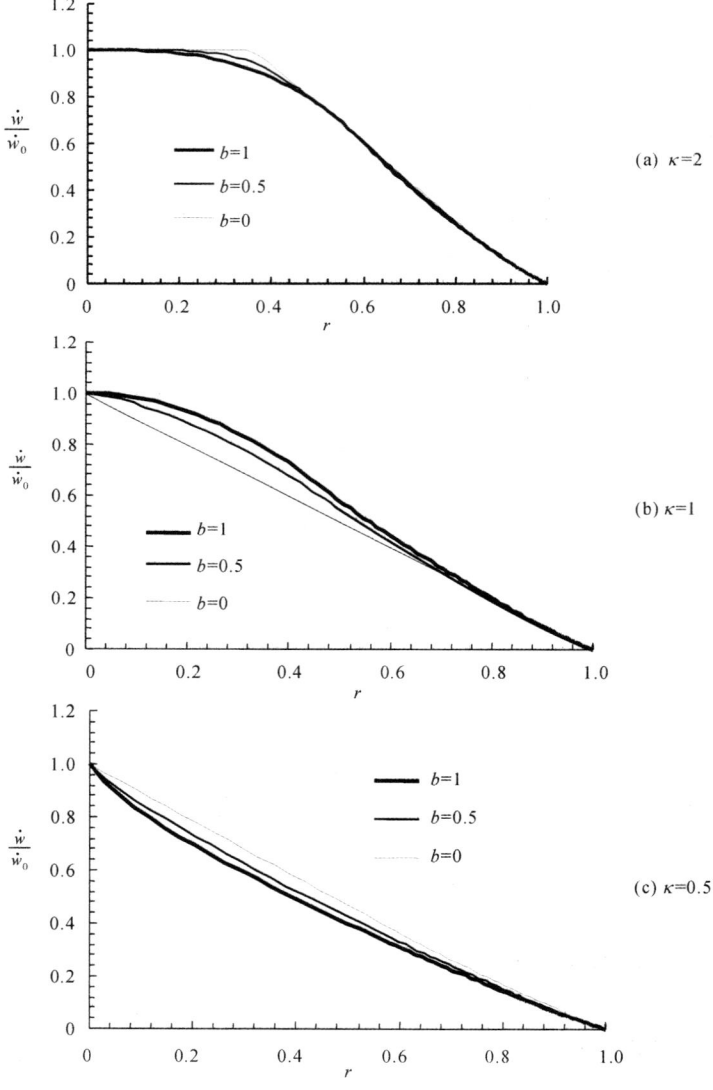

Fig. 13.8. Velocity fields of orthotropic fixed supported circular plate

The curves of plastic limit load and the difference ratio versus the unified yield criterion parameter b for the fixed circular plate are plotted in Fig.13.9(a) with the three different orthotropic ratios $\kappa = 2$, $\kappa = 1$, and $\kappa = 0.5$.

The plastic limit loads again increase monotonically with the increase of the parameter b and are much higher than their counterparts from the simply supported circular plate. The orthotropic ratio influences the plastic limit load very much so that the plastic limit load when $\kappa = 2$ is much higher than the plastic limit loads corresponding to $\kappa = 1$ and $\kappa = 0.5$.

The percentage differences are:

10.8% (from $b = 0$ to $b = 0.5$) to 18.3% (from $b = 0$ to $b = 1$) when $\kappa = 2$;

13.0% (from $b = 0$ to $b = 0.5$) to 21.7% (from $b = 0$ to $b = 1$) when $\kappa = 1$;

16.4% (from $b = 0$ to $b = 0.5$) to 27.3% (from $b = 0$ to $b = 1$) when $\kappa = 0.5$.

They are higher than the counterparts from the simply supported circular plate, which indicates that the yield criterion influences the plastic limit load of fixed circular plates more significantly than for the simply supported circular plate.

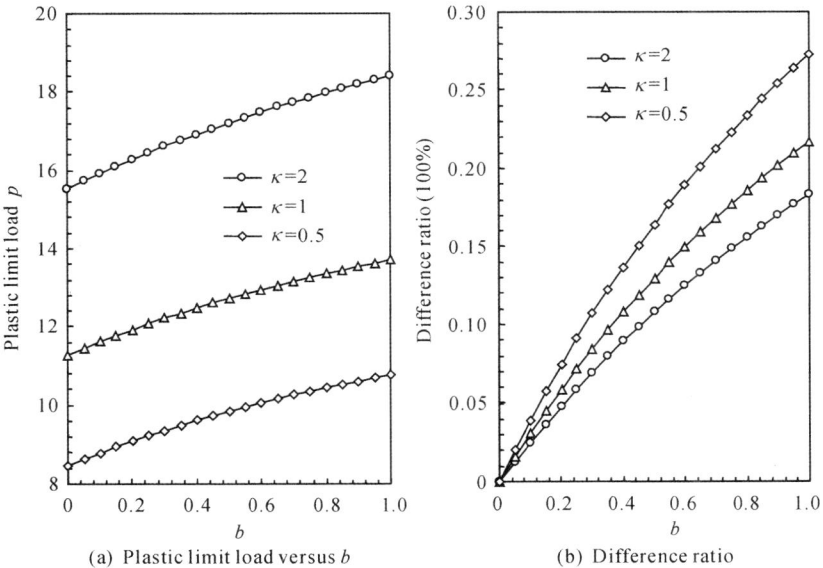

(a) Plastic limit load versus b (b) Difference ratio

Fig. 13.9. Effect of the unified yield criteria parameter b

The effect of a different orthotropic ratio on the plastic limit load is shown in Fig.13.10. The plastic limit load increases again monotonically with the increase of the orthotropic ratio κ.

On the other hand, the difference ratio decreases with the increase of orthotropic ratio. The difference is about 10% and 17% when the value of the unified yield criterion parameter b varies from $b = 0$ to $b = 0.5$ and from $b = 0$ to $b = 1.0$.

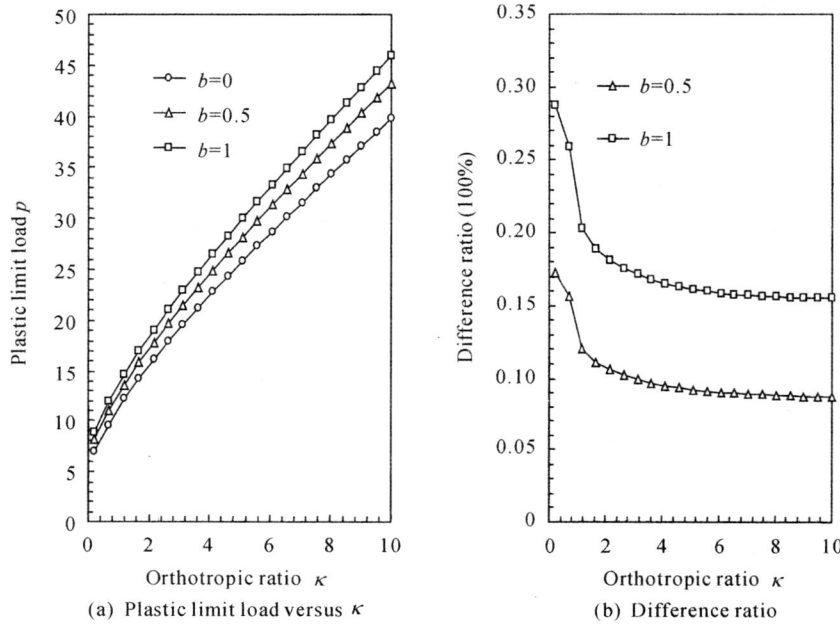

Fig. 13.10. Effect of orthotropic ratio for fixed supported plate

13.6 Summary

A unified piecewise linear orthotropic yield criterion is suggested. A unified solution employing the unified piecewise linear orthotropic yield criterion is derived for orthotropic circular plates in simply and fixed supported edge conditions.

It is found that the plastic limit load increases significantly with the increase of the orthotropic ratio. Different yield criterion affects the plastic limit load, moment and velocity fields.

For the simply supported orthotropic circular plate, the difference ratio depends on the orthotropic ratio and it is 7.1% and 12.6% when $\kappa = 2$, 13.3% and 22.2% when $\kappa = 0.5$ with respect to the modified Mises criterion and the modified Yu criterion respectively. The different ratio is about 7% and 13%

with respect to the modified Mises criterion and the modified Yu criterion when $\kappa > 1$.

For the fixed supported orthotropic circular plate, the difference ratio decreases with the increase of the orthotropic ratio. It is about 10% and 17% with respect to the von Mises criterion and the Yu twin-shear stress criterion respectively, when the orthotropic ratio is greater than 1.

References

Daniel IM, Ishai O (1994) Engineering mechanics of composite materials, Oxford University Press

Hill R (1950) The mathematical theory of plasticity, Clarendon Press, Oxford

Hodge PG (1963) Limit analysis of rotationally symmetric plates and shells, Prentice-Hall, New Jersey

Ma GW, Gama BA, Gillespie JW Jr (2002) Plastic limit analysis of cylindrically orthotropic circular plates, Composite Structures, 55(4):455-466

Ma GW, Hao H, Iwasaki S (1999a) Plastic limit analysis of a clamped circular plate with unified yield criterion. Structural Engineering and Mechanics, 7(5):513-525

Ma GW, Hao H, Iwasaki S (1999b) Unified plastic limit analysis of circular plates under arbitrary load. J. Appl. Mech., ASME, 66(2):568-570

Ma GW, Iwasaki S, Miyamoto Y, Deto H (1998) Plastic limit analysis of circular plates with respect to the unified yield criterion. Int. J. Mech. Sci., 40(10):963-976

Markowitz J, Hu LW (1965) Plastic analysis of orthotropic circular plates. Proc. ASCE, J. Eng. Mech. Div., 90(EM5):251-259

Save MA (1985) Limit analysis of plates and shells: Research over two decades, J. Struct. Mech. 13:343-370

Save MA, Massonnet CE, Saxce G de (1997) Plastic limit analysis of plates, shells and disks, Elsevier, Amsterdam

Tsai SW (1968) Strength theories of filamentary structures. In: Schwartz RT, Schwartz HS (eds.) Fundamental aspects of fiber reinforced plastic composites. Wiley Interscience, New York, 3-11

Yu MH (1961a) General behaviour of isotropic yield function, Res. Report of Xi'an Jiaotong University. Xi'an, China (in Chinese)

Yu MH (1961b) Plastic potential and flow rules associated singular yield criterion, Res. Report of Xi'an Jiaotong University. Xi'an, China (in Chinese)

Yu MH (1983) Twin shear stress yield criterion. Int. J. Mech. Sci., 25(1):71-74

Yu MH (1994) Unified strength theory for geomaterials and its applications. Chinese J. Geotechnical Engineering, 16(2):1-10 (in Chinese, English abstract)

Yu MH (2004) Unified strength theory and its applications. Springer, Berlin

Yu MH, He LN (1991) A new model and theory on yield and failure of materials under the complex stress state. In: Jono M, Inoue T (eds.) Mechanical Behavior of Materials-VI. Pergamon Press, Oxford, 3:841-846

Yu MH, He LN, Liu CY (1992) Generalized twin shear stress yield criterion and its generalization. Chinese Science Bulletin, 37(24):2085-2089

14

Unified Limit Analysis of a Wellbore

14.1 Introduction

A wellbore structure is usually used for underground engineering. The wellbore should be kept stable as it is subjected to earth stress in the mining engineering. A petroleum wellbore sustains the earth stress around the rock as well as the internal pressure of the oil. The stability of the wellbore is of great importance for the successful drilling.

Wellbore stress study in rock and soil engineering usually employs the expansion theory for a thick cylinder. One main aspect of wellbore stability analysis is the selection of an appropriate rock failure criterion, as indicated by Al-Ajmi and Zimmerman (2006), and Al-Ajmi (2006). The commonly used criterion for brittle failure of rocks is the Mohr-Coulomb criterion. This criterion involves only the maximum and minimum principal stresses, σ_1 and σ_3, and therefore assumes that the intermediate principal stress has no influence on rock strength. In contrast to the predictions of the Mohr-Coulomb criterion, much evidence has been accumulating to suggest that intermediate principal stress σ_2 does indeed have a strengthening effect. Wellbore-stability prediction by use of a modified Lade criterion was reported by Ewy (1999). The stability analysis of vertical boreholes using the Mogi-Coulomb failure criterion was presented by Al-Ajmi and Zimmerman (2006). A detailed report is given by Al-Ajmi (2006). Luo and Li (1994) used the twin-shear strength theory (Yu, 1985) to derive the gradually damaged behavior for thick bores in rock and soil. Jian and Shen (1996) used the unified strength theory (Yu, 1991; 1994; 2004) to analyze the expansion trait by considering the strain-softening characteristic of rock and soil. A unified solution for stability analysis of vertical boreholes was derived by Li and Yu (2001; 2002), Xu et al. (2004), and Xu and Hou (2007).

In this chapter the unified strength theory is applied to analyze the stress distribution of rock around wellbore and the limit load of the wellbore.

14.2 Unified Strength Theory

The unified strength theory (Yu et al., 1991; 1992; 2004)

$$F = \sigma_1 - \frac{1}{1+b}(b\sigma_2 + \sigma_3) = \sigma_t \quad \text{when} \quad \sigma_2 \leqslant \frac{\sigma_1 + \alpha\sigma_3}{1+\alpha}, \tag{14.1a}$$

$$F' = \frac{1}{1+b}(\sigma_1 + b\sigma_2) - \sigma_3 = \sigma_t \quad \text{when} \quad \sigma_2 \leqslant \frac{\sigma_1 + \alpha\sigma_3}{1+\alpha}, \tag{14.1b}$$

can be expressed as

$$\begin{aligned} F &= \left[\sigma_1 - \frac{1}{1+b}(b\sigma_2 + \sigma_3)\right] + \left[\sigma_1 + \frac{1}{1+b}(b\sigma_2 + \sigma_3)\right]\sin\varphi_0 \\ &= 2c_0\cos\varphi_0, \\ &\quad \text{when} \quad \sigma_2 \leqslant \frac{\sigma_1 + \sigma_3}{2} + \frac{\sigma_1 - \sigma_3}{2}\sin\varphi_0, \end{aligned} \tag{14.2a}$$

$$\begin{aligned} F' &= \left[\frac{1}{1+b}(\sigma_1 + b\sigma_2) - \sigma_3\right] + \left[\frac{1}{1+b}(\sigma_1 + b\sigma_2) + \sigma_3\right]\sin\varphi_0 \\ &= 2c_0\cos\varphi_0, \\ &\quad \text{when} \quad \sigma_2 \geqslant \frac{\sigma_1 + \sigma_3}{2} + \frac{\sigma_1 - \sigma_3}{2}\sin\varphi_0, \end{aligned} \tag{14.2b}$$

where the unified strength theory parameter b ($0 \leqslant b \leqslant 1$)) is a yield criterion parameter to reflect the relative effect of the intermediate principal stress σ_2. c_0 and φ_0 are the internal cohesion and the angle of internal friction respectively.

The relations of c_0 and φ_0 to other commonly used material parameters are

$$\alpha = \frac{1 - \sin\varphi_0}{1 + \sin\varphi_0}, \quad \sigma_t = \frac{2c_0\cos\varphi_0}{1 + \sin\varphi_0},$$

where α is the tensile and compressible strength ratio of a material, i.e., $\alpha = \sigma_t/\sigma_c$.

For plane strain problems a coefficient m ($0 < m \leqslant 1$) should be introduced. When the considered material is incompressible, m is approximately 1. For simplicity, m is defined as 1 in the following analysis. The yield function can be expressed as

$$\frac{\sigma_1 - \sigma_3}{2} = -\frac{2(1+b)\sin\varphi_0}{2(1+b)+mb(\sin\varphi_0-1)}\frac{\sigma_1+\sigma_3}{2} + \frac{2(1+b)c_0\cos\varphi_0}{2(1+b)+mb(\sin\varphi_0-1)}.$$
$$(14.3a)$$

It can be rewritten as (Yu et al., 1997; 2001)

$$\frac{\sigma_1 - \sigma_3}{2} = -\frac{\sigma_1+\sigma_3}{2}\sin\varphi_{\text{uni}} + C_{\text{uni}}\cos\varphi_{\text{uni}}, \qquad (14.3b)$$

where the unified strength parameters C_{uni} and φ_{uni} were proposed by Yu et al. in 1997 and 2001.

These two unified strength parameters are referred to as the unified effective cohesion and unified effective internal friction angle respectively, with regard to the unified strength theory (UST). Their relations to the material constants C_0 and φ_0 can be written as (Yu et al., 1997; 2001; 2006)

$$\sin\varphi_{\text{uni}} = -\frac{2(1+b)\sin\varphi_0}{2(1+b)+mb(\sin\varphi_0-1)},$$
$$C_{\text{uni}} = \frac{2(1+b)\cos\varphi_0}{2(1+b)+mb(\sin\varphi_0-1)} \cdot \frac{C_0}{\cos\varphi_{\text{uni}}}, \qquad (14.4)$$

where C_{uni} and φ_{uni} are the unified internal cohesion and the unified angle of internal friction, Eq.(14.3) gives the failure criterion for plane strain problems (Yu et al., 1997; 2006).

14.3 Equations and Boundary Conditions for the Wellbore

The plan of a wellbore is shown in Fig.14.1. The cylindrical coordinate system is used, where the z-axis is along the wellbore axis.

Assuming that the wellbore radius is R_0, the internal liquid pressure is p_0, R_∞ ($R_\infty \gg R_0$) represents an infinite radius at which the liquid pressure is p_∞ and the rock lateral pressure is $\sigma_{r\infty}$. The parameter β is the effective void ratio, k is the seepage ratio, E is the elastic modulus, and ν is the Poisson's ratio. D_e and D_p represent the elastic and plastic zones of the surrounding rock respectively. The radius $r = R_d$ gives the boundary of the elastic and plastic zones.

14.3.1 Strength Analysis for Wellbore

The stress state of the rock around the wellbore is plane strain and axisymmetrical. It is assumed that the lateral stress $\sigma_{r\infty}$ around the rock is a constant and the rock is isotropic. The normal stresses σ_r, σ_θ, and σ_z in the

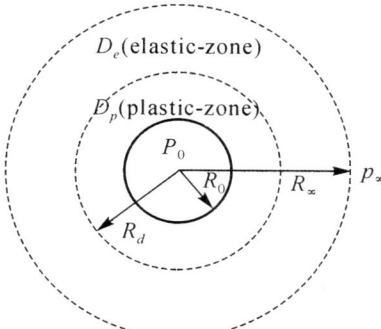

Fig. 14.1. Wellbore subjected to the pore pressure and seepage

radial, circumferential and axial directions are the principal stresses, and the associated shear stress components are zero.

For the present plane strain problem, $\sigma_z = \nu(\sigma_r + \sigma_\theta)$ and there is $\sigma_3 = \sigma_r \leqslant \sigma_z \leqslant \sigma_\theta = \sigma_1$.

When the drilling is finished, the rock around the wellbore is softened. The modulus, internal cohesion, and angle of internal friction will decrease. Denoting that c_1 and φ_1 are the softened internal cohesion and the angle of internal friction respectively, and providing that the rock obeys the failure criterion given in Eq.(14.3), in the initial stage of drilling there is

$$\frac{\sigma_r - \sigma_\theta}{2} = -\frac{\sigma_r + \sigma_\theta}{2}\sin\varphi_{t0} + c_{t0}\cos\varphi_{t0}, \qquad (14.5a)$$

where

$$\sin\varphi_{t0} = \frac{2(1+b)\sin\varphi_0}{2+b+b\sin\varphi_0}, \quad c_{t0} = \frac{2(1+b)c_0\cos\varphi_0}{2+b+b\sin\varphi_0} \cdot \frac{1}{\cos\varphi_{t0}}.$$

The parameters c_{t0} and φ_{t0} represent the effective internal cohesion and the effective angle of internal friction respectively in the original stage with regard to the unified strength theory. When the drilling is finished, the failure condition is

$$\frac{\sigma_r - \sigma_\theta}{2} = -\frac{\sigma_r + \sigma_\theta}{2}\sin\varphi_{t1} + c_{t1}\cos\varphi_{t1}, \qquad (14.5b)$$

where

$$\sin\varphi_{t1} = \frac{2(1+b)\sin\varphi_1}{2+b+b\sin\varphi_1}, \quad c_{t1} = \frac{2(1+b)c_1\cos\varphi_1}{2+b+b\sin\varphi_1} \cdot \frac{1}{\cos\varphi_{t1}},$$

and c_{t1} and φ_{t1} represent the corresponding effective internal cohesion and effective angle of internal friction in the softening stage.

14.3.2 Pore Pressure Analysis

According to the Darcy's law, the pore pressure distribution along the radius is

$$q = \frac{2\pi r k}{\eta} \frac{dp}{dr}, \qquad (14.6)$$

where η is the liquid viscosity, r is the radius of the wellbore, p is the pressure at the inner surface of the wellbore, q is the liquid flux per unit length in the wellbore, and k is the seepage.

The boundary conditions are $p\,|_{r=R_0} = p_0$ and $p\,|_{r=R_\infty} = p_\infty$. The pressure distribution along the radius direction is derived as

$$p = p_0 + (p_0 - p_\infty)\left(\ln\frac{r}{R_0} \Big/ \ln\frac{R_0}{R_\infty}\right), \qquad R_0 \leqslant r \leqslant R_\infty. \qquad (14.7)$$

Equilibrium equation for the rock by considering the seepage effect is

$$\frac{d\sigma_r}{dr} - \chi\frac{dp}{dr} + \frac{\sigma_r - \sigma_\theta}{r} = 0. \qquad (14.8)$$

At the inner surface of the wellbore, the stress boundary condition is

$$\sigma_r|_{r=R_0} = \sigma_{r0} = -p_0(1-\chi). \qquad (14.9a)$$

At R_∞,

$$\sigma_r|_{r=R_\infty} = \sigma_{r\infty} = \sigma_k + \chi p_\infty, \qquad (14.9b)$$

$$\sigma_z|_{r=R_\infty} = \sigma_{z\infty} = p_{\infty b} + \chi p_\infty, \qquad (14.9c)$$

where χ is the effective void ratio, $p_{\infty b}$ and σ_∞ are the vertical pressure and horizontal stress caused by the above rock weight.

14.4 Elastic and Plastic Analysis

14.4.1 Elastic Phase

In the elastic stage the stress distribution can be obtained from Eqs.(14.7), (14.8) and (14.9),

$$\sigma_r = \sigma_{r\infty} + (\sigma_{r\infty} - \sigma_{r0})\frac{R_0^2}{R_\infty^2 - R_0^2}\left(1 - \frac{R_\infty^2}{r^2}\right) - \frac{\chi(p_0 - p_\infty)}{2(1 - \mu)}$$
$$\times\left[\frac{R_0^2}{R_\infty^2 - R_0^2}\left(\frac{R_0^2}{r^2} - 1\right) + \left(\ln\frac{R_0}{r}\right)\bigg/\left(\ln\frac{R_0}{R_\infty}\right)\right], \tag{14.10a}$$

$$\sigma_\theta = \sigma_{r\infty} + (\sigma_{r\infty} - \sigma_{r0})\frac{R_0^2}{R_\infty^2 - R_0^2}\left(1 + \frac{R_\infty^2}{r^2}\right) - \frac{\chi(p_0 - p_\infty)}{2(1 - \nu)}$$
$$\times\left[-\frac{R_0^2}{R_\infty^2 - R_0^2}(\frac{R_0^2}{r^2} + 1) + \left(\ln\frac{R_0}{r} + 1 - 2\nu\right)\bigg/\left(\ln\frac{R_0}{R_\infty}\right)\right]. \tag{14.10b}$$

Based on the plane strain assumption, i.e., $\varepsilon_z = 0$, the following expression can be obtained

$$\sigma_z = \nu(\sigma_r + \sigma_\theta), \tag{14.10c}$$

where σ_r and σ_θ are expressed in Eqs.(14.10a) and (14.10b).

14.4.2 Plastic Limit Pressure

When the pressure p_0 increases to the elastic limit, the rock material around the wellbore falls into the plastic stage. In the plastic zone D_p where $R_0 < r < R_d$, from Eq.(14.5) the relation of σ_r and σ_θ to c_{t1} and φ_{t1} is derived as

$$\frac{\sigma_\theta - \sigma_r}{2} = (c_{t1}\cot\varphi_{t1} - \sigma_r)\frac{\sin\varphi_{t1}}{1 + \sin\varphi_{t1}}. \tag{14.11}$$

From Eqs.(14.11), (14.7), (14.8) and the boundary condition in Eq.(14.9), the pressure at the elastic-plastic boundary can be determined as

$$p_d = p_0 + (p_0 - p_\infty)\ln\frac{R_d}{R_0}\bigg/\ln\frac{R_0}{R_\infty}, \tag{14.12}$$

and the stress distribution in the plastic region D_p can be obtained as

$$\sigma_r = -(1 - \chi)p_0\left(\frac{r}{R_0}\right)^{-\frac{2\sin\varphi_{t1}}{1+\sin\varphi_{t1}}}$$
$$+ \left(c_{t1} - D\frac{2\sin\varphi_{t1}}{1 + \sin\varphi_{t1}}\right)\times\left[1 - \left(\frac{r}{R_0}\right)^{-\frac{2\sin\varphi_{t1}}{1+\sin\varphi_{t1}}}\right], \tag{14.13a}$$

$$\sigma_\theta = \frac{2c_{t1}\cos\varphi_{t1}}{1 + \sin\varphi_{t1}} + \frac{1 - \sin\varphi_{t1}}{1 + \sin\varphi_{t1}}\sigma_r, \tag{14.13b}$$

where

$$D = \chi(p_0 - p_\infty) \Big/ \ln \frac{R_0}{R_\infty}.$$

If R_0 is substituted by R_d in Eq.(14.10a) and Eq.14.10(b), the elastic stress distribution in the elastic zone D_e $(r > R_d)$ can be obtained from Eq.(14.10a) and Eq.14.10(b).

14.4.3 Elastic-plastic Boundary

With $R_\infty \gg R_0$ and $R_\infty \gg R_d$, substituting σ_{r0}, R_0 and p_0 in Eqs.(14.10a) and (14.10b) with σ_{rd}, R_d and p_d, the stress distribution can be deduced from Eqs.(14.12) and (14.13) at the elastic-plastic boundary of $r = R_d$,

$$\sigma_{rd} = -[(1 - \chi)p_0 + c_{t1}\cot\varphi_{t1}]\left(\frac{R_d}{R_0}\right)^{-\frac{2\sin\varphi_{t1}}{1+\sin\varphi_{t1}}} + c_{t1}\cot\varphi_{t1}, \qquad (14.14a)$$

$$\sigma_{\theta d} = 2\sigma_{r\infty} - \sigma_{rd} + \chi\frac{p_0 - p_\infty}{1 - \nu}. \qquad (14.14b)$$

The relation of pressure p_0 at the wellbore surface to the plastic damaged radius R_d is

$$\begin{aligned}
&- [(1 - \chi)p_0 + c_{t1}\cot\varphi_{t1}]\left(\frac{R_d}{R_0}\right)^{-\frac{2\sin\varphi_{t1}}{1+\sin\varphi_{t1}}} + c_{t1}\cot\varphi_{t1} \\
&= (1 + \sin\varphi_{t0})\left[\sigma_{r\infty} + \frac{\chi(p_0 - p_\infty)}{2(1 - \nu)}\right] + c_{t0}\cot\varphi_{t0}.
\end{aligned} \qquad (14.15)$$

When the wellbore surface goes into the plastic yield stage, that is $R_d = R_0$, the maximum radial pressure for retaining the wellbore elastic stabilization can be deduced from Eq.(14.15),

$$p_{e0} = -\frac{(1 + \sin\varphi_{t0})\left[\sigma_{r\infty} + \frac{\chi(p_0-p_\infty)}{2(1-\nu)}\right] + c_{t0}\cot\varphi_{t0}}{1 - \chi\frac{1-\sin\varphi_{t0}-2\nu}{2(1-\nu)}}. \qquad (14.16)$$

When $b = 0$, the elastic limit pressure for the Mohr-Coulomb strength theory can be obtained from Eq.(14.16).

When the rock material is completely in a plastic state, i.e., $R_d \gg R_0$, the maximum radial pressure for the retaining wellbore stabilization can be obtained from Eq.(14.15),

$$p_{p0} = -\frac{2(1 - \mu)(\sigma_{r\infty} + c_{t0}\cos\varphi_{t0} - c_{t1}\cos\varphi_{t1})}{\chi(1 + \sin\varphi_{t0})} + p_\infty. \qquad (14.17)$$

When $b = 1$, the limit load p_{p0} derived from Eq.(14.17) is the limit plastic load based on the Mohr-Coulomb criterion reported by Li (1998). The maximum plastic radius R_d can be obtained from Eqs.(14.15) and (14.17) with reference to the stability of the wellbore.

14.4.4 Example

For an oil drilling wellbore (Li and Li, 1997; Liu et al., 1995) with the radius and the oil pressure of R_0 and p_0, at R_∞ ($R_\infty \gg R_0$), the void pressure p_∞ in the rock is 5 MPa, the radial stress $\sigma_{r\infty}$ is 43.4 MPa, the effective void ratio χ is 25%, the seepage ratio k is $100 \times 10^{-3} \mu m^2$, the elastic modulus E is 1300 MPa, the Poisson's ratio ν is 0.15. The initial yield internal cohesion c_0 and angle of internal friction φ_0 are 0.179 MPa and 31.4° respectively. The softened internal cohesion c_1 and angle of internal friction φ_1 are 0.154 MPa and 25.2°, respectively.

According to the above derivation, the relation of oil pressure on the surface of the wellbore to the plastic radius is shown in Fig.14.2. The elastic and plastic limit pressure for the stability of the oil wellbore are given in Fig.14.3 and Fig.14.4, where $\bar{p}_0 = p_0/p_\infty$, $\bar{p}_{e0} = p_{e0}/p_\infty$, $\bar{p}_{p0} = p_{p0}/p_\infty$.

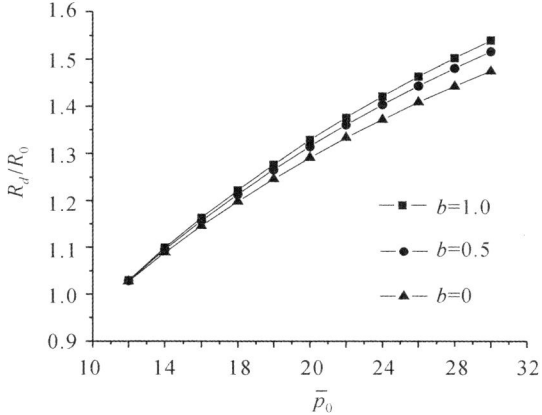

Fig. 14.2. Relation curves of p_0 to R_d

It is shown that the unified strength parameter b influences the plastic radius and the limit pressures. Fig.14.2 shows that the plastic radius increases with the increase of oil pressure in the wellbore, which means that more rock material around the wellbore will enter the plastic phase when the oil pressure on the wellbore surface increases. For a given oil pressure on the wellbore surface, the plastic radius increases with the increase of the parameter b.

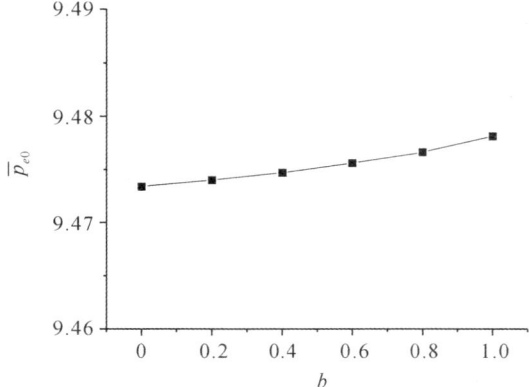

Fig. 14.3. Relation of p_{e0} to the unified strength theory parameter b

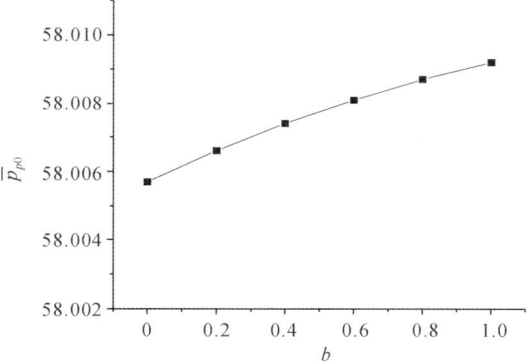

Fig. 14.4. Relation of p_{p0} to the unified strength theory parameter b

Figs.14.3 and 14.4 show the elastic and plastic limit pressures versus the unified strength theory parameter b.

14.4.5 Limit Depth for Stability of a Shaft

Analysis of the stability of a shaft (Fig.14.5) taking into consideration the effect of intermediate principal stress is presented by Xu and Hou (2007).

On the basis of the unified strength theory, a stability analysis of a circular shaft was carried out. The stability formula for the limit depth of the shaft can be expressed as

$$Z_{\max} = \frac{2 + 2b}{2 + b} \cdot \frac{\cos \varphi}{1 - \sin \varphi} \cdot \frac{c}{\gamma}. \tag{14.18}$$

It can be seen from Eq.(14.18) that the influence of the unified strength theory parameter b and friction angle φ (i.e. the intermediate principal stress

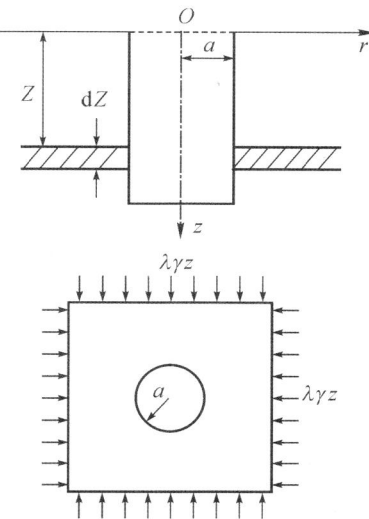

Fig. 14.5. Scheme of a shaft under internal pressures

effect and the strength-differential effect) on the limit depth of the shaft are given. A special case for $b = 0$ can be obtained from Eq.(14.18) as

$$Z_{\max} = \frac{\cos \varphi}{1 - \sin \varphi} \cdot \frac{c}{\gamma}. \tag{14.19}$$

It is the same as the result of the Mohr-Coulomb single-shear theory. The serial results for limit depth $Z_{\max} \gamma / c$ are listed in Table 14.1, which can also be found in Fig.14.6.

Table 14.1. The limit depth of the shaft with the unified strength theory parameter b and φ

$\varphi°$	$b = 0$	$b = 1/4$	$b = 1/2$	$b = 3/4$	$b = 1$
$0°$	1.00	1.11	1.20	1.27	1.33
$15°$	1.30	1.45	1.56	1.66	1.74
$20°$	1.43	1.59	1.71	1.82	1.90
$25°$	2.09	2.13	2.16	2.18	2.20
$30°$	2.53	2.54	2.55	2.55	2.56
$35°$	2.81	2.82	2.83	2.83	2.84

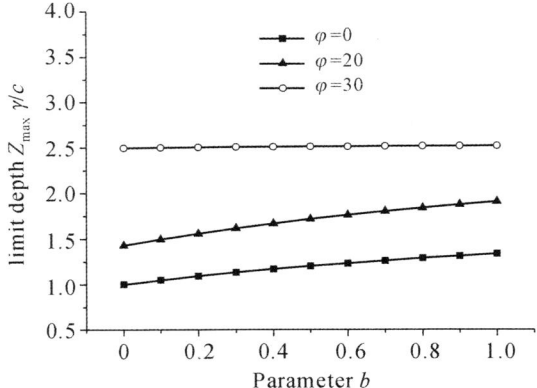

Fig. 14.6. The limit depth of the shaft with the parameters b and φ

The results show that the limit depth of the shaft will increase when the strength-differential effect and the intermediate principal stress effect are considered.

14.5 Summary

Based on the unified strength theory, the elastic and plastic analysis has been carried out for the rock material around the wellbore. The stress distribution of the rock, the elastic and plastic limit loads for the stability of the wellbore and the maximum plastic radius are obtained. The analysis of the stability of a shaft taking into consideration the effect of intermediate principal stress is also discussed.

The analysis results show that the plastic radius increases with the increase of the pressure on the wellbore and the unified strength theory parameter b. It influences the elastic and plastic limit pressures and limit stability depth of the shaft. The analysis results can cover the solutions obtained by other traditional failure conditions, such as the Mohr-Coulomb criterion, the twin shear strength theory.

14.6 Problems

Problem 14.1 Compare the solutions of the elastic and plastic limit pressures of the wellbore.

Problem 14.2 Determine the elastic and plastic limit pressures of the wellbore by using the Mohr-Coulomb criterion ($b = 0$).

Problem 14.3 Determine the elastic limit pressures of the wellbore by using the Mohr-Coulomb criterion ($b = 0$).

Problem 14.4 Determine the elastic limit pressures of the wellbore by using the unified strength theory with $b = 0.5$.

Problem 14.5 Determine the elastic limit pressures of the wellbore by using the unified strength theory with $b = 0.8$.

Problem 14.6 Determine the elastic limit pressures of the wellbore by using the unified strength theory with $b = 1.0$.

Problem 14.7 Determine the plastic limit pressures of the wellbore by using the Mohr-Coulomb criterion ($b = 0$).

Problem 14.8 Determine the plastic limit pressures of the wellbore by using the unified strength theory with $b = 0.5$.

Problem 14.9 Determine the plastic limit pressures of the wellbore by using the unified strength theory with $b = 0.8$.

Problem 14.10 Determine the plastic limit pressures of the wellbore by using the unified strength theory with $b = 1.0$.

Problem 14.11 Compare the solutions of limit depth of the shaft with different criteria.

Problem 14.12 Determine the limit depth of the shaft using the unified strength theory with $b = 0.5$.

Problem 14.13 Determine the limit depth of the shaft using the Mohr-Coulomb criterion ($b = 0$).

Problem 14.14 Determine the limit depth of the shaft using the unified strength theory with $b = 0.8$.

Problem 14.15 Determine the limit depth of the shaft using the unified strength theory with $b = 1.0$.

References

Al-Ajmi AM (2006) Wellbore stability analysis based on a new true-triaxial failure criterion. Doctor Thesis at KTH (Royal Institute of Technology, Sweden), Stockholm, May 2006

Al-Ajmi AM, Zimmerman RW (2006) Stability analysis of vertical boreholes using the Mogi-Coulomb failure criterion. International Journal of Rock Mechanics and Mining Sciences, 43(8):1200-1211

Ewy RT (1999) Wellbore-stability predictions by use of a modified Lade criterion. SPE Drilling Comp. 14(2):85-91

Jiang MJ, Shen ZJ (1996) Unified solution to expansion of cylindrical cavity for geomaterials with strain-softening behavior (by using of the unified strength theory), Rock and Soil Mechanics, 17(1):1-8 (English abstract)

Li JC (1998) Limit analysis of a wellbore based on the twin shear strength theory. In: Yu MH, Fan SC (eds.) Strength Theory: Applications, Developments & Prospects for the 21st Century. Science Press, Beijing, New York, 1103-1108

Li JC, Yu MH, Wang SJ (2001) Unified limit analysis of wellbore under pore pressure and seepage. J. Mech. Strength, 23(3):239-242 (in Chinese)

Li JY, Li ZF (1997) Rock elasto-plastic stresses around a wellbore and wellbore stability under permeation osmosis. Engineering Mechanics, 14(1):131-137 (in Chinese)

Liu YS, Bai JZ, Zhou YH, et al. (1995) Mod density for wellbore stability when formation rock is damaged. Acta Petrolei Sinica,16(3):123-138 (in Chinese)

Luo ZR, Li ZD (1994) Progressive failure of geomaterial thick cylinder (by using the twin shear strength theory of Yu). In: Proceedings of 7th China Conference on Soil Mechanics and Fundamental Engineering. Xi'an, China Civil Engineering Press, 200-203

Xu SQ, He H, Yu MH (2004) Analysis on stability of a shaft with a horizontal soft joint using the unified strength theory. Mine Construction Technology, 25(5):28-31 (in Chinese)

Xu SQ, Hou W (2007) Analysis of stability of a shaft when considering the effect of intermediate principal stress of rock mass. Chinese Journal of Underground Space and Engineering, 3(6):1168-1171

Yu MH (1994) Unified strength theory for geomaterials and its applications. Chinese J. Geotechnical Engineering, 16(2):1-10 (in Chinese, English abstract)

Yu MH (2004) Unified strength theory and its applications. Springer, Berlin

Yu MH, He LN (1991) A new model and theory on yield and failure of materials in the complex stress state. In: Jono M, Inoue T (eds.) Mechanical Behavior of Materials-VI. Pergamon Press, Oxford, 3:841-846

Yu MH, He LN, Liu CY (1992) Generalized twin shear stress yield criterion and its generalization. Chinese Science Bulletin, 37(24):2085-2089

Yu MH, He LN, Song LY (1985) Twin shear stress theory and its generalization. Scientia Sinica (Sciences in China), English ed., Series A, 28(11):1174-1183

Yu MH, Li JC, Zhang YQ (2001) Unified characteristics line theory of spatial axisymmetric plastic problem. Science in China (Series E), English ed., 44(2):207-215; Chinese ed., 44(4):323-331

Yu MH, Ma GW, Qiang HF, et al. (2006) Generalized plasticity. Springer, Berlin

Yu MH, Yang SY, Liu CY, Liu JY (1997) Unified plane-strain slip line theory. China Civil Engineering J., 30(2):14-26 (in Chinese, English abstract)

15

Unified Solution of Shakedown Limit for Thick-walled Cylinder

15.1 Introduction

Correct prediction of the load-bearing capacity of structures is a crucial task in the analysis and design of engineering structures. The plastic limit load of structures from limit analysis or slip-line analysis is usually used as an index of the load-bearing capacity of the structure, subjected to a monotonic loading. When the loading is a repeated loading, the structures fail at a load which is lower than the plastic limit load. This is due to gradual deterioration caused by the alternating plasticity or by the incremental plasticity instead of sudden collapse.

If the load does not exceed the critical value, the structure subjected to the repeated loading may behave plastically at first and then elastically. No further plastic deformation takes place in the structure. The structure shakes down due to the repeated loading. If the load exceeds the critical value the structure does not shake down and fails due to the alternating plasticity or the incremental plasticity. This critical load level is called the shakedown load. The shakedown load is usually regarded as the load-bearing capacity of the structure subjected to the repeated loading.

Many engineering structures or components are subjected to mechanical or other loads varying with time. The shakedown condition should be guaranteed for the safety of such kinds of structures.

Shakedown theory of structures is usually applied for such kinds of problems. A structure in a non-shakedown or inadaptation condition under varying loads may fail by one of two failure modes, namely alternating plasticity or incremental plastic collapse. The structure will shake down if neither of the failure modes occurs (Symonds, 1951; Hodge, 1954; Kachanov, 1971; Martin, 1975; Zyczkowski, 1981; Chakrabarty, 1987, Mroz et al., 1995).

The concept and methods of shakedown analysis were initially addressed in the 1930s and developed in the 1950s. The pioneering works of shakedown include those by Bleich (1932), Melan (1936). Koiter (1956) proved two crucial shakedown theorems, i.e., the static shakedown theorem (Melan's theorem, the first shakedown theorem, or the lower bound shakedown theorem), and the dynamic shakedown theorem (Koiter's theorem, the second shakedown theorem, or the upper bound shakedown theorem), which constitute the fundamentals in the shakedown theory of elasto-plastic structures.

Accordingly, numerous existing methods for shakedown analysis can be divided into two classes, i.e., the static and the dynamic shakedown analysis methods. Shakedown theory has become a well-established branch of plasticity theory.

In recent years shakedown analysis of elasto-plastic structures has increasingly attracted attention from engineers due to the requirements of modern technologies such as in nuclear power plants, chemical industry, the aeronautical and astronautical, electrical and electronic industries. Shakedown theory has been applied with success in a number of engineering problems such as the construction of nuclear reactors, highways and railways and employed as one of the tools for structural design and safety assessment in some design standards, rules, and regulations. A study of the plastic shakedown of structures was made by Polizzotto (1993), and of some issues in shakedown analysis by Maier (2001) and Maier et al. (2000).

Long thick-walled cylinders are very often used as gun barrels and pressure vessels in engineering. They are usually subjected to repeated internal pressure. It is necessary to conduct shakedown analysis in order to determine the shakedown load of the cylinder. The solution to shakedown problem of cylinder is readily available in textbooks of the classical plasticity, and the analytical solution can be found in some published literature and in the Pressure Vessel Code, such as Cases of ASME Boiler and Pressure Vessel Code. However, the solution is based on the Tresca yield criterion, and the analytical solution based on the Huber-von Mises criterion is not readily derivable in most cases due to the nonlinear expression of the criterion. As we have discussed, the Tresca yield criterion considers the effects of only the first and the third principal stresses and ignores the compressive-tensile strength difference (SD) effect of materials. Thus, this classical solution can only be applied to the cylinder made of non-SD materials where the intermediate principal stress effect is negligible. It is of great importance to develop a new approach to cover the SD effect and the intermediate principal stress effect for more general applications. The influence of different strengths in tension and compression for the shakedown of thick-walled cylinders was studied by Feng and Liu (1995). A series of results were given by Feng et al. (1993-1999). An elasto-plastic model incorporating the Yu unified strength theory (UST) was suggested for shakedown analysis of a thick-walled cylinder by Xu and Yu (2005). A closed-form solution of shakedown load for cylinders will be presented in this chapter. The solutions involve the two parameters of Yu unified strength theory m and b and can reflect both the effect of intermediate principal stress and the SD effect in a quantitative manner. It is referred to as the unified solution including serial solutions. By choosing proper values for m and b, the solution is applicable to cylinders made of different materials. In addition, by applying the solution based on Yu unified strength theory, the effects of SD and the intermediate principal stress on the shakedown load of the thick-walled cylinder are evaluated.

15.2 Shakedown Theorem

Many engineering structures or components are subjected to mechanical or other loads varying with time. In many cases only the loading range within which the loads change can be estimated, while the loading path is unknown. It is important to guarantee the shakedown condition for the safety of such kinds of structures.

15.2.1 Static Shakedown Theorem (Melan's Theorem)

The static or Melan's shakedown theorem (Melan, 1936; Kachanov, 1971; Martin, 1975) indicates the necessary condition for the occurrence of shakedown: there exist time-independent fields of residual stresses $\overline{\sigma}_{ij}$ such that the sum $(\overline{\sigma}_{ij} + \sigma_{ij}^e)$ is admissible, where σ_{ij}^e are the elastic components of stresses. It implies that the stress field $(\overline{\sigma}_{ij} + \sigma_{ij}^e)$ is safe if no arbitrary load-variation in the prescribed limits causes the yield surface $f(\overline{\sigma}_{ij} + \sigma_{ij}^e)$ to be reached, i.e.,

$$f(\overline{\sigma}_{ij} + \sigma_{ij}^e) < 0. \tag{15.1}$$

The necessary condition is not obtained if there is no distribution of residual stresses for which $f(\overline{\sigma}_{ij} + \sigma_{ij}^e) < 0$, and so shakedown cannot occur.

On the contrary, shakedown occurs if there is a fictitious residual stress field $\overline{\sigma}_{ij}$ that is independent of time. For any variations of loads within the prescribed limits, the sum of this field with the stress field σ_{ij}^e in a perfectly elastic body is safe (sufficient condition).

The residual stress field is expediently chosen such that the region of admissible load variation is the greatest. Melan's theorem serves as a low bound of the limit load.

15.2.2 Kinematic Shakedown Theorem (Koiter Theorem)

Koiter's theorem (1956), also called the kinematic inadaptation theorem, can be regarded as an extension of the upper bound theorem in limit analysis. The theorem is framed in terms of an admissible plastic strain rate cycle $\dot{\varepsilon}_{ij}^{kp}(s, t)$ for $0 < t < T$. In view of the principle of virtual work, the statement of Koiter's theorem can be interpreted as showing that if the external power of any admissible plastic strain rate cycle $\dot{\varepsilon}_{ij}^{kp}(s,t)$ can be found to exceed the power dissipated in the structure, i.e.,

$$\int_0^T dt \iint_{S_T} p_j \dot{u}_j^k dS > \int_0^T dt \iint_{S_T} \sigma_{ij}^k \dot{\varepsilon}_{ij}^{kp} dV, \tag{15.2}$$

shakedown will not occur, where σ_{ij}^k is the stress field associated with $\dot{\varepsilon}_{ij}^{kp}(s,t)$; $\dot{u}_j^k$ is the velocity field for a cycle by the loads p_j.

It should be noted that the static shakedown theorem and the kinematic non-shakedown theorem determine the lower and the upper bounds to the permissible loading range for the shakedown of a structure.

15.3 Shakedown Analysis for Thick-walled Cylinders

When considering a plane strain thick-walled cylinder under uniform internal pressure p with internal and external radii of r_i and r_e, respectively, for simplicity it is assumed that the material is incompressible and elastic-perfectly plastic with negligible Bauschinger effect. If the pressure p is moderate, the thick-walled cylinder is in an elastic state. The stress field of the cylinder is given by the Lame solutions,

$$\sigma_r = \frac{r_i^2 p}{r_e^2 - r_i^2}(1 - \frac{r_e^2}{r^2}), \tag{15.3a}$$

$$\sigma_\theta = \frac{r_i^2 p}{r_e^2 - r_i^2}(1 + \frac{r_e^2}{r^2}), \tag{15.3b}$$

$$\sigma_z = \frac{r_i^2 p}{r_e^2 - r_i^2}. \tag{15.3c}$$

From Eqs.(15.3a), (15.3b), and (15.3c), σ_θ is the major principal stress, σ_z the intermediate principal stress, σ_r the minor principal stress, and they satisfy

$$\sigma_z \leqslant \frac{m\sigma_\theta + \sigma_r}{m + 1}. \tag{15.4}$$

Therefore, the unified strength theory can be expressed as

$$\sigma_\theta - \frac{1}{m(1+b)}(b\sigma_z + \sigma_r) = \sigma_t, \tag{15.5}$$

where m is the ratio of material strength in compression and in tension, for non-SD materials $m = \sigma_c/\sigma_t = 1$.

From Eq.(15.5) the maximum value for $\sigma_\theta - (b\sigma_z + \sigma_r)/(m+mb)$ occurs on the internal wall of the cylinder. Yielding starts from the internal wall of the cylinder when the internal pressure reaches

$$p_e = \frac{m(1+b)(r_e^2 - r_i^2)}{(m+1+mb)r_e^2 + (m-1)(1+b)r_i^2}\sigma_t, \tag{15.6}$$

where p_e is the elastic limit pressure of the cylinder.

When the internal pressure exceeds p_e, a plastic zone spreads out from the inner radius. If the plastic zone reaches the radius r_p, the cylinder can be divided into two parts: a plastic zone in the range of $r_i \leqslant r \leqslant r_p$, and a elastic zone of $r_p \leqslant r \leqslant r_e$. Using the Lame solution, the boundary condition $\sigma_r = 0$ at $r = r_e$ and at $r = r_p$, the yield condition in Eq.(15.5) is satisfied. The stress components in the elastic zone are derived as

$$\sigma_r = \frac{r_p^2 p_p}{r_e^2 - r_p^2}\left(1 - \frac{r_e^2}{r^2}\right), \tag{15.7a}$$

$$\sigma_\theta = \frac{r_p^2 p_p}{r_e^2 - r_p^2}\left(1 + \frac{r_e^2}{r^2}\right), \tag{15.7b}$$

$$\sigma_z = \frac{r_p^2 p_p}{r_e^2 - r_p^2}, \tag{15.7c}$$

where

$$p_p = \frac{m(1+b)(r_e^2 - r_p^2)}{(m+1+mb)r_e^2 + (m-1)(1+b)r_p^2}\sigma_t$$

is the associated radial pressure on the elasto-plastic interface under the internal pressure p.

According to the equilibrium equation

$$\frac{\mathrm{d}\sigma_r}{\mathrm{d}r} + \frac{\sigma_r - \sigma_\theta}{r} = 0, \tag{15.8}$$

the yield condition in Eq.(15.5), the boundary condition of $\sigma_r = p$ at $r = r_i$, the incompressible condition of materials, then the stress components in the plastic zone are derived as

$$\sigma_r = -\left(p + \frac{m\sigma_t}{m-1}\right)\left(\frac{r_i}{r}\right)^{\frac{2(m-1)(1+b)}{2m+2mb-b}} + \frac{m}{m-1}\sigma_t, \tag{15.9a}$$

$$\sigma_\theta = -\frac{(2+b)}{2m+2mb-b}\left(p + \frac{m\sigma_t}{m-1}\right)\left(\frac{r_i}{r}\right)^{\frac{2(1+b)(m-1)}{2m+2mb-b}} + \frac{m}{m-1}\sigma_t, \tag{15.9b}$$

$$\sigma_z = -\frac{1+m+mb}{2m+2mb-b}\left(p+\frac{m\sigma_t}{m-1}\right)\left(\frac{r_i}{r}\right)^{\frac{2(1+b)(m-1)}{2m+2mb-b}} + \frac{m}{m-1}\sigma_t. \quad (15.9c)$$

With reference to the continuity of σ_r across $r = r_p$, the relationship between the internal pressure p and the radius of the plastic zone r_p is obtained

$$p = \frac{m\sigma_t}{m-1}\left[\frac{(2m+2mb-b)r_e^2}{(m+1+mb)r_e^2+(m-1)(1+b)r_p^2}\left(r_p/r_i\right)^{\frac{2(m-1)(1+b)}{2m+2mb-b}} - 1\right]. \quad (15.10)$$

With the increase of the pressure p, the plastic zone expands further and the elastic-plastic interface moves gradually to the external wall of the cylinder. Setting $r_p = r_e$ in Eq.(15.10), the internal pressure becomes

$$p_s = \frac{m\sigma_t}{m-1}\left[(r_e/r_i)^{\frac{2(m-1)(1+b)}{2m+2mb-b}} - 1\right], \quad (15.11)$$

which is the plastic limit pressure of the cylinder.

If $p_e < p < p_s$, the cylinder is partially plastic. When the cylinder is unloaded there will be residual stress. If p is small the unloading process is purely elastic and the residual stress is derived by superposition of the elastic unloading stress and the elastic-plastic loading stress. The expressions for the residual stresses in the zone adjacent to the internal wall of the cylinder $(r_i \leqslant r \leqslant r_p)$ can be written as

$$\sigma_r^r = -\left(p+\frac{m\sigma_t}{m-1}\right)\left(\frac{r_i}{r}\right)^{\frac{2(m-1)(1+b)}{2m+2mb-b}} + \frac{m}{m-1}\sigma_t$$
$$- \frac{r_i^2 p}{r_e^2-r_i^2}\left(1-\frac{r_e^2}{r^2}\right), \quad (15.12a)$$

$$\sigma_\theta^r = -\frac{(2+b)}{2m+2mb-b}\left(p+\frac{m\sigma_t}{m-1}\right)\left(\frac{r_i}{r}\right)^{\frac{2(m-1)(1+b)}{2m+2mb-b}}$$
$$+ \frac{m}{m-1}\sigma_t - \frac{r_i^2 p}{r_e^2-r_i^2}\left(1+\frac{r_e^2}{r^2}\right), \quad (15.12b)$$

$$\sigma_z^r = -\frac{1+m+mb}{2m+2mb-b}\left(p+\frac{m\sigma_t}{m-1}\right)\left(\frac{r_i}{r}\right)^{\frac{2(m-1)(1+b)}{2m+2mb-b}}$$
$$+ \frac{m}{m-1}\sigma_t - \frac{r_i^2 p}{r_e^2-r_i^2}. \quad (15.12c)$$

Given $r = r_i$, the residual stresses on the internal wall of the cylinder are

$$\sigma_r^r = 0, \tag{15.13a}$$

$$\sigma_\theta^r = -\left[\frac{(2+2b)}{2m+2mb-b} + \frac{r_e^2 + r_i^2}{r_e^2 - r_i^2}\right]p + \frac{(2+2b)m}{2m+2mb-b}\sigma_t, \tag{15.13b}$$

$$\sigma_z^r = -\frac{1}{2}\cdot\left[\frac{(2+2b)}{2m+2mb-b} + \frac{r_e^2 + r_i^2}{r_e^2 - r_i^2}\right]p + \frac{(1+b)m}{2m+2mb-b}\sigma_t. \tag{15.13c}$$

It is seen that σ_r^r, σ_z^r, and σ_θ^r on the internal wall are the major principal stress, the intermediate principal stress, and the minor principal stress respectively, and the intermediate principal stress $\sigma_z^r \geqslant (m\sigma_r^r + \sigma_\theta^r)/(m+1)$. Therefore, the unified strength theory on the internal wall is

$$\frac{1}{1+b}(\sigma_r + b\sigma_z) - \frac{\sigma_\theta}{m} = \sigma_t. \tag{15.14}$$

From Eqs.(15.13) and (15.14), the internal wall of the cylinder yields when the internal pressure reaches

$$p_{max} = \frac{2m(m+1)(1+b)(b+2)/(2m+2mb-b)/(2-mb+2b)}{(2+b)/(2m+2mb-b) + (r_e^2 + r_i^2)/(r_e^2 - r_i^2)}\sigma_t. \tag{15.15}$$

If $p < p_{max}$, a secondary yielding does not take place at the internal wall of the unloaded cylinder. It can be demonstrated that the residual stress induced by the cycle of loading-unloading will not yield any new plastic deformation in the whole cross-section of the cylinder. Therefore the shakedown condition for a thick-walled cylinder under repeated loading and unloading is that the internal pressure p is less than the critical value $p_{shakedown}$ or $p_{plastic}$, i.e.,

$$p_{max,shakedown} =$$
$$\min\left\{\frac{2m(m+1)(1+b)(2+b)/(2m+2mb-b)/(2-mb+2b)}{(2+b)/(2m+2mb-b) + (r_e^2 + r_i^2)/(r_e^2 - r_i^2)}\sigma_t\right\}, \tag{15.16a}$$

$$p_{max,plastic} = \min\left\{\frac{m\sigma_t}{m-1}\left[(r_e/r_i)^{\frac{2(m-1)(1+b)}{2m+2mb-b}} - 1\right]\right\}, \tag{15.16b}$$

which is the shakedown load of the thick-walled cylinder. Setting $m=1$ and $b=0$ in Eq.(15.16b), the shakedown load of the thick-walled cylinder has the form of

$$p_{max} = \min\left\{\sigma_t(1 - r_i^2/r_e^2),\ \sigma_t \ln(r_e/r_i)\right\}, \tag{15.17}$$

which is in agreement with the shakedown load of a cylinder from the classical plasticity based on the Tresca criterion.

The shakedown load given by Eq.(15.16a) is correlated with the compressive tensile strength ratio m, and the unified yield criterion parameter b. It can be said the present approach has the capability to reflect the SD effects and intermediate principal stress on the shakedown load of the cylinder quantitatively, which is ignored in the classical solution.

15.4 Unified Solution of Shakedown Pressure of Thick-walled Cylinders

In order to demonstrate the SD effects and intermediate principal stress on the shakedown load of a thick-walled cylinder, the results from the derived closed-form solution are depicted in Fig.15.1, in which the abscissa denotes the wall ratio of the cylinder r_e/r_i, and the ordinates is the shakedown load p_{max}/σ_t.

From Fig.15.1, the effect of the intermediate principal stress on the shakedown load for non-SD materials ($m=1$) is obvious.

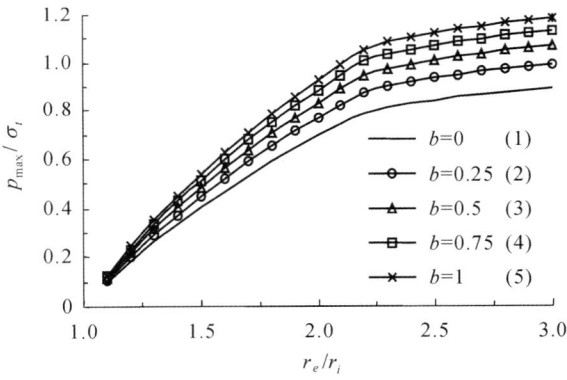

Fig. 15.1. Shakedown load for different values of the unified strength theory parameter b ($m=1.0$)

The curve (1) in Fig.15.1 ($b=0$ and $m=1.0$) is suitable for materials without both the SD and the intermediate principal stress effects, which is exactly the result of the classical solution based on the single-shear yield criterion. The present solution with $m=1$ and $b=0.5$ (curve (3) in Fig.15.1) is a close approximation to the result from the Huber-von Mises criterion. The curve (5) in Fig.15.1 ($b=1.0$ and $m=1$) is the same as the result from the twin-shear stress yield criterion.

Fig.15.2 (m=1.1) and Fig.15.3 (m=1.2) show the shakedown pressure for materials with SD effect. It is seen from these figures that the shakedown load is related to the unified strength theory parameter b which reflects the effect of the intermediate principal stress on material strength. The higher the parameter b, the higher the shakedown load $p_{\max}$. Consequently, for a given compressive-tensile strength ratio m, that of b=0 corresponding to the Tresca criterion or the Mohr-Coulomb criterion gives the lowest value of $p_{\max}$, that of b=1 corresponding to the twin-shear yield criterion or the generalized twin-shear criterion gives the highest value. Therefore, the shakedown load of the cylinder may be underestimated when the effect of the intermediate principal stress of materials is neglected, or an improper yield condition is applied.

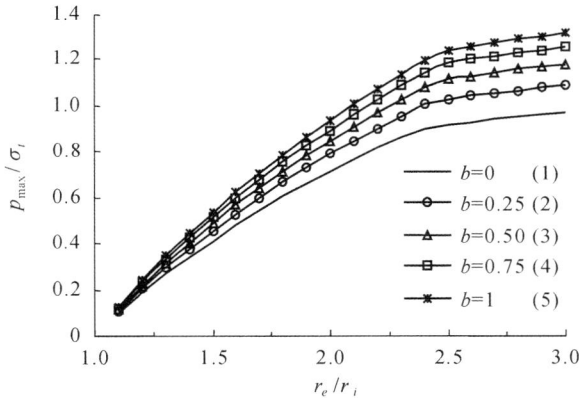

Fig. 15.2. Shakedown load for different values of the unified strength theory parameter b (m=1.1)

Fig.15.4 shows the SD effect on the shakedown load of a cylinder. The results with respect to b=0 are shown in Fig.15.4, which is the same as the result based on the Mohr-Coulomb criterion. It is suitable for materials with negligible intermediate principal stress effect. The curve (1) in Fig.15.4 (b=0 and m=1) is the result of the classical solution based on the Tresca criterion.

Fig.15.5 (corresponding to $b = 0.5$ of the unified strength theory) and Fig.15.6 (corresponding to $b = 1.0$, i.e., the twin-shear strength criterion) are suitable for materials with the intermediate principal stress effect.

From analysis and schematical illustrations of the results, the shakedown load depends on the compressive-tensile strength ratio m and the shakedown load will increase with increasing parameter m. Thus, the shakedown load of the cylinder may be underestimated when the SD effect of materials is ignored. The SD effect of materials on the shakedown load of the cylinder is insignificant when the wall ratio is small, whereas it is prominent when the

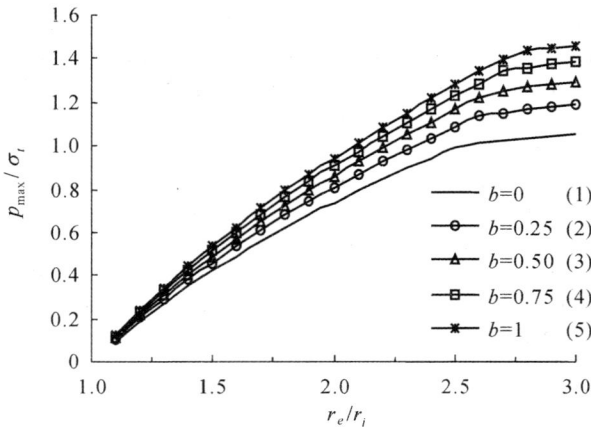

Fig. 15.3. Shakedown load for different values of the unified strength theory parameter b (m=1.2)

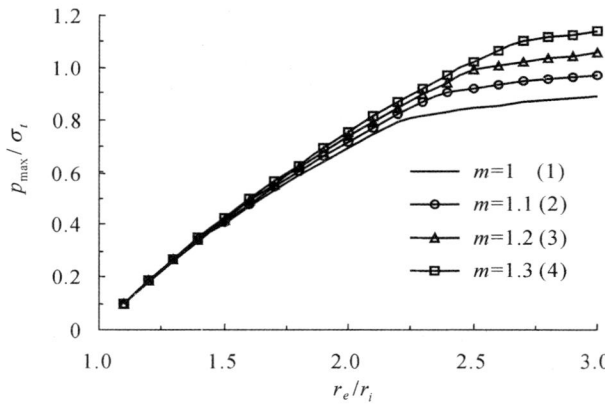

Fig. 15.4. Shakedown load for different values of parameter m (b=0)

wall ratio is high. Therefore, the SD effect of materials should be taken into account in shakedown analysis of the cylinder especially for a high wall ratio of the cylinder.

15.5 Connection between Shakedown Theorem and Limit Load Theorem

Based on the unified strength theory, shakedown analysis of a thick-walled cylinder under internal pressure is carried out and the unified analytical solution of shakedown load for a cylinder is derived in this chapter. This solution

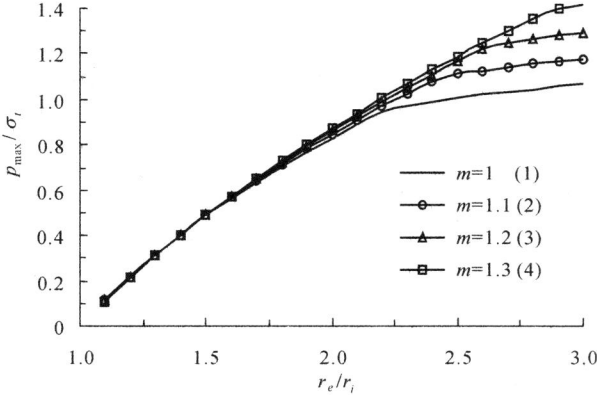

Fig. 15.5. Shakedown load for different values of parameter m (b=0.5)

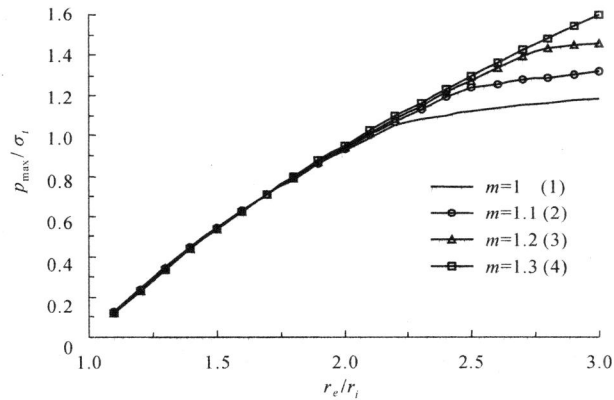

Fig. 15.6. Shakedown load for different values of parameter m (b=1.0)

includes not only the existing classical solution as its special case but gives a series of new results.

It is noted that this solution consists of two parts (Eqs.(15.16a) and (15.16b)): the limit pressure and shakedown pressure (Xu and Yu, 2004a; 2005b), i.e.,

$$p_{\text{max,shakedown}} =$$
$$\min\left\{\frac{2m(m+1)(1+b)(2+b)/(2m+2mb-b)/(2-mb+2b)}{(2+b)/(2m+2mb-b)+(r_e^2+r_i^2)/(r_e^2-r_i^2)}\sigma_t\right\}, \quad (15.18a)$$

$$p_{\text{max,plastic}} = \min\left\{\frac{m\sigma_t}{m-1}\left[(r_e/r_i)^{\frac{2(m-1)(1+b)}{2m+2mb-b}}-1\right]\right\}. \quad (15.18b)$$

The relation between the shakedown pressure and the plastic limit pressure are shown in Fig.15.7 for different parameter b.

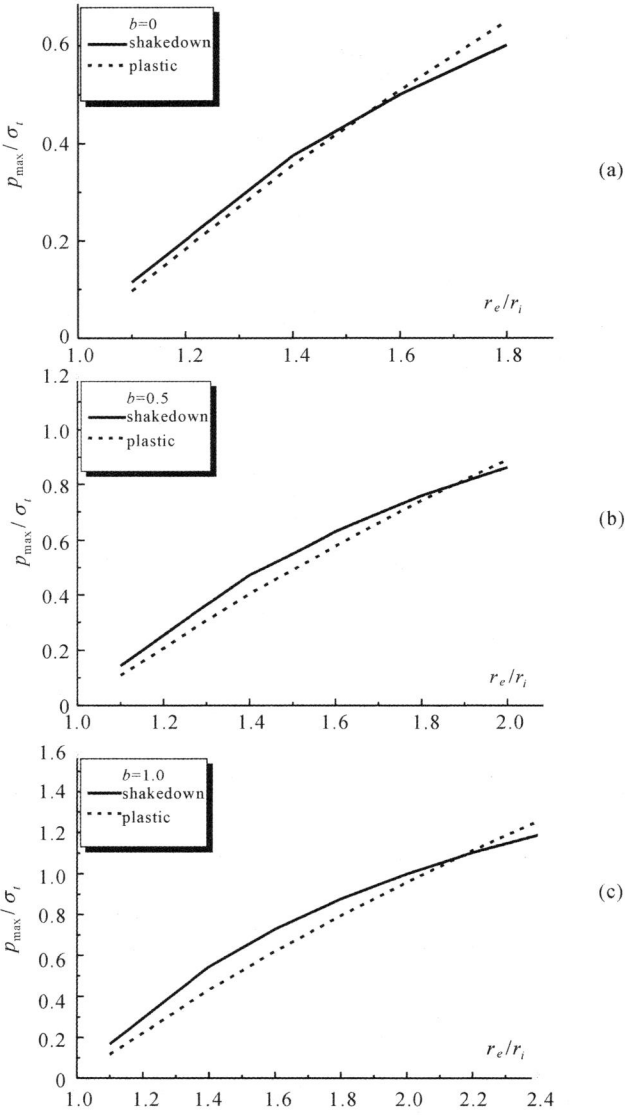

Fig. 15.7. Shakedown load and plastic limit load when $m=1.5$

It is seen that the two curves will intersect when the limit pressure equals the shakedown pressure, i.e.,

$$\frac{m\sigma_t}{m-1}\left[(r_e/r_i)^{\frac{2(m-1)(1+b)}{2m+2mb-b}}-1\right]=$$

$$\frac{2m(m+1)(1+b)(2+b)/(2m+2mb-b)/(2-mb+2b)}{(2+b)/(2m+2mb-b)+(r_e^2+r_i^2)/(r_e^2-r_i^2)}\sigma_t. \quad (15.19)$$

The current unified solution consists of the two parameters m and b to reflect both the SD and the intermediate principal stress effects of materials. With the variation of m and b, the present solution gives a series of values for the shakedown load that can be applied to materials with or without the SD and the intermediate principal stress effects.

In order to demonstrate more clearly the SD effects and intermediate principal stress on the shakedown load, the analytical solution is illustrated schematically. This shows that both the SD and the intermediate principal stress have influences on the shakedown load, and the more pronounced the two effects, the higher the shakedown load. Therefore, for the cylinder made of materials with the SD and/or the intermediate principal stress effect, the classical solution underestimates the shakedown load. It is therefore of significance for the shakedown analysis to take into account their effects.

It is worth mentioning that besides SD and intermediate principal stress, other important properties such as the Bauschinger effect, the strain-hardening effect, etc., should also be considered when their effects are prominent.

15.6 Shakedown Pressure of a Thick-walled Spherical Shell

Shakedown analysis of a thick-walled spherical shell was derived by Liu et al.(1997) using the Mohr-Coulomb criterion and Xu and Yu (2005b) using the UST. The shakedown limit pressure of a thick-walled spherical shell for SD material is

$p_{\text{max,shakedown}}$

$$= \min\left\{\frac{m}{m-1}\left[\left(\frac{r_e}{r_i}\right)^{\frac{2m-2}{m}}-1\right]\sigma_t, \frac{m(m+1)(r_e^3-r_i^3)}{(m-1)r_i^3+(0.5m+1)r_e^3}\sigma_t\right\}.$$

$$(15.20)$$

This result is the same as the solution obtained by using the Mohr-Coulomb strength criterion obtained by Feng and Liu (1995) and Liu et al. (1997) and the twin-shear strength criterion. The relationship of the shakedown limit pressure to the ratio of the strength of material in tension and compression is illustrated in Fig.15.8.

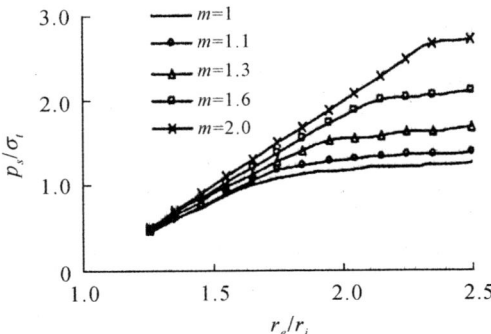

Fig. 15.8. Relationship of shakedown limit pressure to the ratio of material strength in tension and compression

15.7 Summary

The unified strength theory is used to derive unified solutions of the plastic limit and shakedown limit of a thick-walled cylinder. These results are applicable for a wide range of materials and engineering structures.

In the current solutions, the SD effect and the effect of intermediate principal stress acting on the plastic limit loads and shakedown loads of a thick-walled cylinder under uniform internal pressure are presented. By changing the two parameters $m\alpha$ and b (or m and b), a series of values for limit loads and shakedown loads can be obtained from the current solution, which includes both the results from classical plasticity and a series of new results. These solutions are suitable for materials with the SD effect and the intermediate principal stress effect.

Finally, the illustrations of alternatives to the analytical solution are presented to demonstrate graphically to examine the effects of strength difference and intermediate principal stress on the limit loads and the shakedown loads. They show that the limit loads and the shakedown loads depend on both the strength difference in tension and compression and the effect of intermediate principal stress. The limit loads and shakedown loads may be grossly underestimated if these two effects are simply neglected. The unified strength theory gives us a basic theory for use in the strength design of engineering structures. It also provides a tool for estimating accurately the admissible loads with an in-depth understanding of the material strength behavior so that a more economical and optimized design of structures can be achieved.

15.8 Problems

Problem 15.1 Compare the solutions of limit analysis and shakedown analysis.

Problem 15.2 Determine the shakedown load of a pressure cylinder by using the Tresca yield criterion ($m=1$ and $b=0$).

Problem 15.3 Determine the shakedown load of a pressure cylinder by using the Mohr-Coulomb strength theory($b=0$).

Problem 15.4 Determine the shakedown load of a pressure cylinder by using the twin-shear yield criterion ($m=b=1$).

Problem 15.5 Determine the shakedown load of a pressure cylinder by using the twin-shear strength theory ($b=1$).

Problem 15.6 Determine the shakedown load of a cylinder under inter-pressure by using the unified strength theory with $b=0.6$.

Problem 15.7 Determine the shakedown load of a cylinder under tension and inter-pressure by using the twin-shear strength theory ($b=1$).

Problem 15.8 Determine the shakedown load of a cylinder under tension and inter-pressure by using the unified strength theory with $b=0.6$.

References

American Society of Mechanical Engineers (1995) Cases of ASME Boiler and Pressure Vessel Code, Case N-47. ASME, New York

Bleich H (1932) Uber die Bcmessung statich unbcstimmter Stahltragwerke unter Bauingenieur, 13:261-267

Carvelli V, Cen Z, Liu Y, Mater G (1999) Shakedown analysis of defective pressure vessels by a kinematic approach. Arch. Appl. Mech., 69:751-764

Chakrabarty J (1987) Theory of plasticity. McGraw-Hill, New York

Cocchetti G, Maier G (1998) Static shakedown theorems in piecewise linearized poroplasticity. Arch. Appl. Mech., 68:651-661

Cocchetti G, Maier G (2000) Shakedown analysis in poroplasticity by linear programming. Int. J. Numer. Methods Eng., 47:141-168

Feng JJ, Zhang JY, Zhang P, Han JF (2004) Plastic limit load analysis of thick-walled tube based on twin-shear unified strength theory. Acta Mechanica Solid Sinica, 25(2):208-212 (in Chinese)

Feng XQ, Gross D (1999) A global/local shakedown analysis method of elasto-plastic cracked structures. Engineering Fracture Mechanics, 63:179-192

Feng XQ, Liu XS (1993) Factors influencing shakedown of elastoplastic structures. Adv. Mech., 23(2):214-222

Feng XQ, Liu XS (1995) Influence of different strength in tension and compression for shakedown of thick-walled cylinder. Mechanics and Practice, 17(5):28-30 (in Chinese)

Feng XQ, Liu XS (1997) On shakedown of three-dimensional elastoplastic strain-hardening structures. Int. J. Plasticity, 12:1241-1256

Feng XQ, Yu SW (1994) An upper bound on damage of elastic-plastic structures at shakedown. Int. J. Damage Mech, 3:277-289

Feng XQ, Yu SW (1995) Damage and shakedown analysis of structures with strain-hardening. Int. J. Plasticity, 31:247-259

Hachemi A, Weichert D (1992) An extension of the static shakedown theorem to a certain class with damage. Arch. Appl. Mech., 44:491-498

Hamilton R, Boyle JT, Shi J, Mackenzie D (1996) A simple upper-bound method for calculating approximate shakedown loads. J. Pressure Vessel Technol. ASME, 120:195-199

Hodge PG (1959) Plastic analysis of structures. McGraw-Hill, New York

Hodge PG Jr (1954) Shakedown of elastic-plastic structures. In: Osgood WR (ed.) Residual Stresses in Metals and Metal Constructions. Reinhold Pub. Corp., New York

Huang Y, Stein E (1996) Shakedown of a cracked body consisting of kinematic hardening material. Engineering Fracture Mechanics, 54:107-112

Johnson W, Mellor PB (1973) Engineering plasticity. Van Nostrand Reinhold, London

Kachanov LM (1971) Foundations of the theory of plasticity, North Holland Publ. Co., Amsterdam

Kachanov LM (1974) Fundamentals of the theory of plasticity. MIR Publishers, Moscow

Kamenjarzh LA, Weichert D (1992) On kinematic upper bounds of the safety factor in shakedown theory. Int. J. Plasticity, 8:827-837

Kandil A (1996) Analysis of thick-walled cylindrical pressure vessel under the effect of cyclic internal pressure and cyclic temperature. Int. J. Mech. Sci., 38:1319-1332

Koiter WT (1953) Stress-strain relations, uniqueness and variational theorems for elastic-plastic materials with a singular yield surface. Quart. Appl. Math., 11:350-354

Koiter WT (1956) A new general theorem on shakedown of elastic-plastic structures. Proc. K. Ned. Akad. Wet., B59:24-34

Koiter WT (1960) General theorems for elastic-plastic solids. Progress in Solid Mechanics. Sneddon JN, Hill R (eds.) North-Holland Pull. Co., Amsterdam, 1:165-221

König JA (1987) Shakedown of elastic-plastic structures. Elsevier, Amsterdam

König JA, Kleiber M (1978) On a new method of shakedown analysis. Bull. Acad. Pol. Set., Ser. Sci. Tech, 26:165-171

König JA, Maier G (1987) Shakedown analysis of elastic-plastic structures: A review of recent developments. Nucl. Eng. Design, 66:81-95

Liu XQ, Ni XH, Liu YT (1997) The strength difference effect of material on the stable state of thick sphere. J. Theoretical & Applied Mechanics, (1):10-14

Maier G (2001) On some issues in shakedown analysis. J. Appl. Mech., 68:799-808

Maier G, Carvelli V, Coccbetti G (2000) On direct methods for shakedown and limit analysis. Eur. J. Mech. A/Solids (Special issue), 19:S79-S100

Martin JB (1975) Plasticity: fundamentals and general results. The MIT Press

Melan E (1936) Theorie statisch unbestimmter Systeme. In: Prelim. Publ. The Second Congr. Intern. Assoc. Bridge and Structural Eng., Berlin, 43-64

Melan E (1938) Theorie statisch unbestimmter Systeme aus ideal plastischem Baustoff. Sitz. Ber. Akad. Wiss. Wien, II.a, 145:195-218

Mrazik A, Skaloud M, Tochacek M (1987) Plastic design of steel structures. Ellis Horwood, Chichester, New York

Mroz Z, Weichert D, Dorosz S (eds.) (1995) Inelastic behavior of structures under variable loads, Kluwer, Dordrecht

Perry J, Aboudi J (2003) Elasto-plastic stress in thick-walled cylinders. ASME, J. Pressure Vessel Tech., 125:246-252

Pham DC (1997) Evaluation of shakedown loads for plates. Int. J. Mech. Sci., 39(12):1415-1422

Polizzotto C (1982) A unified treatment of shakedown theory and related bounding techniques. Solid Mech. Archives, 7:19-75

Polizzotto C (1993a) On the conditions to prevent plastic shakedown of structures: part 1-Theory. ASME, J. Applied Mechanics, 60:15-19

Polizzotto C (1993b) On the conditions to prevent plastic shakedown of structures: part 2-The plastic shakedown limit load. ASME, J. Appl. Mech., 60:20-25

Polizzotto C (1993c) A study on plastic shakedown of structures: part 1-Basic properties. ASME, J. Appl. Mech., 60:318-323

Polizzotto C (1993d) A study on plastic shakedown of structures: part 2 - Theorems. ASME, J. Applied Mechanics, 60:324-330

Symonds PS (1951) Shakedown in continuous media. J. Appl. Mech., 18(1):85-93

Symonds PS, Neal BG (1951) Recent progress in the plastic methods of structural analysis. J. Franklin Inst., 252:383-407, 469-492

Symonds PS, Prager W (1950) Elastic-plastic analysis of structures subjected to loads varying arbitrarily between prescribed limits. J. Appl. Mech., 17(3):315-323

Weichert D, Maier G (eds.) (2000) Nonelastic analysis of structures under variable repeated loads. Kluwer, Dordrecht

Xu SQ, Yu MH (2004a) Unified analytical solution to shakedown problem of thick-walled cylinder. Chinese J. Mechanical Engineering, 40(9):23-27 (in Chinese, English abstract)

Xu SQ, Yu MH (2004b) Elasto Brittle-plastic carrying capacity analysis for a thick walled cylinder under unified theory criterion. Chinese Quart. Mechanics, 25(4):490-495 (in Chinese, English abstract)

Xu SQ, Yu MH (2005a) Shakedown analysis of thick-walled cylinders subjected to internal pressure with the unified strength criterion. Int. J. Pressure Vessels and Piping, 82(9):706-712

Xu SQ, Yu MH (2005b) Shakedown analysis of thick-walled spherical shell of materials with different strength in tension and compression. Machinery Design and Manufacture, (1):36-37(in Chinese)

Yu MH (1961) General behavior of isotropic yield function. Res. Report of Xi'an Jiaotong University, Xi'an (in Chinese)

Yu MH (1983) Twin shear stress yield criterion. Int. J. Mech. Sci., 25(1):71-74

Yu MH (2002) Advances in strength theories for materials under complex stress state in the 20th century. Applied Mechanics Reviews, ASME, 55(3):169-218

Yu MH (2004) Unified strength theory and its applications: Springer, Berlin

Yu MH, He LN (1991) A new model and theory on yield and failure of materials under complex stress state. Proceedings, Mechanical Behavior of Materials-VI, 3:851-856

Yu MH, He LN, Song LY (1985) Twin shear stress theory and its generalization. Scientia Sinica (Sciences in China), English ed., Series A, 28(11):1174-1183

Zouain N (2001) Bounds to shakedown loads. Int. J. Solids Struct., 38:2249-2266

Zyczkowski M (1981) Combined loadings in the theory of plasticity. Polish Scientific Publishers, PWN and Nijhoff

16

Unified Solution of Shakedown Limit for Circular Plate

16.1 Introduction

The static shakedown theorem (Melan's theorem, the first shakedown theorem, or the lower bound shakedown theorem, Melan, 1936) and the dynamic shakedown theorem (Koiter's theorem, the second shakedown theorem, or the upper bound shakedown theorem, Koiter, 1953; 1956; 1960) and the unified solution of shakedown limit for a thick-walled cylinder have been described in Chapter 15. In this chapter we will deal with the shakedown analysis for a simply supported circular plate and a clamped circular plate. The unified solutions are given for non-SD materials.

Circular plates are used widely in many branches of engineering. They are often subjected to repeated transverse load. Hence it is necessary to conduct the shakedown analysis in order to determine the shakedown load of the plate. Under the varying load the circular plate will deform in elastic and plastic states. The elasto-plastic response of a circular plate to varying loads is a complicated process (Symonds, 1951; König, 1987). The previous shakedown analysis is based on Koiter's upper bound shakedown theorem with the Tresca and Huber-von-Mises yield criteria (Kachanov, 1971; Gokhfeld and Cherniavski, 1980; König, 1978; 1987; Pham, 1996; 1997; 2003). By using numerical methods and based on the static shakedown theorem, the shakedown analysis of perfectly plastic, different kinematic hardening materials is carried out by Stein et al. (1993), Polizzotto (1982; 1993), Ponter and Carter (1997), Maier et al. (2000; 2001).

As discussed by Pham (1997), the shakedown limit will depend on the different yield criteria. From the previous studies we know that the limit load analysis should use different yield criterion for different materials. The Yu unified yield criterion (UYC) will be used in this chapter to investigate the shakedown limit when the circular plates suffer from quasi-static recycle loadings. If the load does not exceed the critical value, the circular plate will behave plastically at first and then elastically and the structure will shakedown due to the repeated loading.

The elastic, plastic and shakedown analysis of a circular plate, which is simply supported or clamped at the edges, will be carried out in this chapter. By choosing proper values for unified yield criterion parameter b, the solution can be applicable to plates made of different materials. In addition, by applying the solution based on the unified yield criterion, the effects of the yield criterion on the shakedown load of the plate are evaluated.

16.2 Unified Solution of Shakedown Limit for Simply Supported Circular Plate

A circular plate with radius a and thickness h is subjected to a uniformly distributed transverse load P, as shown in Fig.16.1, the only non-zero stresses are σ_r, σ_θ and $\tau_{rz} = \tau_{zr}$ in the plate. The generalized stresses can be expressed as

$$M_r = \int_{-h/2}^{h/2} \sigma_r z \mathrm{d}z, \qquad M_\theta = \int_{-h/2}^{h/2} \sigma_\theta z \mathrm{d}z,$$

$$Q_{rz} = \int_{-h/2}^{h/2} \tau_{rz} z \mathrm{d}z, \quad M_0 = \int_{-h/2}^{h/2} \sigma_0 z \mathrm{d}z = \sigma_0 h^2/4, \tag{16.1}$$

where M_r, M_θ and M_0 are the radial, tangential and ultimate (fully plastic) bending moments, respectively, and Q_{rz} is the transverse shear force which is assumed not to influence the plastic yielding.

Defining dimensionless variables, $r = R/a$, $m_r = M_r/M_0$, $m_\theta = M_\theta/M_0$ and $p = Pa^2/M_0$, the equilibrium equation of a circular plate subjected to a constant uniform load is

$$\frac{\mathrm{d}(rm_r)}{\mathrm{d}r} - m_\theta = -\frac{pr^2}{2}. \tag{16.2}$$

When subjected to a uniformly distributed transverse load P, the plate will deform and be in an elastic state, elastic-plastic state and a completely plastic state.

16.2.1 Elastic State

The deformation and the stress state of the plate are in an elastic state when the load p is not big. The dimensionless radial and tangential bending moments m_r and m_θ for a simply supported plate can be written as (Timoshenko and Woinowsky-Krieger, 1959)

$$m_r = \frac{3+\nu}{16}p(1-r^2), \quad m_\theta = \frac{p}{16}\left[(3+\nu) - (1+3\nu)r^2\right]. \tag{16.3}$$

The elastic limit load p_e can be calculated from Eq.(16.3),

$$p_e = \frac{16}{3+\nu}.$$

(16.4)

16.2.2 Elastic-plastic State

The center of the plate ($r = 0$) will firstly go into yield state when $p > p_e$. The plate is in plastic state ranges from 0 to r_e and the plate is in elastic state ranges from r_e to 1. In the plastic zone, if the UYC is chosen as the yield function, the expression of UYC can be written as a piecewise linear function

$$m_\theta = a_i m_r + b_i \qquad (i = 1, ..., 12),$$

(16.5)

where the values of parameters a_i and b_i are shown in Table 5.1.

Substituting Eq.(16.5) into Eq.(16.2) and then integrating Eq.(16.2), m_r falling on the segments L_i (the lines shown in Fig.5.3) is obtained as follows:

$$m_r = \frac{b_i}{1 - a_i} - \frac{pr^2}{2(3 - a_i)} + c_i r^{-1+a_i} \qquad (i = 1, ..., 5),$$

(16.6)

where c_i are the constants and can be derived from the continuous and boundary conditions. For a simply supported circular plate going into a plastic state, the bending moments at every point in the plate are located on the sides AB and BC for the normality requirement of plasticity (Ma et al., 1999). Therefore, i should be 1 and 2 in the plastic zone when the plate is in an elastic-plastic state (Fig.16.1).

In elastic zone, the bending moments can be expressed as (Timoshenko and Woinowsky-Krieger, 1959)

$$m_r = \frac{B}{r^2} - C - \frac{3+\nu}{16} pr^2, \quad m_\theta = -\frac{B}{r^2} - C - \frac{1+3\nu}{16} pr^2,$$

(16.7)

where B and C are the constants. They can be derived from the continuous and boundary conditions.

The boundary and continuous conditions are:

(a) $m_r \ (r = 0) = m_\theta \ (r = 0) = 1$;

(b) $m_r \ (r = r_1)$ and $m_\theta \ (r = r_1)$ are continuous, and $m_r \ (r = r_1) = (1 + b)/(2+b)$, where r_1 is the non-dimensional radius of a ring where the moments correspond to point B in Fig.5.2 where UYC is expressed by generalized stresses;

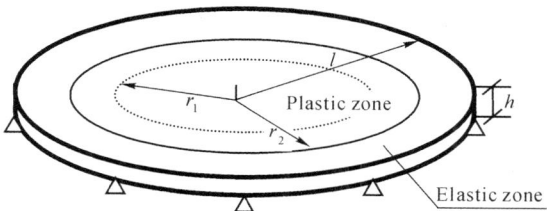

Fig. 16.1. Elastic-plastic state for a simply supported circular plate

(c) m_r $(r = r_2)$ and m_θ $(r = r_2)$ are continuous, where r_2 is the non-dimensional radius of plastic zone;

(d) m_r $(r = 1) = 0$.

The integration coefficients c_i $(i = 1, 2)$ can be derived from Eqs.(16.5) and (16.6) and the boundary and continuous conditions as follows:

$$c_1 = 0, \quad c_2 = \frac{2(1+b)}{3+2b} r_1^{\frac{1}{1+b}}, \tag{16.8}$$

and the relation between load p and r_1 is

$$p = \frac{6+2b}{2+b} \frac{1}{r_1^2}. \tag{16.9}$$

The relations between B, C, r_2 and r_1 are obtained from Eq.(16.5) to Eq.(16.7) and the boundary and continuous conditions (c) and (d) as follows:

$$\begin{aligned}
\frac{B}{r_2^2} - C &= (1+b) + \left(\frac{3+\nu}{8} \cdot \frac{3+b}{1+b} - \frac{3+b}{2+b} \cdot \frac{1+b}{3+2b} \right) \left(\frac{r_2}{r_1} \right)^2 \\
&\quad - \frac{2b(1+b)}{3+2b} \left(\frac{r_1}{r_2} \right)^{\frac{1}{1+b}},
\end{aligned} \tag{16.10}$$

$$B - C = \frac{3+\nu}{8} \cdot \frac{3+b}{2+b} \frac{1}{r_1^2}, \tag{16.11}$$

$$\begin{aligned}
\frac{B}{r_2^2}(1+2b) + C &+ \left[\frac{(1+3\nu)(1+b)(3+b)}{8(2+b)} - \frac{(3+\nu)b(3+b)}{8(1+b)} \right] \left(\frac{r_2}{r_1} \right)^2 \\
&+ (1+b) = 0.
\end{aligned} \tag{16.12}$$

The constants B, C, and plastic zones r_2 and r_1 can be derived from Eq.(16.8) to Eq.(16.11). Then the moment field in elastic and plastic zones can be obtained for a given load p.

16.2.3 Completely Plastic State

If the plate goes completely into a plastic state, r_2 will equal 1, and the plastic limit load p_p derived from Eq.(16.9) is

$$p_p = \frac{6 + 2b}{2 + b} \frac{1}{r_1^2},$$

(16.13)

where r_1 satisfies the following equation:

$$-(3 + 2b)(2 + b) + (3 + b)r_1^{-2} + 2b(2 + b)r_1^{1/(1+b)} = 0.$$

(16.14)

16.2.4 Shakedown Analysis

If the circular plate is unloaded from the initial elastic-plastic state, i.e. $p \to 0$, it will be left with residual stresses. Here we assume that the residual stresses will not produce inverse yielding. In that case the unloading process is purely elastic. So from the elastic solution of the circular plates, we obtain the changes in m_r and m_θ as

$$\Delta m_r = -\frac{p}{16}(3 + \nu)(1 - r^2), \quad \Delta m_\theta = -\frac{p}{16}\left[(3 + \nu) + r^2(1 + 3\nu)\right].$$ (16.15)

The residual stresses in the plate are

$$m_r^r = \frac{b_i}{1 - a_i} - \frac{pr^2}{2(3 - a_i)} + c_i r^{-1+a_i} - \frac{p}{16}(3 + \nu)(1 - r^2),$$

$$m_\theta^r = a_i m_r + b_i - \frac{p}{16}[(3 + \nu) + r^2(1 + 3\nu)], \qquad (i = 1, 2).$$

(16.16)

Observing the residual stress, it can be seen that the reverse yielding would begin first at the center of the plate, that is,

$$m_r^r|_{r=0} = m_\theta^r|_{r=0} = -1.$$

(16.17)

Eqs.(16.16) and (16.17) lead to the minimum uniformly distributed transverse load p_s for reverse yielding to occur in the plate

$$p_s = \frac{32}{1 + \nu} = 2p_e.$$

(16.18)

Evidently, as long as the applied transverse load p does not exceed p_s, the residual stresses will not result in reverse plastic deformation in the circular plate.

When the transverse load p, not exceeding the original value, acts on the circular plate and is then removed, the loading-unloading process will not

result in a new plastic deformation in the plate. From the above analysis
it can been found that if a simply supported circular plate is subjected to
cyclic pressure that ranges from $0 \to p \to 0 \to p \to 0 \to \cdots$, and p does not
exceed p_p (for the first loading from $0 \to p$) and $2p_e$ (for the other loading
from $p \to 0 \to p \to 0 \to \cdots$), yielding will not occur in the circular plate
during the loading-unloading process and the circular plate is safe. When this
happens the circular plate is said to be in shakedown. Hence the shakedown
limit for a circular plate subjected to a load p is

$$p_s = \min\{2p_e, p_p\}. \tag{16.19}$$

16.2.5 Discussion

There is always $2p_e > p_p$ in Eq.(16.19) for ν and parameter b. So p_s is equal to
p_p or $(6+2b)/(2+b)/r_1^2$, where r_1 is satisfied with Eq.(16.14). The parameter
b shows the effect of intermediate principal stress and the difference of various
yield criteria. The influences of parameter b on the shakedown limit p_s and
r_1 are analyzed. Figs.16.2 and 16.3 indicate that the unified yield criterion
parameter b will influence both p_s and r_1. The shakedown limit p_s is the
smallest for b=0 (corresponding to the Tresca criterion) and is the biggest
for b=1.0 (corresponding to the twin-shear yield criterion). The difference of
p_s for these two cases of b=0 and b=1.0 is about 14%. Fig.16.3 shows that
the radius of $m_{\theta max}$ decreases with the increase in the unified yield criterion
parameter b.

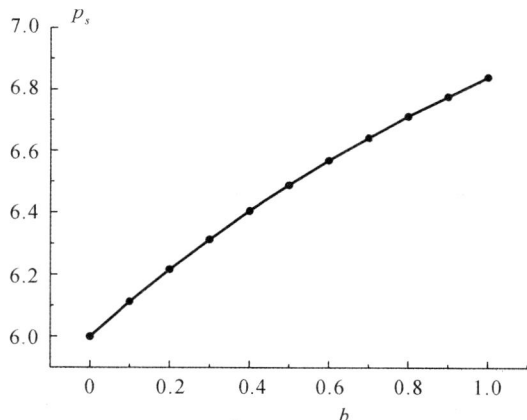

Fig. 16.2. Unified solution of shakedown limit p_s for simply supported circular
plate

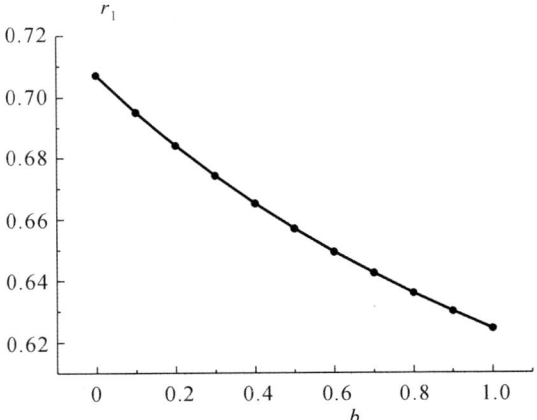

Fig. 16.3. Effect of the unified strength theory parameter b on the radius of $m_{\theta max}$

16.3 Unified Solution of Shakedown Limit for Clamped Circular Plate

16.3.1 Elastic State

In an elastic state the moment fields of a clamped circular plate satisfy $m_r = m_\theta$ at the center of the plate ($r = 0$), $m_\theta = \nu m_r$ at the clamped edge and $m_\theta > m_r$ at other points on the plate. The dimensionless radial and tangential bending moments m_r and m_θ for a clamped plate can be written as (Timoshenko and Woinowsky-Krieger, 1959)

$$m_r = \frac{p}{16}[(1+\nu) - r^2(3+\nu)], \ m_\theta = \frac{p}{16}[(1+\nu) - r^2(1+3\nu)], \qquad (16.20)$$

the elastic limit load p_e can be calculated from Eq.(16.20),

$$p_e = \frac{16}{1+\nu}. \qquad (16.21)$$

16.3.2 Elastic-plastic State

When the plate is in an elastic-plastic state, the boundary and continuous conditions are

(a) $m_r \ (r = 0) = m_\theta \ (r = 0) = 1$;

(b) $m_r \ (r = r_j)$ are continuous, where $j = 1, ..., 4$;

(c) $m_r \ (r = r_e)$ and $m_\theta \ (r = r_e)$ are continuous, where r_e is the dimensionless radius of plastic zone and $r_j < r_e$;

(d) $m_r = m_\theta/\nu$ at $r = 1$.

In the plastic zone $(0 \leqslant r \leqslant r_e)$ the generalized stresses satisfy the UYC and have the same form of Eq.(16.6). The moment fields of the entire clamped plate lie on the five sides corresponding to AB, BC, CD, DE and EF (Ma et al., 1999).

In the elastic zone $(r_e \leqslant r \leqslant 1)$ the generalized stresses can be expressed as the same form of Eq.(16.7), where r ranges from r_e to 1 and the constants B and C can be derived from the continuous and boundary conditions (c) and (d).

16.3.3 Completely Plastic State

When the plate is in a completely plastic state, there is $r_e = 1$ and $m_r(r = 1)$ $= m_\theta / \nu = -(1 + b)/(1 + b - \nu b)$. The load-carrying capacity of a clamped plate in this state has been obtained by Ma et al. (1999),

$$p_p = \frac{6 + 2b}{2 + b} \frac{1}{r_1^2},\tag{16.22}$$

where r_1 can be solved from the boundary and continuous conditions.

16.3.4 Shakedown Analysis

Using the similar analyzing method, the residual stresses can be obtained when the plate is unloaded from the initial load of the elasto-plastic state to zero, i.e. $p \to 0$.

$$m_r^r = \frac{b_i}{1 - a_i} - \frac{pr^2}{2(3 - a_i)} + c_i r^{-1+a_i} - \frac{p}{16}[(1 + \nu) - r^2(3 + \nu)],$$

$$m_\theta^r = a_i m_r + b_i - \frac{p}{16}\left[(1 + \nu) - r^2(1 + 3\nu)\right], \quad (i = 1, ..., 5).\tag{16.23}$$

Observing the residual stress, it can be seen that the reverse yielding would begin first at the center of the plate, that is,

$$m_r^r|_{r=0} = m_\theta^r|_{r=0} = -1.\tag{16.24}$$

Eqs.(16.23) and (16.24) lead to the minimum uniformly distributed transverse load p_s for reverse yielding to occur in the plate

$$p_s = \frac{32}{1 + \nu} = 2p_e.\tag{16.25}$$

So the shakedown limit for a clamped circular plate subjected to a load p is

$$p_s = \min\{2p_e, p_p\}.\tag{16.26}$$

16.3.5 Discussion

There is always $2p_e > p_p$ in Eq.(16.26) for every ν and parameter b. So p_s is equal to p_p or $(6 + 2b)/(2 + b)/r_1^2$. When $b = 0$, the UYC becomes the Tresca criterion and the load-bearing capacity of the circular plate with respect to the Tresca criterion is 11.258 in the case of $\nu=0.25$ which is in good agreement with the analyzing result $p_s=11.26$ ($\nu=0.25$) in the reference (Pham, 1997) with error at approximately 0.018%. In the same case of the Poisson's ratio, the shakedown solution using UYC is 12.23 when $b=0.5$ (near to the von-Mises criterion), while the shakedown result using Mises material (Pham, 1997) is 12.23 with M_0 being substituted by $2/3^{0.5}M_0$. The difference between these two results is only about 7.3%.

The relations between the unified strength theory parameter b with the shakedown limit p_s and the radius r_1 of $m_{\theta\max}$ are illustrated in Fig.16.4 and 16.5. Both the figures show that the parameter b affects the values of shakedown limit p_s and the radius r_1. It can be seen in Fig.16.4 that for a given kind of Poisson's ratio, the shakedown limit p_s is the smallest in the case of $b=0$, and p_s is the biggest in the case of $b=1$. For three kinds of Poisson's ratio, the p_s-b curve increases most slowly when $\nu=0$.

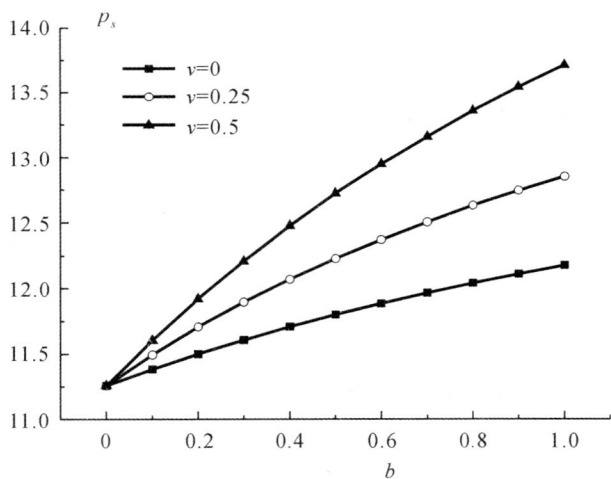

Fig. 16.4. Effect of the unified strength theory parameter b on shakedown limit p_s

When $b=1$, the difference of the shakedown limit p_s for the case of $\nu=0$ and $\nu=0.5$ is about 1.631. For a given Poisson's ratio in Fig.16.5, the radius of $m_{\theta\max}$ decreases with an increase in the unified strength theory parameter b. For a given parameter b, the shakedown limit p_s for $\nu=0.5$ is the biggest, while p_s for $\nu=0$ is the smallest.

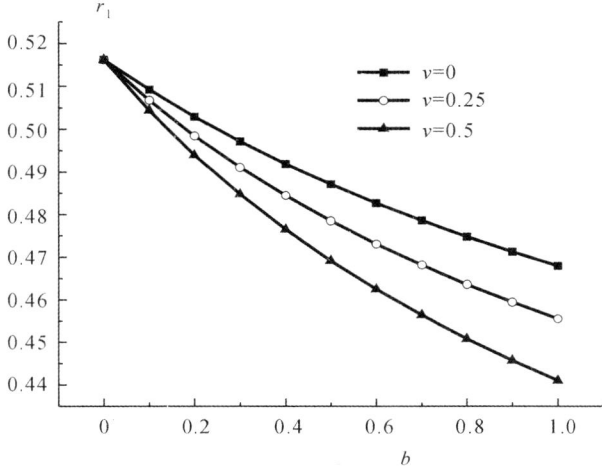

Fig. 16.5. Effect of the unified strength theory parameter b on the radius of $m_{\theta \max}$

16.4 Comparison between Shakedown Solution and Limit Results

Based on the unified yield criterion, a shakedown analysis of a circular plate under a uniformly distributed transverse load is carried out and the unified solution of a shakedown load for a circular plate is derived in this chapter. The solution encompasses the existing classical solution as a special case and a series of new results.

From the above analysis it is noted from Eqs.(16.19) and (16.26) that the shakedown solutions for simply supported and clamped plates are almost the same, that is, both are the minimum of the twice elastic limit load and the plastic limit load,

$$p_s = \min\{2p_e, p_p\}.$$

Because the unified strength parameter b ranges from 0 to 1 and the Poisson's ratio is from 0 to 0.5 no matter what the value of parameter b and the Poisson's ratio, there is always $2p_e > p_p$. Therefore the shakedown solution p_s of a circular plate is equal to its plastic limit load p_p; i.e. $p_s = p_p = (6+2b)/(2+b)/r_1^2$, which is related to the parameter b for the simply supported circular plate and both the parameter b and the Poisson's ratio for the clamped circular plate, as shown in Figs.16.2 and 16.4.

It is also found that the special solutions for shakedown analysis of circular plates when $b=0$ are equal to the plastic limit load of the Tresca criterion; i.e. $p_s=6.0$ for the simply supported circular plate and $p_s=11.258$ for the clamped circular plate. Meanwhile, when $b=0$ the shakedown solution of clamped circular plate is the same as the result in the references. Besides the shakedown solution for $b=0$, the other solutions for different parameter b when it ranges from $b > 0$ to $b=1$ can also be calculated, as can be seen in Fig.16.2 and Fig.16.4. The following table shows the shakedown limit load for both the simply supported and clamped plates when the unified yield criterion parameter b changes.

Table 16.1. Shakedown limit load p_p with the changes in parameter b

b	0	0.3	0.5	0.6	0.8	1.0
Simply supported plate	6.0	6.31	6.49	6.57	6.71	6.8
Clamped plate ($\nu=0.25$)	11.26	11.9	12.23	12.37	12.63	12.85

16.5 Summary

The unified yield criterion is used to analyze the shakedown limit of a circular plate. The results are applicable for a wide range of materials and structures. The shakedown analysis of the circular plate shows the effect of yield criterion on the plastic limit loads and shakedown loads.

For both a simply supported and a clamped circular plate, the shakedown limit p_s increases with the growth of the unified yield criterion parameter b and the shakedown limit p_s for the simply supported plate is smaller than that for the clamped plate. When $b=0$, the analyzed result is in good agreement with the result in the references. The study also shows that the radius of $m_{\theta\mathrm{max}}$ decreases with the growth of the unified yield criterion parameter b. The shakedown limit p_s for the clamped circular plate is influenced by the Poisson's ratio, while p_s for the simply supported circular plate will not be affected by the Poisson's ratio.

By comparison with the limit load, the shakedown solutions p_s of the simply supported and clamped circular plates are both equal to the plastic limit load. p_s of the simply supported circular plate varies only with the unified yield criterion parameter b, while p_s of the clamped circular plate changes with the unified yield criterion parameter b and the Poisson's ratio.

16.6 Problems

Problem 16.1 Compare the solutions of limit analysis and shakedown analysis.

Problem 16.2 Compare the solutions of shakedown analysis of a simply supported and a clamped circular plate.

Problem 16.3 Determine the shakedown load of a simply supported circular plate by using the Tresca yield criterion ($b=0$).

Problem 16.4 Determine the shakedown load of a clamped circular plate by using the Tresca yield criterion ($b=0$).

Problem 16.5 Determine the shakedown load of a simply supported circular plate by using the twin-shear yield criterion ($b=1$).

Problem 16.6 Determine the shakedown load of a clamped supported circular plate by using the twin-shear yield criterion ($b=1$).

Problem 16.7 Determine the shakedown load of a simply supported circular plate by using the unified yield criterion with $b=0.5$.

Problem 16.8 Determine the shakedown load of a clamped circular plate by using the unified yield criterion with $b=0.5$.

Problem 16.9 Determine the shakedown load of a simply supported circular plate by using the unified yield criterion with $b=0.8$.

Problem 16.10 Determine the shakedown load of a clamped circular plate by using the unified yield criterion with $b=0.8$.

References

Gokhfeld DA, Cherniavski OF (1980) Limit analysis of structures at thermal cycling. Sijthoff & Noordhoff

Kachanov LM (1971) Foundations of the theory of plasticity, North-Holland Publ. Co., Amsterdam

Koiter WT (1953) Stress-strain relations, uniqueness and variational theorems for elastic-plastic materials with a singular yield surface. Quart. Appl. Math., 11:350-174

Koiter WT (1956) A new general theorem on shakedown of elastic-plastic structures. Proc. K. Ned. Akad. Wet., B59:24-34

Koiter WT (1960) General theorems for elastic-plastic solids. In: Sneddon JN, Hill R (eds.) Progress in Solid Mechanics. North-Holland Publ. Co., Amsterdam, 1:165-231

König JA (1987) Shakedown of elastic-plastic structures. Elsevier, Amsterdam

König JA, Kleiber M (1978) On a new method of shakedown analysis. Bull. Acad. Pol. Set., Ser. Sci. Tech., 26:165-171

König JA, Maier G (1987) Shakedown analysis of elastic-plastic structures: A review of recent developments. Nucl. Eng. Design, 66:81-95

Ma G, Hao H, Iwasaki S (1999) Plastic limit analysis of a clamped circular plate with unified yield criterion. Structural Engineering and Mechanics, 7(5):513-525

Maier G (2001) On some issues in shakedown analysis. J. Appl. Mech., 68:799-808

Maier G, Carvelli V, Coccbetti G (2000) On direct methods for shakedown and limit analysis. Eur. J. Mech. A/Solids (Special issue), 19:S79-S100

Martin JB (1975) Plasticity: fundamentals and general results. The MIT Press

Melan E (1936a) Theorie statisch unbestimmter Systeme. In: Second Congr. Intern. Assoc. Bridge and. Structural Eng., Berlin, 43-64

Melan E (1936b) Theorie statisch unbestimmter Systeme aus ideal plastischem Baustoff. Sitz. Ber. Akad. Wiss. Wien, II.a, 145:195-218

Neal BG (1956) The plastic method of structural analysis. (2nd ed.) Chapman and Hall, Wiley, New York

Nestor. Zouain (2001) Bounds to shakedown loads. Int. J. Solids Struct., 38:2249-2266

Pham DC (1996) Dynamic shakedown and a reduced kinematic theorem. Int. J. Plasticity, 12:1055-1068

Pham DC (1997) Evaluation of shakedown loads for plates. Int. J. Mech. Sci., 39(12):1416-1422

Pham DC (2003) Plastic collapse of a circular plate under cyclic loads. Int. J. Plasticity, 19(4):547-559

Polizzotto C (1982) A unified treatment of shakedown theory and related bounding techniques. Solid Mech. Archives, 7:19-75

Polizzotto C (1993a) On the conditions to prevent plastic shakedown of structures: part 1-Theory. ASME, J. Appl. Mech., 60:15-19

Polizzotto C (1993b) On the conditions to prevent plastic shakedown of structures: part 2-The plastic shakedown limit load. ASME, J. Appl. Mech., 60:20-25

Polizzotto C (1993c) A study on plastic shakedown of structures: part 1-Basic properties. ASME, J. Appl. Mech., 60:318-323

Polizzotto C (1993d) A study on plastic shakedown of structures: part 2 - Theorems. ASME, J. Appl. Mech., 60:324-330

Ponter ARS, Carter KF (1997) Shakedown state simulation techniques based on linear elastic solutions. Comput. Methods Appl. Mech. Eng., 140:259-279

Stein E, Zhang G, Huang Y (1993) Modeling and computation of shakedown problems for nonlinear hardening materials. Comput. Methods Appl. Mech. Eng., 103:247-272

Symonds PS (1951) Shakedown in continuous media. J. Appl. Mech., 18(1):85-89

Timoshenko S, Woinowsky-Krieger S (1959) Theory of plates and shells. McGraw-Hill, New York

Yu MH (1961a) General behavior of isotropic yield function. Res. Report of Xi'an Jiaotong University, Xi'an (in Chinese)

Yu MH (1961b) Plastic potential and flow rules associated singular yield criterion. Res. Report of Xi'an Jiaotong University, Xi'an (in Chinese)

Yu MH (1983) Twin shear stress yield criterion. Int. J. Mech. Sci., 25(1):71-74

Yu MH (2002a) New system of strength theory. Xi'an Jiaotong University Press, Xi'an

Yu MH (2002b) Advances in strength theories for materials under complex stress state in the 20th century. Applied Mechanics Reviews, ASME, 55(3):169-218

Yu MH (2004) Unified strength theory and its applications. Springer, Berlin

Yu MH, He LN (1991) A new model and theory on yield and failure of materials under complex stress state. Proceedings, Mechanical Behavior of Materials-VI, 3:851-856

Yu MH, He LN, Song LY (1985) Twin shear stress theory and its generalization. Scientia Sinica (Sciences in China), English ed., Series A, 28(11):1174-1183

Shakedown Analysis of Rotating Cylinder and Disc

17.1 Introduction

Rotating cylinders are a commonly-used component in engineering. When they rotate at a constant velocity the centrifugal force is the main loading applied on the cylinders. Elasto-plastic analyses are needed for the cylinders under static loadings and the shakedown of cylinders should be taken into account when subjected to cyclic-variation loading. The shakedown theorem and analysis of structures were described in the literature (Kachanov, 1971; Martin, 1975; Zyczkowski, 1981; König, 1987; Mroz et al., 1995; Weichert et al., 2000). Limits to shakedown loads were discussed by Zouain and Silveira (2001), Feng and Yu (1994) and others.

Elasto-plastic cylinders have been analysed in detail in the classical elasto-plastic theory (Hodge and Balaban 1962; Chakrabarty 1987). These studies are based on the Tresca criterion and have not considered the intermediate principal stress. Ma et al. (1995) employed the unified yield criterion to investigate the effect of intermediate principal stress for the first time and obtained the unified elasto-plastic solution to the rotating disc and cylinder. The unified solution of the limit angular velocity of an annular disc was obtained by Ma and Hao (1999). Limit angular speeds of rotating disc and cylinder for non-SD materials (identical strength both in tension and compression) have been described in Chapter 11.

Shakedown analysis of rotating disks and hollow rotating discs have been presented for non-SD materials (Liu et al., 1988) and (Liu and Wang, 1990). As to materials with different tensile and compressive strengths, Xu and Yu (2004; 2005) used the unified strength theory to discuss the effect of tensile and compressive strength variation as well as the effect of intermediate principal stress on the plasticity and shakedown of cylinders. It is indicated that the unified solution is applicable for both non-SD materials with identical tensile and compressive strength and SD materials with different strengths in compression and in tension.

Elasto-plastic analyses of hollow rotating circular bars, shakedown analyses of hollow rotating circular bars, elasto-plasticity and shakedown of solid rotating circular bars, elasto-plasticity and shakedown analyses of rotating circular bars, elastic and plastic limit analyses of hollow rotating discs by using the Yu unified strength theory are all described in this chapter. The unified solutions of these problems are adopted for both non-SD materials and SD materials.

17.2 Elasto-plastic and Shakedown Analyses of Rotating Cylinder and Disc

The displacement and stress caused by centrifugal force are symmetrical when the cylinders rotate at a constant velocity and σ_r, σ_θ, σ_z are principal stresses. For discs with identical thickness it can be regarded as an axisymmetric plane stress problem and for rotating circular bars it can be solved as an axisymmetric plane strain problem. In this section we will discuss the elasto-plasticity and shakedown of circular bars, and the rotating discs will be studied in the next section. For the sake of simplicity the materials are assumed to be incompressible and elasto-plastic.

When circular bars rotate at a constant velocity, the tangent centrifugal applied on the circular bar is

$$f = \rho\omega^2 r, \tag{17.1}$$

where ρ is the density of materials. The equilibrium equation is

$$\frac{\mathrm{d}\sigma_r}{\mathrm{d}r} + \frac{\sigma_r - \sigma_\theta}{r} + \rho\omega^2 r = 0. \tag{17.2}$$

17.2.1 Elastic Analyses of Hollow Rotating Circular Bars

There is a hollow circular bar with inner radius a and outer radius d. With the increase of rotating velocity, circular bars will transfer to the elasto-plastic state. When the angular velocity is small, the circular bar is in the elastic state and the stress is

$$\sigma_r = c_1 + \frac{c_2}{r^2} - \frac{3 - 2\nu}{8(1 - \nu)}\rho\omega^2 r^2, \tag{17.3a}$$

$$\sigma_\theta = c_1 - \frac{c_2}{r^2} - \frac{1 + 2\nu}{8(1 - \nu)}\rho\omega^2 r^2, \tag{17.3b}$$

$$\sigma_z = c_1 - \frac{1}{4(1 - \nu)}\rho\omega^2 r^2. \tag{17.3c}$$

If $\nu = 0.5$, the general solution for incompressible hollow rotating circular bar stress is

$$\sigma_r = c_1 + \frac{c_2}{r^2} - \frac{1}{2}\rho\omega^2 r^2, \tag{17.4a}$$

$$\sigma_\theta = c_1 - \frac{c_2}{r^2} - \frac{1}{2}\rho\omega^2 r^2, \tag{17.4b}$$

$$\sigma_z = c_1 - \frac{1}{2}\rho\omega^2 r^2, \tag{17.4c}$$

where the integration constants c_1, c_2 are determined by boundary conditions. When both the inner and outer surfaces are free, i.e.,

$$\begin{aligned} \sigma_r &= 0, & r &= a, \\ \sigma_r &= 0, & r &= d, \end{aligned} \tag{17.5}$$

we can obtain

$$\sigma_r = \frac{1}{2}\rho\omega^2(d^2 + a^2 - a^2 d^2/r^2 - r^2), \tag{17.6a}$$

$$\sigma_\theta = \frac{1}{2}\rho\omega^2(d^2 + a^2 + a^2 d^2/r^2 - r^2), \tag{17.6b}$$

$$\sigma_z = \frac{1}{2}\rho\omega^2(d^2 + a^2 - r^2). \tag{17.6c}$$

Apparently σ_r is the minimum principal stress and σ_θ the maximum principal stress. Therefore the yield condition of circular bars is

$$\frac{2m + 2mb - b}{2m(1 + b)}\sigma_\theta - \frac{2 + b}{2m(1 + b)}\sigma_r = \sigma_t. \tag{17.7}$$

For convenience the variables m' and σ_t' are introduced here as

$$m' = \frac{2m + 2mb - b}{2 + b}, \quad \sigma_t' = \frac{2m(1 + b)}{2m + 2mb - b}\sigma_t. \tag{17.8}$$

Hence, Eq.(17.7) can be rewritten as

$$\sigma_\theta - \frac{1}{m'}\sigma_r = \sigma_t'. \tag{17.9}$$

It is seen that when the rotating velocity reaches a certain value, the inner boundary of circular bars will yield first. Substituting $r = a$ into Eqs.(17.4) and (17.9), the limit elastic rotating velocity can be obtained,

$$\omega_{el} = \frac{\sqrt{\sigma_t'/\rho}}{d}. \tag{17.10}$$

Substituting Eq.(17.10) into Eq. (17.6), we can obtain the yield stress of rotating circular bars

$$\sigma_r = \frac{1}{2}\sigma_t'(1 + a^2/d^2 - a^2/r^2 - r^2/d^2), \tag{17.11a}$$

$$\sigma_\theta = \frac{1}{2}\sigma_t'(1 + a^2/d^2 + a^2/r^2 - r^2/d^2), \tag{17.11b}$$

$$\sigma_z = \frac{1}{2}\sigma_t'(1 + a^2/d^2 - r^2/d^2). \tag{17.11c}$$

17.2.2 Elasto-plastic Analyses of Hollow Rotating Circular Bars

When $\omega > \omega_{el}$, the plastic deformation will appear near the inner boundary and then a plastic area is formed. Let c denote the radius of the plastic area, Eqs.(17.2) and (17.9) are solved together with consideration of the boundary condition

$$\begin{aligned}
\sigma_r = &\frac{m'}{m'-1}\left[1 - (a/r)^{\frac{m'-1}{m'}}\right]\sigma_t' \\
&- \frac{m'}{3m'-1}\left[r^2 - a^2(a/r)^{\frac{m'-1}{m'}}\right]\rho\omega^2,
\end{aligned} \tag{17.12a}$$

$$\begin{aligned}
\sigma_\theta = &\frac{1}{m'-1}\left[1 - (a/r)^{\frac{m'-1}{m'}}\right]\sigma_t' \\
&- \frac{1}{3m'-1}\left[r^2 - a^2(a/r)^{\frac{m'-1}{m'}}\right]\rho\omega^2 + \sigma_t',
\end{aligned} \tag{17.12b}$$

$$\begin{aligned}
\sigma_z = &\frac{m'+1}{m'-1}\left[1 - (a/r)^{\frac{m'-1}{m'}}\right]\sigma_t'/2 \\
&- \frac{m'+1}{3m'-1}\left[r^2 - a^2(a/r)^{\frac{m'-1}{m'}}\right]\rho\omega^2 + \sigma_t'/2.
\end{aligned} \tag{17.12c}$$

Based on Eq.(17.4), the stress of circular bars in the elastic area is

$$\sigma_r = c_1' + \frac{c_2'}{r^2} - \frac{1}{2}\rho\omega^2 r^2, \tag{17.13a}$$

$$\sigma_\theta = c_1' - \frac{c_2'}{r^2} - \frac{1}{2}\rho\omega^2 r^2, \tag{17.13b}$$

$$\sigma_z = c_1' - \frac{1}{2}\rho\omega^2 r^2, \tag{17.13c}$$

where c_1' and c_2' are integration constants. According to the continuous conditions of the interface surface between elastic and plastic areas, they should satisfy the following equation:

$$2c_1' = \left[\frac{2m'}{m'-1} - \frac{m'+1}{m'-1}(a/c)^{\frac{m'-1}{m'}} \right] \sigma_t'$$
$$- \frac{m'+1}{3m'-1} \left[c^2 - a^2(a/c)^{\frac{m'-1}{m'}} \right] \rho\omega^2 + \rho\omega^2 c^2,$$

and

$$\frac{2c_2'}{c^2} = -(a/c)^{\frac{m'-1}{m'}} - \frac{m'-1}{3m'-1} \left[c^2 - a^2(a/c)^{\frac{m'-1}{m'}} \right] \rho\omega^2, \tag{17.14}$$

with the boundary condition of outer surface $r = d$, $\sigma_r = 0$, we then have

$$c_1' + \frac{c_2'}{d^2} - \frac{1}{2}\rho\omega^2 d^2 = 0, \tag{17.15}$$

and with the substitution of constants c_1' and c_2' into Eq.(17.15) we obtain the relation between the rotating velocity ω and radius of plastic area c as

$$\left[\frac{2m'}{m'-1} - \frac{m'+1}{m'-1}(a/c)^{\frac{m'-1}{m'}} \right] \sigma_t' - \frac{m'+1}{3m'-1} \left[c^2 - a^2(a/c)^{\frac{m'-1}{m'}} \right] \rho\omega^2$$
$$+ \rho\omega^2 c^2 - (a/c)^{\frac{m'-1}{m'}}\sigma_t' \frac{c^2}{d^2} - \frac{m'-1}{3m'-1} \left[c^2 - a^2(a/c)^{\frac{m'-1}{m'}} \right] \rho\omega^2 \frac{c^2}{d^2} \tag{17.16}$$
$$- \rho\omega^2 d^2 = 0,$$

Apparently the rotating velocity ω and radius of plastic area c can be determined by the above equation. In particular, when $c = d$, the limit plastic rotating velocity can be obtained as

$$\omega_{\text{pl}} = \left[\frac{3m'-1}{m'-1} \frac{1-(a/d)^{\frac{m'-1}{m'}}}{d^2 - a^2(a/d)^{\frac{m'-1}{m'}}} \frac{\sigma_t'}{\rho} \right]^{1/2}. \tag{17.17}$$

Substituting Eq.(17.17) into Eq.(17.12) we can obtain the stress of rotating circular bars in the limit plastic state,

$$\sigma_r = \frac{m'-1}{m'} \left[1 - (a/r)^{\frac{m'-1}{m'}} \right] \sigma_t'$$
$$- \frac{m'-1}{m'} \left[r^2 - a^2(a/r)^{\frac{m'-1}{m'}} \right] \frac{1-(a/d)^{\frac{m'-1}{m'}}}{d^2 - a^2(a/d)^{\frac{m'-1}{m'}}} \sigma_t' \sqrt{b^2 - 4ac}, \tag{17.18a}$$

$$\sigma_\theta = \frac{1}{m'-1}\left[1-(a/r)^{\frac{m'-1}{m'}}\right]\sigma'_t$$
$$-\frac{1}{m'-1}\left[r^2-a^2(a/r)^{\frac{m'-1}{m'}}\right]\frac{1-(a/d)^{\frac{m'-1}{m'}}}{d^2-a^2(a/d)^{\frac{m'-1}{m'}}}\sigma'_t+\sigma'_t,$$

$$(17.18b)$$

$$\sigma_z = \frac{m'+1}{m'-1}\left[1-(a/r)^{\frac{m'-1}{m'}}\right]\sigma'_t/2$$
$$-\frac{m'+1}{m'-1}\left[r^2-a^2(a/r)^{\frac{m'-1}{m'}}\right]\frac{1-(a/d)^{\frac{m'-1}{m'}}}{d^2-a^2(a/d)^{\frac{m'-1}{m'}}}\frac{\sigma'_t}{2}+\frac{\sigma'_t}{2}.$$

$$(17.18c)$$

17.2.3 Shakedown Analyses of Hollow Rotating Circular Bars

When the rotating velocity is the range of $\omega_{\mathrm{el}} < \omega < \omega_{\mathrm{pl}}$, the circular bar is in the elasto-plastic state, and then if the loading is unloaded completely, there will appear residual stress in the circular bars. The unloaded stress can be calculated according to the elastic solution when the unloading is considered elastic. Based on Eq.(17.6), we can find the stress variation of rotating circular bars in the unloading

$$\Delta\sigma_r = -\frac{1}{2}\rho\omega^2\left(d^2+a^2-a^2d^2/r^2-r^2\right)\frac{-b\pm\sqrt{b^2-4ac}}{2a}, \qquad (17.19a)$$

$$\Delta\sigma_\theta = -\frac{1}{2}\rho\omega^2\left(d^2+a^2+a^2d^2/r^2-r^2\right), \qquad (17.19b)$$

$$\Delta\sigma_z = -\frac{1}{2}\rho\omega^2\left(d^2+a^2-r^2\right). \qquad (17.19c)$$

The residual stress is the superposition of loading stress and unloading stress. For elastic loading there is no residual stress after unloading; i.e., the rotating circular bar is in the zero stress state. For elasto-plastic loading there exists residual stress in the circular bars. Summating Eqs.(17.19) and (17.12) with Eq.(17.13) respectively, the residual stress of the plastic area is obtained as

$$\sigma_r^r = \frac{m}{m'-1}\left[1-(a/r)^{\frac{m-1}{m}}\right]\sigma'_t-\frac{m}{3m'-1}\left[r^2-a^2(a/r)^{\frac{m-1}{m}}\right]\rho\omega^2$$
$$-\frac{1}{2}\rho\omega^2\left(d^2+a^2-a^2d^2/r^2-r^2\right),$$

$$(17.20a)$$

$$\sigma_\theta^r = \frac{1}{m'-1}\left[1-(a/r)^{\frac{m-1}{m}}\right]\sigma'_t-\frac{1}{3m'-1}\left[r^2-a^2(a/r)^{\frac{m-1}{m}}\right]\rho\omega^2+\sigma'_t$$
$$-\frac{1}{2}\rho\omega^2\left(d^2+a^2+a^2d^2/r^2-r^2\right),$$

$$(17.20b)$$

$$\sigma_z^r = \frac{m+1}{m'-1}\left[1 - (a/r)^{\frac{m-1}{m}}\right]\sigma_t'/2 - \frac{m+1}{3m'-1}\left[r^2 - a^2(a/r)^{\frac{m-1}{m}}\right]\frac{\rho\omega^2}{2} + \frac{\sigma_t'}{2}$$
$$- \frac{1}{2}\rho\omega^2\left(d^2 + a^2 - r^2\right),$$

(17.20c)

and the residual stress of the elastic area is

$$\sigma_r^r = c_1' + \frac{c_2'}{r^2} - \frac{1}{2}\rho\omega^2 r^2 - \frac{1}{2}\rho\omega^2\left(d^2 + a^2 - a^2 d^2/r^2 - r^2\right), \quad (17.21a)$$

$$\sigma_\theta^r = c_1' - \frac{c_2'}{r^2} - \frac{1}{2}\rho\omega^2 r^2 - \frac{1}{2}\rho\omega^2\left(d^2 + a^2 + a^2 d^2/r^2 - r^2\right), \quad (17.21b)$$

$$\sigma_z^r = c_1' - \frac{1}{2}\rho\omega^2 r^2 - \frac{1}{2}\rho\omega^2\left(d^2 + a^2 - r^2\right). \quad (17.21c)$$

It is seen that when the rotating velocity reaches a certain value, the inner surface will appear as an inverse yield first. From Eq.(17.20) the residual stress at $r = a$ should be

$$\sigma_r^r = 0, \quad (17.22a)$$
$$\sigma_\theta^r = \sigma_t' - \rho\omega^2 d^2, \quad (17.22b)$$
$$\sigma_z^r = (\sigma_t' - \rho\omega^2 d^2)/2. \quad (17.22c)$$

In order to ensure that no inverse yield will appear when the circular bar is unloaded, the residual stress at the inner surface should satisfy

$$\sigma_r^r - \frac{1}{m'}\sigma_\theta^r \leqslant \sigma_t'. \quad (17.23)$$

Substituting the residual stress at the inner surface into Eq.(17.23), the maximum rotating velocity ω_s which will not cause inverse yield is

$$\omega_s = \frac{1}{d}\sqrt{\frac{m'+1}{m'}\frac{\sigma_t'}{\rho}}. \quad (17.24)$$

It is seen that if the rotating velocity $\omega \leqslant [(m'+1)\sigma_t'/(m'\rho)]^{1/2}/d$ satisfies the first loading, no plastic deformation will come into being in the opposite direction when unloaded completely. If loaded again after that, new plastic deformation will not appear until $\omega > [(m'+1)\sigma_t'/(m'\rho)]^{1/2}/d$ due to the existence of residual stress. It can be concluded that when the rotating velocity varies in the range of $0 \sim [(m'+1)\sigma_t'/(m'\rho)]^{1/2}$, the circular bar will be in the elastic state, i.e., shakedown state, except for one plastic deformation. However it should be noted here that only centrifugal force caused by velocities is taken into account, neglecting that caused by accelerations. In addition, the velocity in Eq.(17.24) cannot exceed the limit plastic rotating velocity of circular bars. Therefore, to keep the circular bars in the shakedown state when subject to repeated loads, the rotating velocity should satisfy

$$\omega < \omega_{\mathrm{s}} = \left(\frac{1}{d} \sqrt{\frac{m'+1}{m'} \frac{\sigma_t'}{\rho}}, \ \omega_{\mathrm{pl}} \right), \tag{17.25}$$

where ω_s is the limit rotating velocity of circular bars in shakedown state.

17.2.4 Elasto-plasticity and Shakedown of Solid Rotating Circular Bars

In this section the solid rotating circular bars will be studied. The general solution of stress for circular bars in the elastic stage is as shown in Eq.(17.3). Thus, considering the boundary conditions at $r = 0$ and $r = d$ for solid rotating circular bars, we can obtain the stress solution of solid rotating circular bars

$$\sigma_r = \frac{3 - 2\nu}{8(1 - \nu)} \rho\omega^2 (d^2 - r^2), \tag{17.26a}$$

$$\sigma_\theta = \frac{3 - 2\nu}{8(1 - \nu)} \rho\omega^2 \left(d^2 - \frac{1 + 2\nu}{3 - 2\nu} r^2\right), \tag{17.26b}$$

$$\sigma_z = \frac{3 - 2\nu}{8(1 - \nu)} \rho\omega^2 \left(d^2 - \frac{2}{3 - 2\nu} r^2\right). \tag{17.26c}$$

For incompressible circular bars we have

$$\sigma_r = \frac{1}{2} \rho\omega^2 (d^2 - r^2), \tag{17.27a}$$

$$\sigma_\theta = \frac{1}{2} \rho\omega^2 (d^2 - r^2), \tag{17.27b}$$

$$\sigma_z = \frac{1}{2} \rho\omega^2 (d^2 - r^2). \tag{17.27c}$$

Apparently, σ_r is the minimum principal stress and σ_θ the maximum principal stress. It is known that the centre of circular bars will yield first when the rotating velocity reach a certain value. Substituting $r = 0$ into Eq.(17.27) and then Eq.(17.9), the limit elastic rotating velocity of circular bars is obtained as

$$\omega_{\mathrm{el}} = \frac{\sqrt{2\sigma_t' m'/(m' - 1)/\rho}}{d}. \tag{17.28}$$

When $\omega > \omega_{el}$, the plastic deformation will appear near the inner boundary and then a plastic area is formed. Let c denote the radius of the plastic area the Eqs.(17.2) and (17.9) are solved together with consideration of the boundary condition at $r = 0$, and then the stress of the plastic area is obtained as

$$\sigma_r = \frac{m'}{m'-1}\sigma_t' - \frac{m'}{3m'-1}\rho\omega^2 r^2, \tag{17.29a}$$

$$\sigma_\theta = \frac{1}{m'-1}\sigma_t' - \frac{1}{3m'-1}\rho\omega^2 r^2 + \sigma_t', \tag{17.29b}$$

$$\sigma_z = \frac{m'+1}{m'-1}\sigma_t'/2 - \frac{m'+1}{3m'-1}\rho\omega^2 r^2/2 + \sigma_t'/2. \tag{17.29c}$$

Based on Eq.(17.4), the stress of elastic area is

$$\sigma_r = c_1' + \frac{c_2'}{r^2} - \frac{1}{2}\rho\omega^2 r^2, \tag{17.30a}$$

$$\sigma_\theta = c_1' - \frac{c_2'}{r^2} - \frac{1}{2}\rho\omega^2 r^2, \tag{17.30b}$$

$$\sigma_z = c_1' - \frac{1}{2}\rho\omega^2 r^2. \tag{17.30c}$$

According to the continuous conditions of the interface surface between elastic and plastic areas, they should satisfy the following equation:

$$c_1' = \frac{m'\sigma_t'}{m'-1} + \frac{m'-1}{3m'-1}\rho\omega^2 c^2, \quad c_2' = \frac{1-m'}{3m'-1}\rho\omega^2 c^2.$$

Further noting the boundary condition of the outer surface $r = d$, $\sigma_r = 0$, we have

$$\frac{m'}{m'-1}\sigma_t' d^2 + \frac{m'-1}{3m'-1}\rho\omega^2 c^2 d^2 - \frac{m'-1}{2(3m'-1)}\rho\omega^2 c^4 - \frac{1}{2}\rho\omega^2 d^4 = 0. \tag{17.31}$$

The rotating velocity and radius of plastic area c can be determined by the above equation. In particular, when $c = d$ the limit plastic rotating velocity can be obtained as

$$\omega_{pl} = \sqrt{\frac{(3m'-1)\sigma_t'}{(m'-1)\rho d^2}}. \tag{17.32}$$

Substituting Eq.(17.32) into Eq.(17.29), we can get the stress of rotating circular bars in the limit plastic state,

$$\sigma_r = \frac{m'}{m'-1}\sigma'_t(1 - r^2/d^2), \tag{17.33a}$$

$$\sigma_\theta = \frac{1}{m'-1}\sigma'_t(1 - r^2/d^2) + \sigma'_t, \tag{17.33b}$$

$$\sigma_z = \frac{1}{2}\frac{m'+1}{m'-1}\sigma'_t(1 - r^2/d^2) + \sigma'_t/2. \tag{17.33c}$$

The shakedown of rotating circular bars is investigated as follows. It is assumed that the circular bar is unloaded in the plastic limit state and the unloading is elastic. According to Eqs.(17.32) and (17.27) the unloaded stress is obtained as

$$\Delta\sigma_r = -\frac{1}{2}\frac{3m'-1}{m'-1}\sigma'_t(1 - r^2/d^2), \tag{17.34a}$$

$$\Delta\sigma_\theta = -\frac{1}{2}\frac{3m'-1}{m'-1}\sigma'_t(1 - r^2/d^2), \tag{17.34b}$$

$$\Delta\sigma_z = -\frac{1}{2}\frac{3m'-1}{m'-1}\sigma'_t(1 - r^2/d^2). \tag{17.34c}$$

Summating Eq.(17.34) with Eq.(17.33), the residual stress is obtained as

$$\sigma^r_r = -\sigma'_t + \sigma'_t r^2/d^2, \tag{17.35a}$$

$$\sigma^r_\theta = \sigma'_t r^2/d^2, \tag{17.35b}$$

$$\sigma^r_z = \sigma'_t r^2/d^2 - \sigma'_t/2. \tag{17.35c}$$

Apparently the residual stress satisfies

$$\sigma^r_\theta - \frac{1}{m'}\sigma^r_r \leqslant \sigma'_t. \tag{17.36}$$

Hence, when the solid rotating circular bar is unloaded from the plastic limit state, there is no inverse yield. It is indicated that the solid rotating circular bar is in shakedown within the range of the plastic limit.

17.3 Summary of Elasto-plasticity and Shakedown Analyses of Rotating Circular Bars

From the above elasto-plastic and shakedown analyses for the rotating circular bars with a constant velocity, the calculation expression for the stress, radius of plastic area and elastic limit rotating velocity as well as shakedown limit rotating velocity are presented. Fig.17.1 shows the relation of the non-dimensional elastic limit rotating velocity of hollow rotating circular bars $\Omega_{\mathrm{el}} = \omega_{\mathrm{el}}/[\sigma_t/(\rho d_2)]^{1/2}$ with respect to the unified strength theory parameter b at different values of the ratio of material strength in compression and in tension m. Fig.17.2 gives the relation of non-dimensional plastic limit rotating velocity of hollow rotating circular bars $\Omega_{\text{pl}} = \omega_{\text{pl}}/[\sigma_t/(\rho d_2)]^{1/2}$ with respect to the unified strength theory parameter b at different values of m. Fig.17.3 presents the relation of non-dimensional shakedown limit rotating velocity $\Omega_s = \omega_s/[\sigma_t/(\rho d_2)]^{1/2}$ with respect to the thickness d/a at different values of m. Fig.17.4 indicates the relation of shakedown limit rotating velocity Ω_s with respect to the thickness d/a at different values of the unified strength theory parameter b.

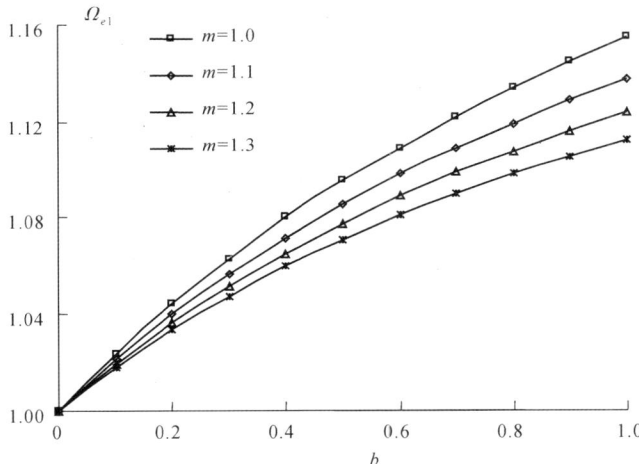

Fig. 17.1. Relation of elastic limit rotating velocity of hollow rotating circular bars with respect to unified strength theory parameter b and m

It is seen from these figures that the loading capability of hollow circular bars will increase when the effect of compressive and tensile strength variation and intermediate principal stress are taken into account. And a similar conclusion can be drawn for the solid circular bars.

17.4 Elasto-plastic and Shakedown Analyses of Rotating Disc

Rotating discs will be studied in this section. Similar to rotating circular bars, rotating discs are axisymmetric as well and σ_r, σ_θ, σ_z are principal stresses. But the rotating discs are in the state of plane stress, i.e., $\sigma_z = 0$. The unified strength theory will be employed to investigate the elasto-plasticity and shakedown of rotating discs.

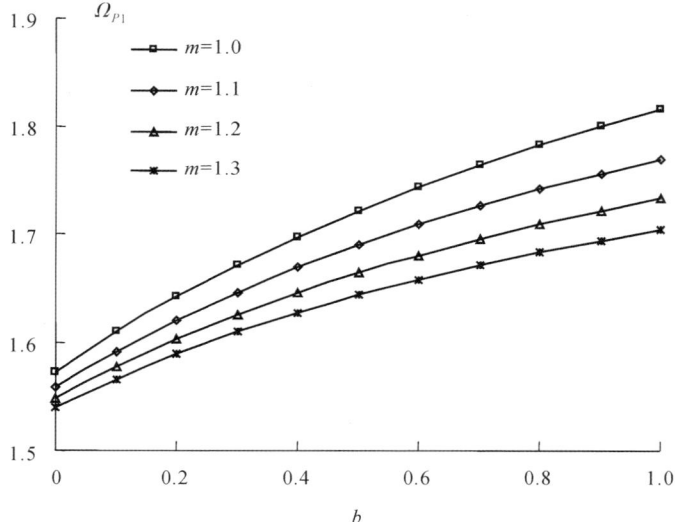

Fig. 17.2. Relation of plastic limit rotating velocity of hollow rotating circular bars with respect to unified strength theory parameter b and m

17.4.1 Elastic Analyses of Hollow Rotating Discs

It is assumed that the hollow discs with inner radius a and outer radius d rotate with the angel velocity ω. Several non-dimensional variables are cited for the sake of convenience.

$$R = \frac{r}{d}, \quad \overline{\sigma}_r = \frac{\sigma_r}{\sigma_t}, \quad \overline{\sigma}_\theta = \frac{\sigma_\theta}{\sigma_t}, \quad \Omega^2 = \frac{\rho \omega^2 d^2}{\sigma_t}, \quad \beta = \frac{a}{d}.$$

The equilibrium equation of rotating discs is the same as that of rotating circular bars, i.e., Eq.(17.2). Substituting the above non-dimensional variables into Eq.(17.2), we have

$$\frac{d\overline{\sigma}_r}{dR} + \frac{\overline{\sigma}_r - \overline{\sigma}_\theta}{R} + \Omega^2 R = 0. \tag{17.37}$$

The elastic stress solutions of hollow rotating discs are

$$\overline{\sigma}_r = c_1 + \frac{c_2}{R^2} - \frac{7}{16}\Omega^2 R^2, \tag{17.38a}$$

$$\overline{\sigma}_\theta = c_1 - \frac{c_2}{R^2} - \frac{5}{16}\Omega^2 R^2, \tag{17.38b}$$

where c_1; c_2 are integration constants. For the discs with free inner and outer boundaries we have

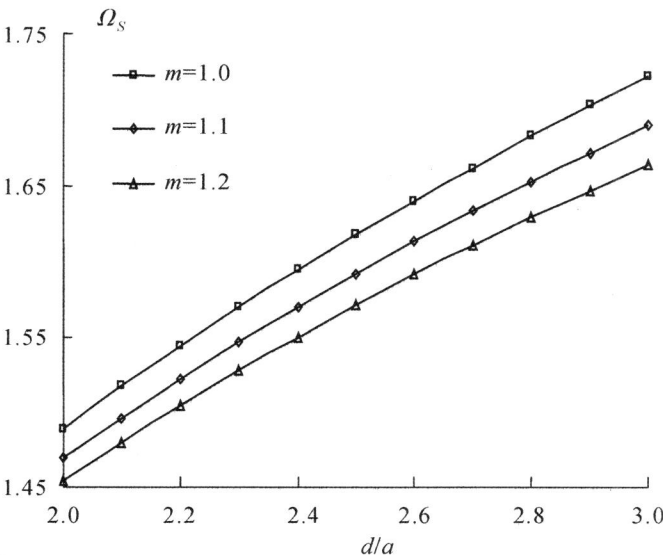

Fig. 17.3. Relation of shakedown limit rotating velocity of hollow rotating circular bars with respect to $m(b = 0.5)$

$$\overline{\sigma}_r = \frac{7}{16}\Omega^2 \left(1 + \beta^2 - \beta^2/R^2 - R^2\right), \tag{17.39a}$$

$$\overline{\sigma}_\theta = \frac{7}{16}\Omega^2 \left(1 + \beta^2 + \beta^2/R^2 - 5R^2/7\right). \tag{17.39b}$$

The yield condition of rotating discs is

$$\overline{\sigma}_\theta = a_i\overline{\sigma}_r + b_i, \tag{17.40}$$

where a_i, b_i are constants. Fig.17.5 presents the yield loci of the unified strength theory in the $\overline{\sigma}_r$-$\overline{\sigma}_\theta$ stress state.

The stresses of rotating discs should satisfy $\overline{\sigma}_\theta > \overline{\sigma}_r \geqslant \overline{\sigma}_z = 0$, therefore the curves corresponding to the stress of the plastic area are lines AB and BC in Fig.17.5. The function of line AB is

$$\overline{\sigma}_\theta = a_1\overline{\sigma}_r + b_1, \tag{17.41a}$$

where $a_1 = -b$, $b_1 = 1 + b$.

The function of line BC is

$$\overline{\sigma}_\theta = a_2\overline{\sigma}_r + b_2, \tag{17.41b}$$

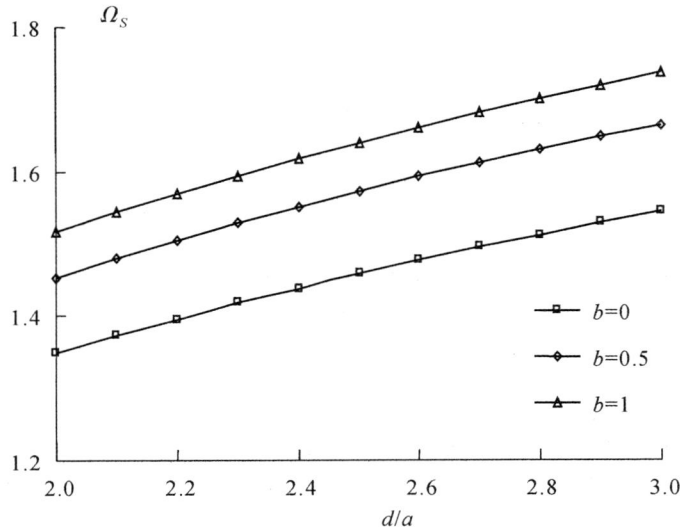

Fig. 17.4. Relation of shakedown limit rotating velocity of hollow rotating circular bars with the unified strength theory parameter b ($m = 1.2$)

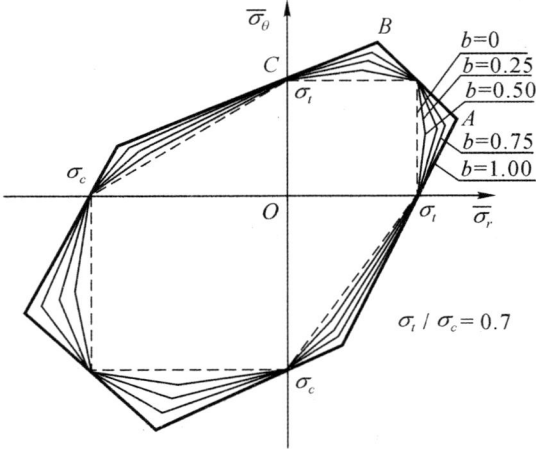

Fig. 17.5. Yield loci of the unified strength theory in plane stress state

where $a_2 = b/(m + mb)$, $b_2 = 1$. It is seen that the inner boundary of discs will yield first when the rotating velocity reaches a certain value. Substituting $r = a$ into Eqs.(17.39) and (17.41b), the limit elastic rotating velocity can be obtained

$$\Omega_{\mathrm{el}} = \sqrt{8/(7 + \beta^2)}. \tag{17.42}$$

Thus the stresses of discs in the elastic state are

$$\bar{\sigma}_r = \frac{7}{2(7 + \beta^2)} \left(1 + \beta^2 - \beta^2/R^2 - R^2\right), \tag{17.43a}$$

$$\bar{\sigma}_\theta = \frac{7}{2(7 + \beta^2)} \left(1 + \beta^2 + \beta^2/R^2 - 5R^2/7\right). \tag{17.43b}$$

17.4.2 Plastic Limit Analyses of Hollow Rotating Discs

The inner surface arrives at the plastic state when the rotating velocity $\Omega = \Omega_{\mathrm{el}}$, and the plastic area expands with the increase in velocity. When $\Omega = \Omega_{\mathrm{pl}}$, the complete area of discs is in the state of the plastic area, namely the plastic limit state, and is called the plastic limit rotating velocity of discs.

When the discs are in the plastic limit state, the stresses still satisfy $\bar{\sigma}_r > \bar{\sigma}_\theta \geqslant \bar{\sigma}_z = 0$. According to Fig.17.5, there are two possible cases for the plastic limit stress state of hollow discs: (1) the stress state is in line BC; (2) the stress state is in lines AB and BC. Both cases are discussed in detail as follows.

Case 1

When the ratio of the inner and outer radius β is over a critical value β_{cr}, the plastic stress state of discs corresponds to line BC. Solving Eq.(17.38) and yield condition Eq.(17.41b) together and considering the boundary conditions $R = 1$, $\bar{\sigma}_r = 0$, the stress distributions of discs are obtained

$$\bar{\sigma}_r = \frac{m + mb}{m + mb - b} \left(1 - R^{-\frac{m+mb-b}{m+mb}}\right) - \frac{(m + mb)\Omega_{\mathrm{pl}}^2}{3m + 3mb - b} \left(R^2 - R^{-\frac{m+mb-b}{m+mb}}\right), \tag{17.44a}$$

$$\bar{\sigma}_\theta = \frac{b}{m + mb - b} \left(1 - R^{-\frac{m+mb-b}{m+mb}}\right) - \frac{b\Omega_{\mathrm{pl}}^2}{3m + 3mb - b} \left(R^2 - R^{-\frac{m+mb-b}{m+mb}}\right) + 1. \tag{17.44b}$$

When $R = \beta$, $\bar{\sigma}_r = 0$. Hence the plastic limit rotating velocity of discs in Case 1 is obtained

$$\Omega_{\mathrm{pl}} = \left[\frac{(3m + 3mb - b)(1 - \beta^{-\frac{m+mb-b}{m+mb}})}{(m + mb - b)(\beta^2 - \beta^{-\frac{m+mb-b}{m+mb}})}\right]^{1/2}. \tag{17.45}$$

Assuming that at $R = R_0$ the radial stress $\bar{\sigma}_r$ reaches the maximum, then we have

$$\frac{\mathrm{d}\bar{\sigma}_r}{\mathrm{d}R}(R = R_0) = 0, \tag{17.46}$$

where $\bar{\sigma}_r \ (R = R_0) \leqslant (m + mb)/(m + mb + 1)$ and we obtain

$$\left(1 - \frac{m + mb - b}{3m + 3mb - b}\Omega_{\mathrm{pl}}^2\right) R_0^{-\frac{2m + 2mb - b}{m + mb}} - \frac{2m + 2mb}{3m + 3mb - b}\Omega_{\mathrm{pl}}^2 R_0 = 0. \quad (17.47)$$

The stress state of the maximum radial stress should correspond to point B in Fig.17.5 when $\beta = \beta_{\mathrm{cr}}$ and we have

$$\bar{\sigma}_r(R = R_0;\ \beta = \beta_{\mathrm{cr}}) = \frac{m + mb}{m + mb + 1}. \quad (17.48)$$

Solving Eq.(17.44) to Eq.(17.48) together, the critical value β_{cr} can be determined.

Case 2

When β is smaller than the critical value β_{cr}, the stress state of discs is in lines AB and BC of Fig.17.5. Based on the yield condition and boundary conditions, the discs can be divided into three areas: plastic area I ($\beta \leqslant R \leqslant R_1$); plastic area II ($R_1 \leqslant R \leqslant R_2$); plastic area III ($R_2 \leqslant R \leqslant 1$) (refer to Fig. 17.6), where R_1 and R_2 are the radii of plastic areas I and II, respectively. The stress state of plastic areas I and III corresponds to line BC in Fig.17.5, and the stress state of plastic area II corresponds to line AB.

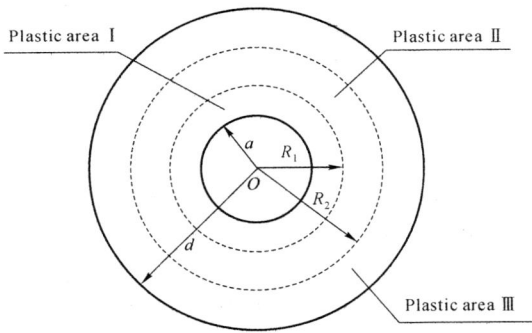

Fig. 17.6. Distribution of plastic areas of hollow rotating discs

Solving Eq.(17.38) and yield condition Eq.(17.41), the stress solution of three plastic areas can be derived. Here the radial stresses of the plastic area are listed as,

Plastic area I:

$$\bar{\sigma}_r = \frac{b_2}{1 - a_2} - \frac{\Omega_{\mathrm{pl}}^2}{3 - a_2}R^2 + d_1 R^{-1 + a_2}. \quad (17.49)$$

Plastic area II:

$$\overline{\sigma}_r = \frac{b_1}{1 - a_1} - \frac{\Omega_{\mathrm{pl}}^2}{3 - a_1} R^2 + d_2 R^{-1+a_1}, \tag{17.50}$$

Plastic area III:

$$\overline{\sigma}_r = \frac{b_2}{1 - a_2} - \frac{\Omega_{\mathrm{pl}}^2}{3 - a_2} R^2 + d_3 R^{-1+a_2}, \tag{17.51}$$

where d_1, d_2 and d_3 are integration constants. With consideration of boundary conditions and continuous conditions, the relations of d_1, d_2, d_3 and R_1, R_2, Ω_{pl} are as follows:

$$\frac{b_2}{1 - a_2} - \frac{\Omega_{\mathrm{pl}}^2}{3 - a_2} \beta^2 + d_1 \beta^{-1+a_2} = 0, \tag{17.52}$$

$$\frac{b_2}{1 - a_2} - \frac{\Omega_{\mathrm{pl}}^2}{3 - a_2} R_1^2 + d_1 R_1^{-1+a_2} = \frac{m + mb}{m + mb + b}, \tag{17.53}$$

$$\frac{b_1}{1 - a_1} - \frac{\Omega_{\mathrm{pl}}^2}{3 - a_1} R_1^2 + d_2 R_1^{-1+a_1} = \frac{m + mb}{m + mb + b}, \tag{17.54}$$

$$\frac{b_1}{1 - a_1} - \frac{\Omega_{\mathrm{pl}}^2}{3 - a_1} R_2^2 + d_2 R_2^{-1+a_1} = \frac{m + mb}{m + mb + b}, \tag{17.55}$$

$$\frac{b_2}{1 - a_2} - \frac{\Omega_{\mathrm{pl}}^2}{3 - a_2} R_2^2 + d_3 R_2^{-1+a_2} = \frac{m + mb}{m + mb + b}, \tag{17.56}$$

$$\frac{b_2}{1 - a_2} - \frac{\Omega_{\mathrm{pl}}^2}{3 - a_2} + d_3 = 0. \tag{17.57}$$

Substituting the radial stresses of the plastic area into yield condition Eq.(17.41), the tangent stresses can be determined.

17.4.3 Shakedown of Hollow Rotating Discs

The shakedown of rotating hollow discs subject to repeated loadings will be studied in this section. For convenience b is taken as zero and the effect of the tensile and compressive strength vibration of materials on the shakedown of rotating discs is mainly discussed here.

On the basis of the above analyses, the plastic limit rotating velocity of hollow rotating discs when $b = 0$ is easily obtained as

$$\Omega_{\mathrm{pl}} = \sqrt{3/(1 + \beta + \beta^2)}, \tag{17.58}$$

and the elastic limit rotating velocity is still expressed by Eq.(17.42). When the loading is increased to $\Omega_{\mathrm{el}} < \Omega < \Omega_{\mathrm{pl}}$, the disc is in the elasto-plastic state, and its internal part is a plastic area while its external part is an elastic area. Assuming that the radius of the plastic area is c corresponding to a certain value of Ω, then the yield condition can be simplified as

$$\overline{\sigma}_\theta = 1. \tag{17.59}$$

Solving Eqs.(17.2) and (17.57) together, and considering boundary condition $R = \beta$, $\overline{\sigma}_r = 0$, the stresses of the plastic area are obtained

$$\overline{\sigma}_r = 1 - \beta/R - \frac{1}{3}\Omega^2 R^2 \left[1 - (\beta/R)^3 \right], \tag{17.60a}$$

$$\overline{\sigma}_\theta = 1. \tag{17.60b}$$

According to Eq.(17.36), the stresses of the elastic area are

$$\overline{\sigma}_r = c_1' + \frac{c_2'}{R^2} - \frac{7}{16}\Omega^2 R^2, \tag{17.61a}$$

$$\overline{\sigma}_\theta = c_1' - \frac{c_2'}{R^2} - \frac{5}{16}\Omega^2 R^2, \tag{17.61b}$$

where c_1' and c_2' are integration constants. Based on the stress continuous conditions on the interface between elastic and plastic areas, we obtain

$$2c_1' = 2 - \frac{\beta}{c} + \frac{5}{12}\Omega^2 c^2 + \frac{1}{3}\Omega^2\beta^3/c, \tag{17.62a}$$

$$\frac{2c_2'}{c^2} = -\frac{\beta}{c} - \frac{5}{24}\Omega^2 c^2 + \frac{1}{3}\Omega^2\beta^3/c. \tag{17.62b}$$

Taking into account the boundary condition of the outer surface $R = 1$, $\overline{\sigma}_r = 0$, we have

$$c_1' + c_2' - \frac{7}{16}\Omega^2 = 0, \tag{17.63}$$

Substituting integration constants c_1' and c_2' into Eq.(17.63), the relation between rotating velocity Ω and the radius of the plastic area c is obtained as

$$\Omega^2 = \frac{24(2c - \beta c^2 - \beta)}{21c - 8\beta^3 c^2 - 10c^3 + 5c^5 - 8\beta^3}. \tag{17.64}$$

If $c = \beta$, $c = 1$, we can obtain easily Eqs.(17.42) and (17.58). When the rotating velocity decreases to zero, i.e., unloaded completely, there will be residual stress in the discs. Assuming that the unloading is elastic, the stress variation of discs can be derived on the basis of Eq.(17.39),

$$\Delta \bar{\sigma}_r = -\frac{7}{16}\Omega^2 \left(1 + \beta^2 - \beta^2/R^2 - R^2\right), \tag{17.65a}$$

$$\Delta \bar{\sigma}_\theta = -\frac{7}{16}\Omega^2 \left(1 + \beta^2 + \beta^2/R^2 - 5R^2/7\right). \tag{17.65b}$$

Summating the above equations with Eqs.(17.60) and (17.61) respectively, the residual stresses of the plastic area after unloading are obtained as

$$\bar{\sigma}_r^r = 1 - \beta/R - \frac{1}{3}\Omega^2 R^2 \left[1 - (\beta/R)^3\right] - \frac{7}{16}\Omega^2 \left(1 + \beta^2 - \beta^2/R^2 - R^2\right), \tag{17.66a}$$

$$\bar{\sigma}_\theta^r = 1 - \frac{7}{16}\Omega^2 \left(1 + \beta^2 + \beta^2/R^2 - 5R^2/7\right), \tag{17.66b}$$

and the residual stresses of the elastic area are

$$\bar{\sigma}_r^r = c_1' + \frac{c_2'}{R^2} - \frac{7}{16}\Omega^2 R^2 - \frac{7}{16}\Omega^2 \left(1 + \beta^2 - \beta^2/R^2 - R^2\right), \tag{17.67a}$$

$$\bar{\sigma}_\theta^r = c_1' - \frac{c_2'}{R^2} - \frac{5}{16}\Omega^2 R^2 - \frac{7}{16}\Omega^2 \left(1 + \beta^2 + \beta^2/R^2 - 5R^2/7\right). \tag{17.67b}$$

It is seen that when the rotating velocity reaches a certain value, the residual stresses of the discs will result in the inverse yield of discs. It will appear at the inner surface first. It is known from Eq.(17.66) that $\bar{\sigma}_r^r = 0$, $\bar{\sigma}_\theta^r = 1 - \Omega^2 \Omega_{\text{el}}^2 < 0$ at the inner surface. Therefore the yield condition of the inner surface is

$$\bar{\sigma}_r^r - \frac{\bar{\sigma}_\theta^r}{m} = \sigma_t. \tag{17.68}$$

Substituting the stress at the inner surface into the above equation, the maximum rotating velocity which will not cause the inverse yield when the discs are unloaded is obtained as

$$\Omega_s = \sqrt{8(1+m)/(7+\beta^2)}. \tag{17.69}$$

Takeing into account the consideration that the maximum rotating velocity of discs should not exceed the plastic limit rotating velocity, the shakedown limit velocity is thus obtained

$$\Omega_s = \min\left(\sqrt{8(1+m)/(7+\beta^2)},\ \sqrt{3/(1+\beta+\beta^2)}\right). \tag{17.70}$$

17.5 Elasto-plastic and Shakedown Analyses of Solid Rotating Discs

Similarly, the elasto-plastic and shakedown analyses of solid rotating discs can be carried out in the same way. Here we neglect the detailed derivation and give the main solution for solid rotating discs.

Elastic stresses

$$\bar{\sigma}_r = \frac{7}{16}\Omega^2 \left(1 - R^2\right), \tag{17.71a}$$

$$\bar{\sigma}_\theta = \frac{7}{16}\Omega^2 - \frac{5}{16}\Omega^2 R^2, \tag{17.71b}$$

and the elastic limit rotating velocity

$$\Omega_{\mathrm{el}} = \sqrt{7/16}. \tag{17.72}$$

The plastic areas of the plastic limit state can be divided into two parts: the plastic area I in the central part $(0 \leqslant R < R_1)$ and the plastic area II in the surrounding part $(R_1 \leqslant R < 1)$ (refer to Fig.17.7). The stress state of plastic area I corresponds to line AB in Fig.17.5 and the stress state of plastic area II corresponds to line BC, where R_1 is the critical radius between plastic areas I and II. The stresses of plastic area I are

$$\bar{\sigma}_r = \frac{b_1}{1 - a_1} - \frac{\Omega_{\mathrm{pl}}^2}{3 - a_1}R^2, \tag{17.73a}$$

$$\bar{\sigma}_\theta = a_1\bar{\sigma}_r + b_1. \tag{17.73b}$$

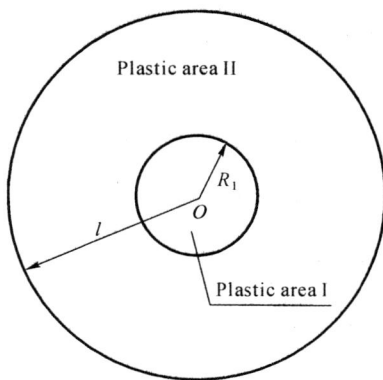

Fig. 17.7. Distribution of plastic areas for solid rotating discs

The stresses of plastic area II are

$$\overline{\sigma}_r = \frac{b_2}{1 - a_2} - \frac{\Omega_{\mathrm{pl}}^2}{3 - a_2} R^2 + d_1 R^{-1+a_1}, \qquad (17.74a)$$

$$\overline{\sigma}_\theta = a_2 \overline{\sigma}_r + b_2, \qquad (17.74b)$$

where d_1, Ω_{pl} and R_1 are determined by the following equation:

$$\frac{b_1}{1 - a_1} - \frac{\Omega_{\mathrm{pl}}^2}{3 - a_1} R_1^2 = \frac{m + mb}{m + mb + b}, \qquad (17.75)$$

$$\frac{b_2}{1 - a_2} - \frac{\Omega_{\mathrm{pl}}^2}{3 - a_2} R_1^2 + d_1 R_1^{-1+a_2} = \frac{m + mb}{m + mb + b}, \qquad (17.76)$$

$$\frac{b_2}{1 - a_2} - \frac{\Omega_{\mathrm{pl}}^2}{3 - a_2} + d_1 = 0. \qquad (17.77)$$

17.6 Summary of Elastic and Plastic Analyses of Rotating Discs

Elasto-plastic and shakedown analyses have been conducted for the rotating discs. It is seen that the elastic limit rotating velocity is neither related to the effects of the intermediate principal stress of materials, nor to the tensile and compressive strength variation, while the plastic limit rotating velocity is related to them.

Fig.17.8 gives the relation of plastic limit rotating velocity Ω_{pl} with respect to the ratio of compressive and tensile strength m as well as the unified strength theory parameter b. Parameter b is also the parameter for the effect of the intermediate principal stress.

It is seen from Fig.17.8 that the plastic limit rotating velocity will increase with the increase in b and the effect of m on Ω_{pl} is related to the value of b. When $b = 0$ (the Mohr-Coulomb strength theory), m has no effect on plastic limit rotating velocity Ω_{pl}, and with the increase in b the influence of m is more and more obvious.

Fig.17.9 presents the relation of shakedown rotating velocity Ω_s with respect to the ratio of inner and outer radii of the discs a/d at various values of m. It is indicated that the effect of m on the shakedown rotating velocity Ω_s is related to a/d. When a/d is large, m has no effect on Ω_s, and when a/d is small, and m has a significant effect on Ω_s.

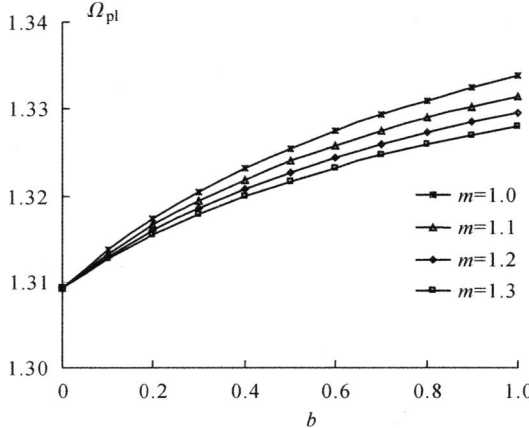

Fig. 17.8. Relation of plastic limit rotating velocity of hollow rotating discs with respect to unified strength theory parameter b and m ($\beta = 0.5$)

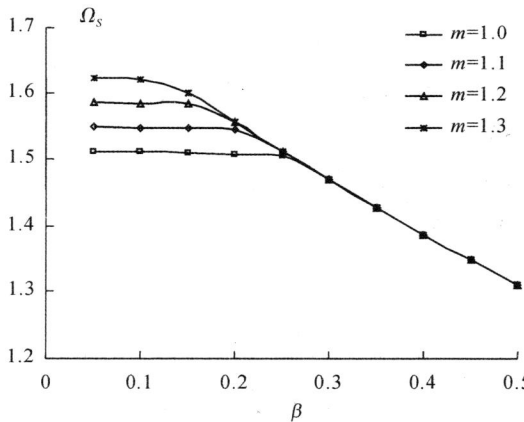

Fig. 17.9. Relation of shakedown limit velocity of hollow rotating discs with respect to β and m

17.7 Summary

In this chapter the elasto-plastic and shakedown analyses for the rotating circular bars and discs are carried out using the unified strength theory. The closed solution of the elasto-plastic stress, elastic limit rotating velocity, plastic limit rotating velocity and shakedown limit rotating velocity of rotating circular bars and discs are derived. With these solutions, the effects of tensile and compressive strength variation and intermediate principal stress on those limit rotating velocities are researched. The unified solution obtained by using the unified strength theory gives a series of results. The following conclusions are drawn:

(1) The elastic limit rotating velocity, plastic limit rotating velocity and shakedown limit rotating velocity of rotating circular bars increase with the increase of intermediate principal stress coefficient, and the effect of tensile and compressive strength variation is related to the value of the intermediate principal stress.

(2) The elastic limit rotating velocity of rotating discs is not related to the tensile and compressive strength variation. For the hollow rotating discs, the plastic limit rotating velocity increases with the intermediate principal stress coefficient and the ratio of compressive and tensile strength. The effect of the ratio of compressive and tensile strength m on the shakedown limit velocity of discs Ω_s is related to the ratio of the inner and outer radii a/d. When a/d is large m has no effect on Ω_s and when a/d is small m has a significant effect on Ω_s.

The unified strength theory has been applied to various problems of structural plasticity. A series of new results are presented. Comments on the unified strength theory are given by Shen (2004), Teodorescu (2006), and Altenbach and Kolupaev (2008). Applications of the unified strength theory in plastic limit analysis and shakedown analysis for different structures are still developing. It is expected that the unified strength theory will have more and more applications in the future in addition to the plastic analysis of structures. We hope that the Chinese idiom "throwing out a brick to attract a piece of jade" becomes real and this book can serve as a solid brick.

Applications of the unified strength theory in computational plasticity are still to be explored. We will study them in the next monograph: *Computational Plasticity: with Emphasis on the Application of the Unified Strength Theory.*

References

Altenbach H, Kolupaev VA (2008) Remarks on Model of Yu MH. In: Zhang TY, Wang B, Feng XQ (eds.) The Eighth International Conference on Fundamentals of Fracture (ICFF VIII), 270-271

Chakrabarty J (1987) Theory of plasticity. McGraw-Hill New York

Feng XQ, Yu SW (1994) An upper bound on damage of elastic-plastic structures at shakedown. Int. J. Damage Mechanics, 3:277-289

Guven (1992) Elastic-plastic stresses in a rotating annular disk of variable thickness and variable density. Int. J. Mech. Sci., 34:133-138

Hodge DG, Balaban M (1962) Elastic-plastic analysis rotating cylinder. Int. J. Mech. Sci., 4:465

Kachanov LM (1971) Foundation of the theory of plasticity. North-Holland Publ. Co., Amsterdam

König JA (1987) Shakedown of elastic-plastic Structures. Elsevier, Amsterdam

Liu XS, Xu BY, Liu Y (1988) Shakedown analysis of rotating disks. Journal of Ching Hua University, 28:84-92 (in Chinese)

Liu Z, Wang RG (1990). Shakedown analysis of hollow rotating disc. Chinese J. Mech. Eng., 26(5):23-28 (in Chinese)

Ma G, Hao H (1999) Investigation on limit angular velocity of annular disc. In: Wang CM, Lee K H, Ang KK (eds.) Computational Mechanics for the Next Millennium, Elsevier Science Ltd., Amsterdam, 1:399-1015

Ma GW, Yu MH, Iwaski S, et al. (1995) Unified elasto-plastic solution to rotating disc and cylinder. J. Struc. Eng., 41:79-85

Martin JB (1975) Plasticity: fundamentals and general results. The MIT Press

Shen ZJ (2004) Reviews to "*Unified Strength Theory and Its Applications*". Adv. Mech., 34(4):562-563 (in Chinese)

Teodorescu PP (2006) Review of "Unified Strength Theory and Its Applications. Springer, Berlin, 2004". Zentralblatt MATH 2006, Cited in Zbl. Reviews, 1059.74002 (02115115)

Timoshenko S, Goodier JN (1970) Theory of elasticity. (3rd ed.) McGraw-Hill, New York

Xu SQ, Yu MH (2004) Unified analytical solution to shakedown problem of thick-walled cylinder. Chinese J. Mechanical Engineering, 40(9):23-27 (in Chinese)

Xu SQ, Yu MH (2005a) Shakedown analysis of thick-walled cylinders subjected to internal pressure with the unified strength criterion. Int. J. Pressure Vessels and Piping, 82(9):706-712

Xu SQ, Yu MH (2005b) Shakedown analysis of thick spherical shell of material with different strength in tension and compression. Machinery Design & Manufacture, 1:36-37

Yu MH (1992) A new system of strength theory. Xian Jiaotong University Press, Xi'an (In Chinese)

Yu MH (2004) Unified strength theory and its applications. Springer, Berlin

Yu MH, He LN (1991) A new model and theory on yield and failure of materials under complex stress state. In: Mechanical Behavior of Materials-VI, Pergamon Press, Oxford, 3:841-846

Yu MH, Yang SY, Fan SC, et al. (1999) Unified elasto-plastic associated and non-associated constitutive model and its engineering applications. Computers and Structures, 17:627-636

Yu MH, Yang SY, Liu CY (1997) Unified plane-strain slip line theory. China Civil Engineering. J., 30(2):14-26 (in Chinese)

Zouain N and Silveira JL (2001) Bounds to shakedown loads. Int. J. Solids Struct., 38:2249-2266

Zyczkowski M (1981) Combined loadings in the theory of plasticity. Polish Scientific Publishers, PWN and Nijhoff

Index